COMPUTER FORENSICS

COMPUTER FORENSICS

Cybercriminals, Laws, and Evidence

MARIE-HELEN MARAS, PhD

Assistant Professor of Criminal Justice
State University of New York—Farmingdale

JONES & BARTLETT
LEARNING

World Headquarters
Jones & Bartlett Learning
40 Tall Pine Drive
Sudbury, MA 01776
978-443-5000
info@jblearning.com
www.jblearning.com

Jones & Bartlett Learning
Canada
6339 Ormindale Way
Mississauga, Ontario L5V 1J2
Canada

Jones & Bartlett Learning
International
Barb House, Barb Mews
London W6 7PA
United Kingdom

Jones & Bartlett Learning books and products are available through most bookstores and online booksellers. To contact Jones & Bartlett Learning directly, call 800-832-0034, fax 978-443-8000, or visit our website, www.jblearning.com.

Substantial discounts on bulk quantities of Jones & Bartlett Learning publications are available to corporations, professional associations, and other qualified organizations. For details and specific discount information, contact the special sales department at Jones & Bartlett Learning via the above contact information or send an email to specialsales@jblearning.com.

Production Credits
Publisher, Higher Education: Cathleen Sether
Acquisitions Editor: Sean Connelly
Senior Associate Editor: Megan R. Turner
Production Manager: Jenny L. Corriveau
Associate Production Editor: Jill Morton
Associate Marketing Manager: Lindsay White
Manufacturing and Inventory Control Supervisor: Amy Bacus
Composition: DataStream Content Solutions, LLC
Cover Design: Kristin E. Parker
Photo Research and Permissions Supervisor: Christine Myaskovsky
Cover Image: Abstract of fingerprint on monitor, © Saniphoto/Dreamstime.com; Abstract of human figures with numbers, © Kts/Dreamstime.com
Chapter Opener Image: © Pixel 4 Images/ShutterStock, Inc.
Printing and Binding: Malloy Incorporated
Cover Printing: Malloy Incorporated

Library of Congress Cataloging-in-Publication Data
Maras, Marie-Helen, 1979–
    Computer forensics : cybercriminals, laws, and evidence / by Marie-Helen Maras.
        p. cm.
    Includes bibliographical references and index.
    ISBN-13: 978-1-4496-0072-3
    ISBN-10: 1-4496-0072-7
    1. Electronic evidence—United States.   2. Computer crimes—Investigation—United States.   I. Title.
    KF8947.5.M37 2011
    363.25'9680973—dc22

                                                                                                2010050880

6048

Printed in the United States of America
15 14 13 12 11   10 9 8 7 6 5 4 3 2

DEDICATION

*Χαίροις της Αιγύπτου θείος βλαστός, χαίροις Γερανείων θεοδώρητος θησαυρός,
Χαίροις Λουτρακίου και πάσης Κορινθίας, αντίληψις και κλέος, Άγιε Πατάπιε.*

~

In loving memory of my father, Pete Maras (Petoulis).
Thank you for the best 28 years of my life—of unconditional love, laughter, and adventure.

BRIEF CONTENTS

CONTENTS

Chapter 12   Mobile Phones and PDAs in Computer Forensics Investigations   293

Chapter 13   The Pretrial and Courtroom Experiences of a Computer Forensics Investigator   321

PREFACE

Computer forensics cannot be divorced from the law. A computer forensics investigator needs knowledge of the law to effectively do his or her job. Meanwhile, legal professionals working on cybercrimes must have knowledge of the hardware, software, and technology involved in computer forensics to effectively do their jobs.

The available textbooks on computer forensics are either too technical, placing too much emphasis on the hardware and software used, or too thick in legal analysis, to the extent that a comprehensive background in law is required for their review. There is currently no textbook in the market that falls somewhere in between these two extremes— a book that is tailored to, and can be used by, the individual who does not have a comprehensive legal and technical background. This textbook seeks to fill this void in the literature.

This book is intended to appeal to a wide range of groups. By steering away from both a thicket of legal terms and reams of technical analysis, it seeks to interest a much broader audience of writers and researchers working on computer forensics. Moreover, by providing a concise yet sufficiently detailed account of the most significant and current developments in computer forensics and their implications for a number of different fields (e.g., computer science, law, public policy and administration, security, and criminology), it is likely to prove an extremely useful resource for academics, practitioners, and graduate and undergraduate students in these areas. Criminal justice and socio–legal scholars and professionals should also find food for thought in this work.

Given that this textbook covers the technology and software currently used in the field, it will be of interest to law enforcement agencies and professionals working as computer forensics specialists.

Specifically, this book is intended for the following audiences:

- Law enforcement agents seeking to expand their knowledge of investigations to the field of computers

- Students and professionals seeking a career in computer forensics investigations

- Computer forensics specialists concerned about the legality of searches and evidence seizure, storage, transport, and evaluation

- Legal professionals seeking to understand computer forensics investigations, rules concerning electronic evidence, and the admissibility of this evidence in court

- Computer specialists in the private sector who may be required by courts or law enforcement agents—sometime in the future—to search, restore, transmit, copy, or store electronic data during a computer forensics investigation

Anyone interested in learning about computer forensics, investigations, and electronic evidence will also benefit from this textbook.

The computer forensics field is gaining prominence because of the current worldwide media coverage of cybercrimes and cybercriminals. This textbook will also be relevant to civil liberties groups and professional associations, as news of extradition of cybercriminals and the acceptance of the use of hearsay evidence in computer crime cases is becoming a more common practice.

Finally, this textbook is intended for students in computer forensics courses. It is also intended for students in legal courses who are seeking an introduction to the technology involved in computer forensics investigations and the technical and legal difficulties involved in searching, extracting, maintaining, and storing electronic evidence, while simultaneously looking at the legal implications of such investigations and the rules of legal procedure relevant to electronic evidence.

Supplements

This textbook is accompanied by a series of valuable supplements. An instructor's manual (with Microsoft® PowerPoint® slides) is available to assist instructors in teaching computer forensics and other cybercrime courses. Additionally, a TestBank containing discussion questions and practical exercises is provided to stimulate the critical thinking skills.

ACKNOWLEDGMENTS

I would like to warmly thank Sean Connelly, Megan Turner, and Jill Morton at Jones & Bartlett Learning for all of their direction and assistance during the development and production of this textbook. Additionally, I am grateful to my former thesis supervisor at the University of Oxford, Dr. Lucia Zedner, for her continuous support and encouragement. I thank you for making my experience at Oxford truly memorable and for continuing to cheer me on long after graduation. I would also like to thank my former professors at the Center for Criminology at the University of Oxford for making my learning experience unforgettable. Last and by no means least, I am especially grateful to John Kostanoski, Chair of the Criminal Justice Department at Farmingdale State College, State University of New York, for his guidance toward my professional development.

On behalf of Jones & Bartlett Learning, we would like to thank the following people for their valuable insight in the review of the text:

Qinghai Gao, Farmingdale State College at SUNY
Camille Gibson, Praire View A&M University
Robert Haack, Suffolk County Community College
Raymond Hsieh, California University of Pennsylvania
Alexander Muentz, Temple University

Chapter 1

Entering the World of Cybercrime

During 2006 and 2007, inmates at the Plymouth County, Massachusetts, correctional facility were provided with computer privileges to conduct legal research. Stringent security measures were put in place that prevented inmates from accessing e-mail, the Internet, and other computer programs—at least that is what prison officials thought.

Francis G. Janosko, an inmate at the correctional facility at that time, managed to hack into the computer network. Specifically, he gained unauthorized access to the computer system to send e-mails and provide inmates in the facility with access to the personal information (names, Social Security numbers, home addresses, and telephone numbers, among other items) of more than 1000 current and former correctional facility workers.[1]

Janosko's actions put the lives of these employees and their families in harm's way. This case is but one example of how the technology and information age has provided criminals with the means to cause catastrophic harm or damage with just a few presses on a keyboard—harm or damage that one could accurately say would not have been possible without the existence of such technology. It also raises a troubling question: Which other types of crimes have the Internet, computers, and related technologies made possible? To answer this question, this chapter focuses on what cybercrime is, how it differs from traditional forms of crime, which types of cybercrimes are distinguished, and which crimes are considered "cybercrimes."

Cybercrime: Defined

The exponential expansion of computer technologies and the Internet have spawned a variety of new criminal behaviors and provided criminals with a new environment within which to operate. **Cybercrime** involves the use of the Internet, computers, and related technologies in the commission of a crime. It includes technologically specific crimes that would not be possible without the use of computer technology as well as traditional crimes committed with the assistance of a computer.[2] The range of criminal activities has also been increasing as a result of the advent of cybercrime, as many more crimes were created and have been the exclusive product of technology and the Internet.

Cybercrime Versus Traditional Crime

Cybercrime differs from traditional crime in several ways. One important difference is that cybercrime knows no physical, geographic boundaries because the Internet provides criminals with access to people, institutions, and businesses around the globe. Consider the crime of fraud. Normally, fraud involves face-to-face communication with the victim or lengthy conversations over the phone to gain the target's trust. In today's world, however, fraudulent e-mails and websites can be used to con victims worldwide. For instance, from 2004 to 2009, Icarus Dakota Ferris manufactured and sold counterfeit postage stamps online by claiming that they were discontinued. He made approximately $345,000 in profits by defrauding victims globally.[3] As this example suggests, cybercrimes can be committed on a far broader scale than their traditional, real-world counterparts.

The Internet has augmented the ease and speed with which criminal activities are conducted. Prior to the advent of the Internet, if someone wanted to steal money from the bank, he or she would either rob the bank during its daily operation or steal the money after business hours. Either way, the thief would have to physically remove the money from the bank. As such, bank robbers were restricted to taking as much money as they could possibly carry outside the bank. In the online environment, such physical restrictions no longer apply. Larger monetary rewards can be gained without expending any physical energy. Billions of dollars can be stolen from a bank in the online environment within minutes.

The Internet also affords perpetrators with the opportunity to expend less effort in defrauding someone. Consider mail fraud, which has changed a great deal since the advent of computers. The amount of correspondence between individuals has increased exponentially as a result of technology because it takes significantly less time to send and receive a letter. Instead of individuals sending out fake letters through regular mail, criminals now do so electronically. In the past, regular mail might contain vital information, such as credit card and financial information, which nefarious individuals might steal. Now, criminals are using electronic mail (e-mail) to obtain the same kind of information. Criminals can also send bulk amounts of e-mail without paying; in the practice known as spamming, huge numbers of e-mails are sent out to multiple recipients almost instantaneously.

Another crime affected by computer and information technology is the theft of proprietary information and trade secrets from businesses. In the past, businesses might use spies, tap into phone wires, or set up cameras or voice recorders to obtain information on their rivals. They would steal paperwork or sift through the trash for valuable documents that might have been discarded carelessly. Often, stealing this information was relatively difficult for the criminal. With the advent of computers, however, once someone has gained unauthorized access to a computer system, he or she has all of the information desired at their fingertips.

Social networking sites such as Facebook, MySpace, and Twitter have also made the burglar's job much easier. Today, burglars no longer need to stake out someone's home to determine whether the individual (target) is at home. Instead, they can simply request to be the target's friend on Facebook or MySpace or a follower of the subject on Twitter. Given that many people accept friend and follower requests from strangers, such a request is very likely to give the would-be thief access to the subject's profile page. On these sites, the targets are usually more than happy to share with the rest of the world their comings and goings—to such an extent that they normally declare when they are leaving their homes and where they are going. Individuals also like to share information about their homes and purchases. In California, a group of teenagers—mostly girls—were accused of burglarizing the homes of celebrities such as Orlando Bloom, Brian Austin Green, Megan Fox, Paris Hilton, and Lindsay Lohan using this information. More than $3 million in jewelry, clothes, and accessories was stolen from these celebrities.[4] The group tracked their victims' whereabouts online through sites such as Twitter to determine when the celebrities would be away from their homes.

Perhaps the National Academy of Sciences said it best: "The modern thief can steal more with a computer than with a gun. Tomorrow's terrorist may be able to do more damage with a keyboard than with a bomb."[5]

New Crimes, New Tactics

Crimes that would not have been possible without the use of technology include threats against software and networks such as **hacking** (defined as unauthorized intrusion into computers) and **malware** (malicious software), which includes computer viruses, worms, and Trojan horses (each of which is explored in this chapter in further detail).

Cyberterrorism is another example.[6] **Cyberterrorism** may be defined "as the politically motivated use of computers as weapons or as targets, by sub-national groups or clandestine agents intent on violence, to influence an audience or cause a government to change its policies."[7] A cyberterrorist may hack into U.S. critical infrastructure in an attempt to cause grave harm such as loss of life or significant economic damage. Such attacks are aimed at wreaking havoc on information technology systems that are an integral part of public safety, traffic control, medical and emergency services, and public works. While this type of wide-scale disruption has not yet come to fruition in real life, film has depicted it (for example, *Live Free or Die Hard*) and academicians, practitioners, researchers, law enforcement agencies, and politicians have entertained the possibility of such events occurring. Additionally, computer security experts have created mock cyberterrorism attacks to expose weaknesses in the United States' critical infrastructure during a war game titled "Digital Pearl Harbor" hosted by the U.S. Naval War College in 2002.[8] These attacks illustrated that the most vulnerable systems were the Internet and computer infrastructure systems of financial institutions.

Old Crimes, New Tactics

Cybercrime also includes crimes that put a twist on traditional crimes. Extortion can occur online. **Cyberextortion** occurs when someone uses a computer to attack or threaten to attack an individual, business, or organization if money is not provided to prevent or stop the attack. In Long Island, for example, two teenagers attempted to extort MySpace by threatening to post a method for stealing MySpace users' personal information online, unless the site's operators paid the teenagers $150,000. In 2010, Anthony Digati tried to extort money from a New York–based life insurance company. In particular, Digati, via e-mail, threatened to damage the reputation of the insurance company and cost it millions of dollars in revenue if the company did not pay him approximately $200,000.[9]

Crimes of **vandalism** can also occur online, although they take a different form from physical vandalism, such as graffiti on the walls. Online vandalism can occur by defacing websites, for example. Web defacement involves the unauthorized access to a website and the alteration or replacement of its content without causing permanent damage. A group of U.S. hackers (known as 'Team Spl0it') broke into government websites during the conflict in Kosovo in 1999 and posted antiwar messages.[10] A more recent example concerns the Red Eye Crew. In January 2010, this group of hackers (allegedly from Brazil) defaced more than 30 websites owned by various U.S. House of Representatives and House Committee members. The hackers left an offensive message on the compromised websites that was aimed at the President of the United States, Barrack Obama.[11]

Even certain public order crimes, which include—among other things—victimless crimes that threaten the general well-being of society and challenge its accepted moral principles, can be committed online. One such example is prostitution. Prostitutes provide a range of sexual behaviors (e.g., sadism, masochism, and sexual intercourse) in exchange for remuneration. They may provide their services through brothels or escort services or search for potential customers on the streets. Nowadays, their services can be offered and arranged over the Internet. Specifically, prostitutes (or their pimps) may set up websites through which customers may solicit sexual services or post ads on online forums, wait for clients to answer the ads, and arrange meetings with their customers in hotels or other locations. This has come to be known as **cyberprostitution**. There have been many cases on Craigslist (a website that provides, among other things, forums for housing, jobs, personals, services, and events) where prostitutes have been soliciting sex in the "Casual Encounters" section (other sections as well). In fact, in 2009, undercover police officers in Worcester, Massachusetts, posted false offers of prostitution on Craigslist's "Casual Encounters" section that resulted in the arrests of 50 people.[12]

Cybercrime Categories

Computers can be an incidental aspect of the commission of the crime and may contain information about the crime. Computers may be the source of evidence for certain crimes (for example, drug dealing and child pornography). Drug dealers may sell their illegal substances through encrypted e-mail, arrange meetings with clients to distribute their drugs, and swap recipes for drugs in restricted-access chat rooms. Profits and sales are sometimes recorded and kept in files in the dealer's computer. In *United States v. Carey*,[13] police officers searched for evidence of the sale and possession of drugs on the suspect's computer and found child pornography.

Serious criminals, such as those who engage in **terrorism** and **organized crime**, may store evidence of their crimes on computers. Consider the coordinated terrorist attacks on July 7, 2005, in London, which targeted the public transportation system. The investigation of the London bombings revealed that the perpetrators (Mohammad Sidique Khan, Lindsay Germaine, Shehzad Tanweer, and Hasib Hussain) had documents and manuals downloaded on their computers that contained information on how to construct bombs. In fact, various electronic journals (e.g., *Mu'askar al-Battar*) and manuals (e.g., the *Anarchist Cookbook* and *Terrorist Handbook*) are available online that provide logistical information on targets and offer instructions on everything from bomb making, construction of crude chemical weapons, and sniper training to guidelines for establishing secret training camps and safe houses.[14] According to the U.S. Department of Justice (DOJ) website, a DOJ Committee member accessed a website that housed more than 110 different texts on bomb making, containing titles such as the following:[15]

- "How to Make a CO_2 Bomb"
- "Mail Grenade"
- "Calcium Carbide Bomb"
- "Jug Bomb"
- "Chemical Fire Bottle"

Additionally, a website that promotes al-Qaeda's positions even offers its visitors courses on how to prepare explosives.[16] What is apparent is that terrorists and their supporters are

> Building a massive and dynamic online library of training materials . . . covering such varied subjects as how to mix ricin poison, how to make a bomb from commercial chemicals, how to pose as a fisherman and sneak through Syria into Iraq, how to shoot at a U.S. soldier, and how to navigate by the stars while running through a night-shrouded desert.[17]

The Internet has also been (and is being) used by terrorists for training purposes. For instance, one website, which is mostly in Arabic, provides video and audio clips that include "detailed advice on physical training, surveillance of targets, and various operational tactics."[18] Scheuer has argued that the perpetrators of the London bombings "may well have profited from the urban-warfare training al-Qaeda has made readily available on the Internet."[19]

The determination as to whether a particular crime is a cybercrime depends on the role of the computer in the crime. To be considered a cybercrime, computers must be either the target of the crime or the tools used to commit the crime.

Category 1: The Computer as Target of the Crime

Cybercrime consists of crimes where the target is the computer. For crimes where the computer is the target, a perpetrator:

- Attempts to break into a computer and/or
- Steals information from the computer and/or
- Bombards the computer or launches an attack from the outside and/or
- Causes damage to the computer.

Examples of crimes in which the computer is the target include, but are not limited to, hacking, cracking, denial of service (DoS) attacks, distributed denial of service (DDoS) attacks, and malicious software dissemination (computer viruses, worms, Trojan horses, logic bombs, sniffers, and botnets). Each of these crimes is explored in this section.

Hacking and Cracking

A hacker can be defined as a computer user who seeks to gain unauthorized access to a computer system.[20] Consider hacker David C. Kernell. Kernell gained unauthorized access to Alaska Governor Sarah Palin's private e-mail account. Subsequently, Kernell released screen shots of Palin's e-mail account and information about the contents of her e-mails online.[21] The tools with which to hack a computer are readily available online for anyone to use; even persons who have only limited computer knowledge can learn how to access a computer without authorization. Typing the words "how to hack a computer" on YouTube.com will provide you with many hits of videos walking you through this process (not all work, however).

Cracking occurs when a perpetrator seeks to gain "unauthorized access to a computer system in order to commit another crime such as destroying information contained in that system."[22] Consider the case of LifeGift Organ Donation Center. The Center's former director of information technology, Danielle Duann, gained unauthorized access to its computer network and deleted software applications along with their backups and organ donation database records.[23] Crackers, both advanced and novices (also known as "script-kiddies"), seek to gain unauthorized "root"-level access to a computer system so that they have the same power as a legitimate administrator and can modify the system as they please.

According to Sinrod and Reilly, a cracker is basically a hacker with criminal intent: Cracking occurs when a person maliciously sabotages a computer (or computers) or seeks to disrupt a computer network (or networks).[24] A case in point is *United States v. Peterson*,[25] in which Peterson hacked into a credit reporting service, stole financial information from its database, and used that information to create fraudulent credit cards. During the Gulf War, Dutch hackers stole information about the movements of U.S. troops from U.S. Defense Department computers and tried to sell it to the Iraqis (who turned the offer down, thinking that it was a hoax).[26] On numerous occasions, individuals have hacked into telephone companies; one such notorious hacker was Jonathan Bosanac, also known as "The Gatsby." In 2004, an individual hacked into T-Mobile's customer information for at least seven months, accessing information (such as names and Social Security numbers) belonging to approximately 400 customers.[27]

Denial of Service and Distributed Denial of Service Attacks

With a **denial of service** (DoS) attack, the perpetrator seeks to prevent users of a particular service from effectively using that service. Typically, a network server is bombarded with more (phony) authentication requests than it can handle. The attack overwhelms the resources of the target computer or computers, causing them to deny server access to other computers making legitimate requests. Ultimately, the server shuts down.

An example of a DoS occurred in the Scott Dennis case.[28] Dennis launched three DoS attacks against a private mail list server of the U.S District Court for the Eastern District of New York (Judsys). The Eastern District's server was overwhelmed with e-mail messages. Dennis claimed that he engaged in such an attack to prove that the server was vulnerable to outside attacks.

Distributed denial of service (DDoS) attacks occur when a perpetrator seeks to gain control over multiple computers and then uses these computers to launch an attack against a specific target or targets.[29] Consider the case of the 15-year-old Canadian boy who was known as "Mafiaboy." Mafiaboy launched DDoS attacks against major websites, such as CNN.com, Yahoo.com, and eBay.com. The Federal Bureau of Investigation (FBI) was able to track Mafiaboy down by reviewing online chat room sessions where he had discussed with other individuals which websites he should target; among them were those sites that Mafiaboy actually attacked (e.g., CNN.com).

Another example of a DDoS attack involved Brian Thomas Mettenbrink, who targeted websites of the Church of Scientology in January 2008.[30] His attack was carried out when Mettenbrink downloaded a piece of software from the website of an underground group known as Anonymous, which had a history of launching DDoS attacks against the Church of Scientology's sites. When launched, this software sent out large amounts of illegitimate traffic to several Scientology websites. Due to the amount of traffic, the target site could not process all of the information. Accordingly, all the resources of these servers were sucked up and legitimate traffic was denied.

Malicious Software Dissemination

Malware (or malicious software) causes damage to a computer or invades a computer to steal information from it. Some forms of malware include computer viruses, worms, Trojan horses, spyware, and botnets.

Computer Virus

A **computer virus** is a software program that is designed to spread itself to other computers and to damage or disrupt a computer, such as interrupting communications by overwhelming a computer's resources. Sometimes a computer virus may be relatively benign and may simply be designed to annoy a computer user—for instance, by displaying a message at a certain time on a user's computer screen. Consider the Elk Clone virus, which spread in 1982 through floppy disks on computers running the Apple II operating system.[31] This virus did not damage the computer, but rather caused an infected computer to display the following poem:

> Elk Cloner: The program with a personality
> It will get on all your disks
> It will infiltrate your chips
> Yes it's Cloner!
> It will stick to you like glue
> It will modify RAM too
> Send in the Cloner!

Computer viruses are designed to attach themselves (or piggyback) on files or programs. They are passed on by a computer user's activity, such as opening up an e-mail attachment infected by the virus.

The first person ever prosecuted for writing a computer virus was David L. Smith, who created the Melissa virus. The Melissa virus spread through e-mails containing an infected Word document attachment. This e-mail was disguised as an important message from a friend or colleague whose computer had been infected by the virus. The Melissa virus was designed to be passed on when the e-mail attachment that was infected by the virus was downloaded and opened. Once opened, an infected e-mail was then sent to the first 50 e-mail addresses listed in the computer user's Microsoft Office address book. The Melissa virus disrupted computer networks all over the world, resulting in millions of dollars worth of damage.

Worms

Worms are similar to computer viruses. Unlike a computer virus, however, a worm does not need to piggyback on a file or program to replicate itself. Also, unlike a computer virus, a worm does not need user activity to make copies of itself and spread.

In 1988, Robert Tappan Morris, who was then a graduate student at Cornell University, created and unleashed a worm (the Morris worm) that spread and multiplied to the point that it eventually caused computers at numerous government and military sites, leading

Jurisdiction Issues

Given the universal reach of cybercrimes, nations will not be able to effectively deal with them unless uniform cybercrime laws are enacted worldwide. Several cybercrime cases have illustrated the need for harmonized criminal law.

Consider the case of hacker/cracker Gary McKinnon. McKinnon, a resident of the United Kingdom, hacked into U.S. Department of Defense and National Aeronautics and Space Administration (NASA) computers and caused more than $700,000 in damage by deleting critical data and often rendering the computers he attacked unusable for an extended period of time. The U.K. Crown Prosecution Service (CPS) could not effectively charge McKinney for his crime. According to CPS, McKinnon could only be charged under Section 2 of the Computer Misuse Act 1990 for unauthorized access with the intent to commit or facilitate the commission of further offenses, which carried a maximum penalty of five years and unlimited fines. By comparison, in the United States, McKinnon might have received a sentence of as much as 60 years in a maximum security prison.[32]

Needless to say, McKinnon's defense team fought vigorously to keep him in the United Kingdom to be tried under the Computer Misuse Act. McKinnon lost his appeal on July 31, 2009, and was to be extradited to the United States to be tried for his crimes pursuant to the U.S./U.K. Extradition Treaty of 2003.[33] Had such a treaty not existed, the result might have been far different for McKinnon.

The problems associated with the lack of cybercrime laws (or these laws' inefficiencies in dealing with the threat of cybercrime) in certain countries are clearly illustrated by the case of the "I LOVE YOU" computer virus. This computer virus was created by Onel de Guzman, a resident of the Philippines. The virus hid itself within an e-mail message titled "I LOVE YOU." When an individual opened the e-mail, the virus spread by sending itself to every name in the individual's address book if the computer ran Microsoft Outlook's computer program. For those users who opened the e-mail, the computer virus destroyed stored music and picture files on the user's computer, replaced each with a copy of the virus, and rerouted users seeking to access websites via Internet Explorer to predetermined sites containing a program that scanned the individual's computer for usernames and passwords. In May 2000, the I LOVE YOU virus spread across the United States, Europe, and Asia and caused billions of dollars' worth of damages. While other countries affected by the computer virus (such as the United States) had adequate computer crime laws to prosecute an individual for such an offense, the Philippines did not. Under that country's criminal law statutes at that time, Guzman had not violated the law. What Guzman did was a cybercrime in other countries, however, even if in his own country it was not considered a violation of criminal law.

Practical Exercise

Search online for an international cybercrime case where the lack of a cybercrime law or its inefficiency played a role in the cybercriminal's sentence. Report the findings of your search, making sure to include the jurisdiction issues that arose in the case in your analysis. If the criminal was not charged with an offense, indicate why this occurred.

distinguished educational institutions, and medical research facilities to crash.[34] Another example of a worm was Code Red, a network worm that surfaced in 2001 and took advantage of a flaw in Microsoft software. The only thing that was required for this worm to spread was a network connection. This worm also defaced Web pages with the text "Hacked by Chinese."

In 2003, another worm made the headlines of the news: the Blaster worm, which similarly took advantage of a flaw in Microsoft software. Eighteen-year-old Jeffrey Parson released a "B variant Blaster worm" from his home near Minneapolis, Minnesota. The "B variant Blaster worm" infected approximately 7000 individual Internet-connected computers, which it then used as drones to launch an attack.[35] The attack caused a significant amount of damage to Microsoft and users with Microsoft software.

Trojan Horses

A **Trojan horse** is a type of malware named after a "gift" the Greeks gave to the citizens of Troy during the Trojan War. The Greeks, who were unable to enter the city of Troy despite a long siege, devised a strategy to trick the Trojans into letting their soldiers into the city. The Greeks delivered what appeared to be a peace offering—a huge wooden horse—making the Trojans think that the Greeks had left. In reality, Greek soldiers were hiding in this horse. When the Trojans moved the horse into their city, the soldiers hiding in the horse crept out at night and opened the gates to the rest of the Greek army, which was waiting outside the city walls. Subsequently, the Greek army conquered Troy.

In similar fashion, in the world of computers, a Trojan horse tricks the computer user into thinking that it is legitimate software, when it actually contains hidden functions. When the computer user downloads and installs the program, these hidden functions are executed along with the software. These features can damage a computer by, for example, erasing files on a user's computer system or can steal personal information. During 2006 and 2007, Aleksey Volynskiy used Trojan horses to hack into victims' brokerage accounts at Charles Schwab and capture personal account information, make unauthorized sales of securities and launder more than $246,000 through the accounts (unbeknownst to the account holders).[36] Not all Trojan horses are harmful, however, they may simply annoy a user by turning the text on documents upside down). Unlike worms and viruses, Trojan horses do not self-replicate or reproduce by infecting other files or programs.

Trojan horses can also place logic bombs in a computer system. A **logic bomb** does not self-replicate. Instead, when triggered by an event such as a predetermined date and time or when a particular program is run a specific number of times, it executes a previously dormant program. For instance, a logic bomb that has been inserted into a user's computer may be activated to delete the hard drive of the computer when the computer is restarted. One of the first cases prosecuted where a logic bomb was used was in 1987, after Donald Burleson was fired from the insurance company where he worked for being confrontational. Burleson had planted a logic bomb on the computer system (two years before he was fired), which was designed to erase more than 100,000 vital commission records (used to prepare the monthly payroll) from the company's computers. He triggered the logic bomb the day he was fired.[37]

Spyware

Spyware consists of software that enables the remote monitoring of a computer user's activities or information on an individual's computer where this software has been installed. For instance, in 2008, Omar Khan gained unauthorized access to his school computers and uploaded spyware to remotely connect to a secure network of Tesoro High School in Orange County, California.[38]

Spyware can also secretly gather information on users (e.g., passwords, credit card details) without their knowledge and relay these data to interested third parties. An example of such software is a **keylogger**; once installed, it records every keystroke of the user and reports this information back to its source. In a recent case, a NASA employee opened an e-mail attachment (from an individual whom she had met at an online dating site) containing spyware, which then infected her government-issued computer. In another case, JuJu Jiang, a 24-year-old male, installed keylogging software on the computer stations at Kinkos stores throughout Manhattan, New York.[39] Through this software, he was able to obtain bank account information, usernames and passwords, and other important personal information. Jiang used this information to access bank accounts and open fraudulent bank accounts online. He was fully aware that the keylogging software could damage the computer stations and openly admitted to fraudulently using the information obtained with the software.

Sniffers are programs that are used to monitor and analyze networks. They also can be used to collect individuals' usernames, passwords, and other personal information. In 2000, Ikenna Iffih, a 28-year-old male, was charged with gaining unauthorized access to a NASA computer Web server and installing a sniffer program there. This program was designed to save logon names and passwords of users.[40] In another case, 25-year-old Stephen Watt, a resident of New York, created a sniffer program that was stored on servers in California, Illinois, New Jersey, Ukraine, Latvia, and the Netherlands. Watt used this software to monitor and gather millions of credit and debit card numbers.[41]

Botnets

Computer hijacking occurs when an individual takes control over a computer (or computers), with malicious code known as "bot code." Bot code allows for the remote control of computers that are infected by it without the knowledge of the computer user. A **botnet** is a network of zombie (bot-infected) computers; those who control botnets are known as **botherders**. In addition to engaging in whatever activities the botherder wants the botnet to perform, the botherder may harvest data from the bot-infected computers, which can include personal information, credit card information, e-mail account data, usernames, and passwords.

A case in point involved Christopher Maxwell. Maxwell, a 20-year-old male, who created bot code that searched through the Internet for computers with low security. Maxwell's bot code infected these computers, thereby creating a botnet. Maxwell (the botherder), along with two co-conspirators, installed adware (and obtained commission for doing so) into the computers that made up the botnet without the computer owners' knowledge of the installation.[42]

The first botherder to receive a lengthy federal prison sentence (57 months) was Jeanson James Ancheta from Downey, California. Ancheta created botnets to launch attacks and install adware using unsuspecting computer owners' machines.[43] Another case involved Jesse William McGraw (also known as "Ghost Exodus"), a 25-year-old contract security guard at a hospital in Texas. McGraw posted a video on YouTube.com of himself compromising the hospital computer system with what he titled "botnet infiltration."[44] Bot code was found on his laptop at home. McGraw claimed that the botnet would be used to launch a denial of service attack against the website of a rival hacker group.

Category 2: The Computer as Tool Used to Commit Crime

In this type of cybercrime, computers are tools with which to commit a crime. Crimes that fall within this category include copyright infringement, embezzlement, phishing, cyberharassment, cyberstalking cyberbullying, illegal online sales of prescription drugs and controlled substances, and Internet gambling.[45]

Copyright Infringement

Copyright law gives the creator of an original work a limited monopoly to reproduce or distribute his or her work. **Copyright infringement** (otherwise known as piracy) is the unlawful copying of movies, TV shows, music, software, literature, and videogames. It occurs when part or all of the creator's work is copied without authorization or when an individual has enabled others to make such copies. For example, Gilberto Sanchez violated federal copyright laws by "unlawfully distributing a pirated copy" of the film *X-Men Origins: Wolverine* online at www.Megaupload.com.[46] Numerous other cases have been prosecuted, including the following:[47]

- Owen Moody uploaded *Slumdog Millionaire* to an Internet site (piratebay.com) where other individuals could download the movie.

- Derek Hawthorne uploaded *Australia* and *The Curious Case of Benjamin Button* to websites where users could download the movies.

- Jake Yates uploaded a pre-screen release of *The Love Guru,* making the film available to Internet users before its showing in theaters.

Music piracy cases also abound. For example, the music piracy group known as Rabid Neurosis (RNS) was charged with copyright infringement related to music. From 1999 to 2007, the group illegally uploaded thousands of copyright-protected music files online, which were subsequently reproduced and distributed by online users hundreds of thousands of times.[48]

Unauthorized sales of copyrighted material can result in significant illicit gains. Dashiell Ponce de Leon, for example, made more than $1 million through illegal Internet sales of software and video games from websites he ran.[49]

Prior to the advent of computers, pirates would copy VHS cassette tapes and sell them on the streets. While this activity still happens, the Internet and computers have made it possible to illegally download movies and music online within minutes. Peer-to-peer (P2P) file sharing programs (e.g., Kazaa) and torrent downloading sites (e.g., Pirate Bay) allow people to quickly upload and download movies, music, books, and software without paying for them. In addition, programs such as "AnyDVD" allow individuals to remove copyright protection from movies so that they can copy the DVDs that they borrow or rent. Many online sites exist where individuals can go and view TV shows and movies (even those that are being played in cinemas and theaters worldwide), including the following sites:

- www.surfthechannel.com
- www.fastpasstv.com
- www.flickpeek.com
- www.ovguide.com
- www.tv-links.cc
- www.movie-forumz.org
- www.planetmoviez.com
- www.tvduck.com

Computers are not just used to download and upload music, movies, literature, and software online without permission. The CD/DVD/Blu-ray ROMs in the computers can be used to rip, copy, and burn music CDs, DVD/Blu-ray movies, and software without permission from the manufacturer or creator. In particular, computer programs such as Windows Media Player (**Figure 1–1**) provide users with the ability to copy the contents of music CDs to their computers. Users may subsequently burn the songs copied onto their computers on CDs and distribute them (either for profit or not for profit)—in so doing, they violate copyright laws.

Embezzlement

Embezzlement is a type of theft in which an employee of a company transfers the company's financial assets to his or her own account, often using a computer. In one case, accounts supervisor Diana Simon stole more than $3 million from her employer, Kaneka Texas (a chemical company), by wiring money from her employer's bank account to her own account at a Houston bank.[50] Another embezzlement case involved the vice president of finance of the Columbian Retirement Home. Thom Randle embezzled approximately $700,000 from this company by opening unauthorized accounts on behalf of the

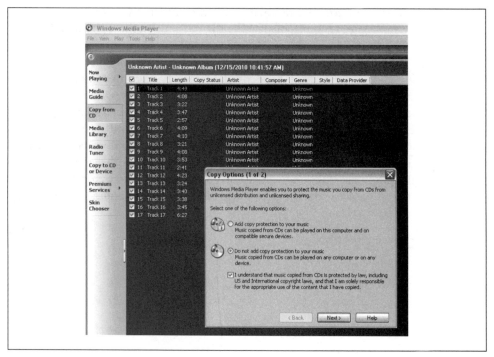

Figure 1–1 Windows Media Player

retirement home, transferring money into them, and subsequently transferring money out of these accounts and into his own personal account.[51] In yet another example of computer-enabled embezzlement, Jessica Quitugua Sabathia, an employee of North Bay Health Care Group, used her work computer to access her employer's accounting software without authorization and issue approximately 127 checks payable to herself and others.[52]

Phishing

Computers can be used as tools to commit fraud, including phishing. **Phishing** occurs when individuals deceive others by posing as legitimately established enterprises so as to steal communication users' valuable personal information, such as account data and credit card information.[53] Usually, the cybercriminals use a legitimate look-a-like e-mail from a bank that asks the victim to open a link to verify his or her information. After the user clicks on the link, he or she is directed to a website that looks identical to the actual bank's website. Once the victim types in the personal information and clicks the Submit button, the information goes straight to the criminal. The user is subsequently redirected to the actual bank's website and asked to verify the same information (to log on again).

In the case prosecuted as *United States v. Carr*,[54] Helen Carr sent fakes e-mails to America Online (AOL) users advising them that they needed to update their personal and credit card information in their accounts. One of the phishing e-mails Carr sent went to an FBI agent who happened to be a cybercrime expert. After further investigation, George Patterson and Helen Carr were found to be the ringleaders of this phishing scheme.

In 2004, Zachary Keith Hill was sentenced to four years in prison for his elaborate phishing scheme, in which he stole personal information, masqueraded as other entities (AOL and PayPal), and then used the information garnered in this scheme to make his own purchases.[55] His method of phishing was one that is commonly employed—namely, sending an e-mail to the customers of the targeted companies, AOL and PayPal, and claiming that their financial and personal information had to be updated. The e-mail further stated that if the consumers did not "update" their information, then their accounts would be reviewed and suspended. Hill provided a link to what appeared to be AOL/PayPal, which requested Social Security numbers, bank account details, names, and home addresses. Needless to say, this scheme tricked many people into giving the data to Hill because they believed that this information was ultimately being provided to either AOL or PayPal. At the time Hill was sentenced, he had already amassed 473 credit card numbers, 152 sets of bank account numbers and routing numbers, and 566 sets of usernames and passwords for Internet services accounts.

Some phishing schemes feed off of individuals' fears and vulnerabilities. In 2009, the Centers for Disease Control and Prevention (CDC) provided information about a new phishing scheme aimed at taking advantage of citizens who were frightened by the H1N1 (swine flu) virus.[56] The phishing scheme involved an e-mail, supposedly sent by the CDC, claiming that a state vaccination program was being implemented. According to this e-mail, users had to create a personal H1N1 vaccination profile on the cdc.gov website. A link was provided that would supposedly bring targeted individuals to the vaccination profile page to provide their personal information, including their names, home addresses, and Social Security numbers.

Cyberharassment

Cyberharassment is a crime that occurs when someone uses the Internet, e-mail, or other forms of communication to intentionally annoy, attack, alarm or otherwise bother another person.[57] It can include obscene and derogatory comments that focus on, for instance, gender, race, sexual orientation, nationality, or religion. In one case, two individuals from Andover, Massachusetts, were arrested on charges of harassing their neighbors. Their harassing behavior included the use of Craigslist to place fake advertisements, e-mails, phone calls, and visits from a social worker concerning false claims that the couple's 14-year-old son was abusing a female at school.[58] In 2007, Sam Sloan filed a lawsuit against two members of the United States Chess Federation board—Susan Polgar and her husband, Paul Truong—claiming Internet harassment. In particular, Sloan stated that Polgar and Truong had posted thousands of obscene and defamatory remarks on public Internet bulletin boards to win the election to the United States Chess Federation board.[59] Sloan also ran for reelection to the board at that time, albeit unsuccessfully.

Cyberstalking

Cyberstalking occurs when an individual repeatedly harasses or threatens another person through the use of the Internet, e-mail, or other electronic communications devices.[60] When engaging in cyberstalking, an individual may harass or threaten someone to the point where the target (the person being stalked) fears for his or her well-being. Shannon Marie Mitchell, a 17-year-old senior at Vanguard High School in Marion County, Florida, was arrested and charged with aggravated cyberstalking.[61] Mitchell was accused of posting photographs, contact information, and sexually explicit information about her boyfriend's ex-girlfriend (unnamed, 15 years old) on a pornographic website. Her actions resulted in the 15-year-old girl receiving phone calls and text messages from individuals seeking sexual encounters.

Sometimes victims of this crime do not know they are being stalked. Consider the case of Gary Dellapenta. After meeting a 28-year-old woman in church, Dellapenta placed ads in her name on AOL and other sites describing the victim's supposed fantasies of being gang-raped. Certain individuals started responding to these ads. Dellapenta, posing as the woman, corresponded with these individuals through e-mail and in chat rooms and provided them with details of the woman's physical appearance, her home address, her phone number and a means of bypassing her home security unit. On at least six different occasions, men knocked on the woman's front door saying they wanted to rape her.

In some cases, cyberstalking evolves into physical stalking. The situation can thus take a turn for the worse and lead to physical harm or even death of the victim.[62]

Cyberbullying

Cyberbullying is a crime that involves the use of telecommunication and electronic communication technologies to harass, insult, or humiliate a child, pre-teen, or teen.[63] Consider the following case: A New York teenager, Denise Finkel, who is enrolled in State University of New York in Albany, is suing four of her former Oceanside High School classmates (Michael Dauber, Jeffrey Schwartz, Leah Herz, and Melinda Danowitz) and their parents for cyberbullying. Finkel is also suing the social networking site, Facebook, where the cyberbullying took place. It is alleged that the four high school students named in the lawsuit created a password-protected chat room on Facebook dedicated to ostracizing, ridiculing, and humiliating Finkel.

Illegal Online Sales of Prescription Drugs and Controlled Substances

Traditional, licensed pharmacies require a prescription from a doctor before dispensing prescription-only drugs. Some online pharmacies

> Use state-licensed pharmacists and require consumers to obtain valid prescriptions from licensed physicians before ordering drugs online. They verify that a

licensed physician actually has issued the prescription to the patient before they dispense any drugs. These sites . . . are covered by a comprehensive regulatory scheme . . . , established under existing laws, [which] has created a safety net to protect the American public from injuries resulting from unsafe drugs, counterfeit drugs, and improper prescribing and dispensing practices.[64]

The same cannot be said about other pharmacies in the online environment—which leads to a discussion of **illegal prescription drug sales**. Some unscrupulous pharmacies allow individuals to purchase drugs that would normally require prescription without one. Individuals who have been denied certain drugs by doctors either for health reasons or because of suspected misuse may be able to obtain these drugs via such sites. Other online pharmacies will "diagnose" a patient based on an online form and questionnaire the individual fills out and prescribe drugs based on the "diagnosis" given. Some doctors have claimed that such operations spend only a few minutes (sometimes even seconds) reviewing these orders.[65] Additionally, state medical boards have questioned the ability of such questionnaires to meet the requisite standard of care required in practicing medicine.[66] Despite the lack of reliability of such questionnaires, many online sites use them to prescribe the prescription drugs that are sold by their pharmacies.

An example of such a site was Affpower, an online pharmaceutical distribution network, which sold controlled and non-controlled prescription drugs online.[67] These drugs were provided to individuals who did not have a prescription for them. Doctors were employed by Affpower to supply the prescriptions for these drugs. However, these prescriptions were provided without a physical examination of the customer and without any contact with the customer. In fact, one doctor working for Affpower, Dr. Subramanya K. Prasad,

> Admitted that he and other Affpower doctors: had no contact with Affpower customers and lacked any physician–patient relationship with them; were not issuing prescriptions in the usual course of a professional practice; and were not issuing the prescriptions for a legitimate medical purpose, but simply to make money.[68]

Several studies have been conducted on the significant health risks associated with providing online diagnoses and prescriptions of medications. Medical examinations should always be performed before any diagnosis is made, but especially when the end result is a prescription for one of those drugs that are highly sought after online. One drug that is often bought online to avoid the personal embarrassment associated with requesting it from physicians in person is Viagra, a medication used to treat erectile dysfunction. Viagra has been known to cause heart attacks. Accordingly, before prescribing Viagra, a cardiac evaluation of the patient should be conducted—men are at high risk of cardiac arrest if they use this product without medical oversight.[69]

Many of the online pharmacies that do not require prescriptions before dispensing medications offer controlled substances such as Valium, Oxycontin, and Percocet.[70]

Controlled substances are drugs whose manufacture, distribution, possession, and use are tightly regulated by the government. Controlled substances are divided into schedules under the Comprehensive Drug Abuse and Control Act of 1970; these schedules classify drugs according to their medical use (if any), potential for abuse, and level of harmfulness.[71] Types of controlled substances sold online include narcotics, stimulants, depressants, and anabolic steroids.

In 2007, the FBI conducted an undercover operation focusing on the illegal sale of anabolic steroids (a type of performance enhancement drug) and other controlled substances online, known as Operation Phony Pharm.[72] Several individuals were charged with selling anabolic steroids online that they had manufactured from raw materials purchased from China. These individuals used MySpace profiles and websites, such as www.anabolic-superstore.com and www.sell.com, to distribute these controlled substances. Some online sites even offer Rohypnol (flunitrazepam), a so-called date rape drug.[73] The manufacture, distribution, possession, and use of this drug was completely banned in the United States in 2000 with the Samantha Reid Date-Rape Drug Prohibition Act.

Another danger associated with online sales is the inability of a customer to know where the drugs were manufactured and under what conditions. Customers are also unable to establish the legitimacy of the company selling prescription drugs. Moreover, several instances of the sale of counterfeit drugs have been reported.[74] One such example involved High-Tech Pharmaceuticals. High-Tech Pharmaceuticals imported and distributed a variety of counterfeit prescription drugs online:[75]

- Xanax
- Valium
- Zoloft
- Viagra
- Ambien
- Cialis

Customers could purchase these drugs without a prescription. It was also discovered that these counterfeit drugs were being manufactured in unsanitary conditions in Belize.

It is important to note that prescription drugs and controlled substances are not the only items illegally sold online. Firearms are also sold illegally online to bypass existing rules and regulations for purchasing them. Furthermore, there have been numerous instances where women, children, and human organs have been sold through this medium.

Internet Gambling

Numerous Internet real-money gambling sites exist online. An individual from anywhere in the world can call a number or access a website to place a wager on a sporting event, for example. All that is required is an account number, a password (to be given when the number is dialed or the website is accessed), and the terms of the bet.[76]

The transmission of bets or wagers via the Internet is federally prohibited in the United States.[77] As such, Internet gambling sites have taken steps to conceal their true identities from U.S. authorities. Bank transactions involving **illegal gambling** proceeds have also been concealed. Specifically, certain illegal gambling sites used the Automated Clearing House (ACH) system to receive money owed by customers. The ACH system "allows money to be electronically transferred from a gambler's U.S. checking account to an Internet gambling company simply by the gambler going to the Internet gambling company's website and entering his bank account information."[78] To disguise this unlawful activity, ACH transactions are made to shell companies that the Internet gambling operators have created with names unrelated to gambling.

Many Internet gambling sites allow gamblers to use credit cards to make payments. In an attempt to limit Internet gambling, U.S. credit card companies have blocked the use of their cards on such sites. Nevertheless, these sites have found ways to bypass this restriction by disguising the payments made or retrieved from such cards. For example, one international Internet gambling company, PartyGaming, took steps to disguise payments of the winnings of U.S. customers to circumvent such limits.[79]

The use of such sites by U.S. residents and the offering of the services of these sites to U.S. residents is strictly prohibited by law.[80] The detection of such sites is extremely difficult, however, as they take significant steps to conceal their identities from law enforcement authorities. For instance, many of these sites have been set up with their servers located offshore. Consequently, U.S. law enforcement agencies have had significant difficulty in obtaining records of gambling activities from these offshore locations.

Combating Cybercrime

The proliferation of cybercrime has created a need for new laws and enforcement processes to attack this problem. Given that technology and crimes committed via this technology are evolving very rapidly, effective enforcement of cybercrime is becoming particularly challenging. To effectively combat cybercrime, one must innovate faster than cybercriminals "and must be smarter in thinking ahead and making counter-moves—especially those that take offenders time, effort and expense to combat."[81] Cybercriminals constantly seek new opportunities to exploit new victims. As such, authorities need to develop the capacity to anticipate the features of computer, information, and communication technology that might open up new opportunities to cybercriminals to engage in illicit activity.

Chapter Summary

This chapter focused on cybercrime and the ways it differs from crimes committed before the advent of the Internet and computer technologies. Cybercrimes can be committed on a far broader scale than their traditional, real-world counterparts. Using the Internet and the right technology, a criminal can commit theft, extort victims, plan a robbery, vandalize property, solicit prostitutes, and bully, harass, and stalk a target in the privacy of his or her own home. Additionally, a criminal can produce exact digital copies of movies, music, and software and distribute them worldwide within a matter of minutes (maybe even seconds). Online pharmacies have been known to illegally sell prescription drugs and controlled substances. Instances of illegal online gambling have also increased exponentially. The reach of the threat has no geographic boundaries and, as a result, the targeting of cybercriminals poses unique challenges for authorities worldwide.

Key Terms

Botherder	Embezzlement
Botnet	Hacking
Computer virus	Illegal gambling
Controlled substance	Illegal prescription drug sales
Copyright infringement	Keylogger
Cracking	Logic bomb
Cyberbullying	Malware
Cybercrime	Organized crime
Cyberextortion	Phishing
Cyberharassment	Sniffer
Cyberprostitution	Spyware
Cyberstalking	Terrorism
Cyberterrorism	Trojan horse
Denial of service	Vandalism
Distributed denial of service	Worm

Practical Exercise

Can you think of a crime that falls into one of the categories included in this chapter? Look at the section on "Computer Crime and Intellectual Property" on the U.S. Department of Justice's website. Find a case that is not mentioned in this chapter, which relates to one of the categories. In your answer, explain why you think your chosen crime belongs in that category.

Critical Thinking Question

Are there any crimes left that have not been touched at all by information, communication, and computer technologies?

Review Questions

1. What is cybercrime?

2. How does computer crime differ from traditional crime?

3. Identify and describe two or three traditional crimes in which computers are now used as an instrument. How has the nature of these crimes changed as a result of the technology?

4. What are the two main categories of cybercrime? Provide a few examples of each.

5. How can vandalism occur online?

6. What is malware? Provide a few examples of it.

7. What is a botnet? How does it work?

8. What is embezzlement?

9. How does copyright infringement occur?

10. What are the dangers associated with online sales of prescription drugs?

11. Which problems does cybercrime pose to authorities seeking to investigate it?

Footnotes

[1]For further information, see United States Attorney's Office, District of Massachusetts. (2009, December 22). Press release: Former inmate sentenced for hacking prison computer. US Department of Justice, Computer Crime & Intellectual Property Section. Retrieved from http://www.justice.gov/criminal/cybercrime/janoskoSent.pdf

[2]For the purpose of this analysis, traditional crimes are considered those that were committed prior to the advent of computer technologies. Traditional crimes include, but are not limited to, fraud, theft, distribution of child pornography, harassment, stalking, embezzlement, blackmail, and espionage.

[3]United States Attorney's Office, Western District of New York. (2009, November 24). Man admits selling $345,000 of counterfeit stamps. US Department of Justice, Computer Crime & Intellectual Property Section. Retrieved from http://www.justice.gov/criminal/cybercrime/ferrisPlea.pdf

[4]Massarella, L., Venezia, T., & Dahl, J. (2009, October 25). Teen gang busted in celeb burglaries. *New York Post* [online]. Retrieved from http://www.nypost.com/p/news/national/ teen_gang_busted_in_celeb_burglaries_VSy1DgvGuvoxqR2nmdVDOK

[5]National Academy of Sciences. (1990). *Computers at risk: Safe computing in the information age.* Washington, DC: National Academies Press, p. 7.

[6]Cyberterrorism is explored in further detail in Chapter 5.

[7]Wilson, C. (2003, October 17). Computer attack and cyber terrorism: Vulnerabilities and policy issues for Congress. *US Congressional Research Report* RL32114, p. 4. Retrieved from http://www.fas.org/irp/crs/RL32114.pdf

[8]Wilson, C. (2008, January 29). Botnets, cybercrime, and cyberterrorism: Vulnerabilities and policy issues for Congress. *US Congressional Research Report* RL32114, p. 20. Retrieved from http://www.fas.org/sgp/crs/terror/RL32114.pdf

[9]United States Attorney's Office, Southern District of New York. (2010, March 8). Manhattan U.S. Attorney charges California man with cyber-extortion of New York–based insurance company. US Department of Justice, Computer Crime & Intellectual Property Section. Retrieved from http://www.justice.gov/criminal/cybercrime/digatiChar.pdf

[10]Denning, D. (2001). Cyberwarriors: Activists and terrorists turn to cyberspace. *Harvard International Review, 23*(2), 70–75.

[11]For further information, see Constantin, L. (2010, January 28). Mass defacement hits Congressional Websites. *Softpedia* [online]. Retrieved from http://news.softpedia.com/news/Mass-Defacement-Hits-Congressional-Websites-133436.shtml

[12]Croteau, S. J. (2009, April 19). 50 arrested in prostitution ring. *Worcester Telegram and Gazette* [online]. Retrieved from http://www.telegram.com/article/20090419/FRONTPAGENEWS/904190287

[13]172 F.3d 1268.

[14]See, more generally, Lia, B. (2005). *Globalisation and the future of terrorism: Patterns and predictions.* New York: Routledge; Emerson, S. (2006). *Jihad Incorporated: A guide to militant Islam.* New York: Prometheus Books, p. 472.

[15]US Department of Justice. (1997, April). Report on the availability of bombmaking information. Submitted to the US House of Representatives and the US Senate. Retrieved from http://www.cybercrime.gov/bombmakinginfo.html

[16]See http://al3dad.jeeran.com; Lia, B. (2005). *Globalisation and the future of terrorism: Patterns and predictions.* New York; Routledge, p. 176.

[17]Coll, S., & Glasser, S. B. (2005, August 7). Terrorists turn to the Web as base of operations. *Washington Post* [online], p. A01. Retrieved from http://www.washingtonpost.com/wp-dyn/content/article/2005/08/05/AR2005080501138_pf.html

[18]Emerson, S. (2006). *Jihad Incorporated: A guide to militant Islam.* New York: Prometheus Books, p. 471.

[19]Scheuer, M. (2005, August 5). Assessing London and Sharm al-Sheikh: The role of Internet intelligence and urban warfare training. Retrieved from http://jamestown.org/terrorism/news/article.php?articleid=2369764

[20]Sinrod, E. J., & Reilly, W. P. (2000). Cyber-crimes: A practical approach to the application of federal computer laws. *Santa Clara Computer and High-Technology Law Journal, 16*(2), p. 181; Brenner, S. W. (2006). Defining cybercrime: A review of state and federal law. In R. D. Clifford (Ed.), *Cybercrime: The investigation, prosecution and defense of a computer-related crime* (2nd ed.). North Carolina: Carolina Academic Press, p. 15.

[21]For further information on this case, see *United States v. Kernell*, E.D. Tenn., October 7, 2008, No. 3:08-CR-142.

[22]Brenner, S. W. (2006). Defining cybercrime: A review of state and federal law. In R. D. Clifford (Ed.), *Cybercrime: The investigation, prosecution and defense of a computer-related crime* (2nd ed.). North Carolina: Carolina Academic Press, pp. 15–16.

[23]For further information, see United States Attorney's Office, Southern District of Texas. (2009, May 1). Computer administrator pleads guilty to hacking former employer's computer system. US Department of Justice, Computer Crime & Intellectual Property Section. Retrieved from http://www.justice.gov/criminal/cybercrime/duannPlea.pdf

[24]Sinrod, E. J., & Reilly, W. P. (2000). Cyber-crimes: A practical approach to the application of federal computer laws. *Santa Clara Computer and High-Technology Law Journal, 16*(2), p. 182.

[25]*United States v. Peterson,* 98 F.3d 502, 504 (9th Cir. 1996).

[26]Christensen, J. (1999, April 6). Bracing for guerrilla warfare in cyberspace. *CNN Interactive.* Retrieved from http://www.cnn.com/TECH/specials/hackers/cyberterror/Interactive

[27]Schneier, B. (2005, February 14). T-Mobile hac. *Schneier on Security Blog.* Retrieved from http://www.schneier.com/blog/archives/2005/02/tmobile_hack_1.html

[28]United States Attorney's Office, District of Alaska. (2000, April 19). Alaska man indicted for alleged attack on United States court computer system. US Department of Justice, Computer Crime & Intellectual Property Section. Retrieved from http://www.justice.gov/criminal/cybercrime/dennis.htm

[29]Sinrod, E. J., & Reilly, W. P. (2000). Cyber-crimes: A practical approach to the application of federal computer laws. *Santa Clara Computer and High-Technology Law Journal, 16*(2), p. 189.

[30]United States Attorney's Office, Central District of California. (2010, January 25). Nebraska man agrees to plead guilty in attack of Scientology websites orchestrated by 'Anonymous.' US Department of Justice, Computer Crime & Intellectual Property Section. Retrieved from http://www.justice.gov/criminal/cybercrime/mettenbrinkPlea.pdf

[31]Jesdanun, A. (2007, September 2). High school prank began era of computer virus. *USA Today* [online]. Retrieved from http://www.usatoday.com/tech/news/computersecurity/wormsviruses/2007-09-02-computer-virus-anniversary_N.htm

[32]Culf, A. (2009, July 31). Hacker Gary McKinnon loses appeal against extradition to U.S. *The Guardian* [online]. Retrieved from http://www.guardian.co.uk/world/2009/jul/31/gary-mckinnon-loses-extradition-appeal

[33]The U.S./U.K. Extradition Treaty does not require the United States to present evidence to secure the extradition of a U.K. citizen, thus no evidence was required to be shown to the U.K. authorities to have McKinnon extradited to the United States. The U.K. Home Secretary is reviewing the Extradition Act of 2003 and the U.S./U.K. Extradition Treaty of 2003 to determine whether it is being applied even-handedly. The U.K. government has yet to declare whether McKinnon's extradition will be postponed until after the review is completed (i.e., summer 2011).

[34]See *United States v. Morris,* 928 F.2d 504 (1991).

[35]United States Attorney's Office, Western District of Washington. (2003, August 29). Minneapolis, Minnesota 18 year old arrested for developing and releasing B variant of Blaster computer worm. US Department of Justice, Computer Crime & Intellectual Property Section. Retrieved from http://www.justice.gov/criminal/cybercrime/parsonArrest.htm

[36]United States Attorney's Office, Southern District of New York. (2010, April 7). Computer hacker sentenced to 37 months in prison in Manhattan federal court for scheme to steal and launder money from brokerage accounts. US Department of Justice, Computer Crime & Intellectual Property Section. Retrieved from http://www.justice.gov/criminal/cybercrime/bhararaSent.pdf

[37]For further information, see Icove, D., Seger, K., & VonStorch, W. (2002). Fighting computer crime. *Computer Crime Research Center.* Retrieved from http://www.crime-research.org/library/crime1.htm

[38]Ayres, C. (2008, June 19). Schoolboy hacker Omar Khan who upped his grades faces 38 years in jail. *The Times* [online]. Retrieved from http://www.timesonline.co.uk/tol/news/world/us_and_americas/article4168112.ece

[39]United States Attorney's Office, Southern District of New York. (2005, February 28). Queens man sentenced to 27 months imprisonment on federal charges of computer damage, access device fraud and software. US Department of Justice, Computer Crime & Intellectual Property Section. Retrieved from http://www.justice.gov/criminal/cybercrime/jiangSent.htm

[40]US Department of Justice. (2000, February 23). Boston computer hacker charged with illegal access and use of state's government and private systems. Retrieved from http://www.cybercrime.gov/iffih.htm

[41]Zetter, K. (2009, August 28). Accused TJX hacker agrees to guilty plea—faces 15 to 25 years. *Wired* [online]. Retrieved from http://www.wired.com/threatlevel/tag/carding/'/page/3/

[42]United States Attorney's Office, Western District of Washington. (2006, May 4). California man pleads guilty in "botnet" attack that impacted Seattle hospital and Defense Department. US Department of Justice, Computer Crime & Intellectual Property Section. Retrieved from http://www.justice.gov/criminal/cybercrime/maxwellPlea.htm

[43]United States Attorney's Office, Central District of California. (2006, May 8). "Botherder" dealt record prison sentence for selling and spreading malicious computer code. US Department of Justice, Computer Crime & Intellectual Property Section. Retrieved from http://www.justice .gov/criminal/cybercrime/anchetaSent.htm

[44]United States Attorney's Office, Northern District of Texas. (2010, January 25). Arlington security guard, who hacked into hospital's computer system, pleads guilty to federal charges. US Department of Justice, Computer Crime & Intellectual Property Section. Retrieved from http://www.justice.gov/criminal/cybercrime/mcgraw.pdf

[45]This list is by no means exhaustive. Throughout this book, numerous cybercrimes, both included and not included on this list, will be explored in further detail.

[46]United States Attorney's Office, Central District of California. (2009, December 16). New York man arrested on federal charges of illegally distributing copy of "Wolverine" movie. Federal Bureau of Investigation, Los Angeles. Retrieved from http://losangeles.fbi.gov/dojpressrel/ pressrel09/la121609.htm

[47]United States Attorney's Office, Central District of California. (2009, February 20). Three charged in movie piracy cases involving illegal posting of theatrical films on Internet: Screeners of "Slumdog Millionaire" and "Benjamin Button" were made available. Federal Bureau of Investigation, Los Angeles. Retrieved from http://losangeles.fbi.gov/dojpressrel/pressrel09/ la022009a.htm

[48]US Department of Justice. (2009, September 9). Four members of alleged Internet music piracy group charged with copyright infringement conspiracy. Office of Public Affairs. Retrieved from http://www.justice.gov/opa/pr/2009/September/09-crm-940.html

[49]United States Attorney's Office, District of Columbia. (2005, October 6). Texas man sentenced to 46 months in jail and ordered to pay $1.54 million in restitution for illegally selling copyright protected software and video games over the Internet. US Department of Justice, Computer Crime & Intellectual Property Section. Retrieved from http://www.justice.gov/criminal/ cybercrime/poncedeleonSent.htm

[50]United States Attorney's Office, Southern District of Texas. (2010, February 22). Former accounts supervisor of area chemical company sentenced to prison for embezzling millions of dollars. Federal Bureau of Investigation, Houston. Retrieved from http://houston.fbi.gov/dojpressrel/ pressrel10/ho022210.htm

[51]United States Attorney's Office, Southern District of Texas. (2010, May 14). Chico man pleads guilty to embezzling $693,000 from charity. US Department of Justice, Computer Crime & Intellectual Property Section. Retrieved from http://www.justice.gov/criminal/cybercrime/ randlePlea.pdf

[52]United States Attorney's Office, Eastern District of California. (2004, July 28). Vallejo woman admits to embezzling more than $875,035. US Department of Justice, Computer Crime & Intellectual Property Section. Retrieved from http://www.justice.gov/criminal/cybercrime/ sabathiaPlea.htm

[53]For more information on these threats, see Schneier, B. (2004). *Secret and lies: Digital security in a networked world.* Indianapolis, IN: Wiley.

[54]Ed VA (2003).

[55]US Department of Justice. (2004, May 18). Fraudster sentenced to nearly four years in prison in Internet ePhishing case. Retrieved from http://www.cybercrime.gov/hillSent.htm

[56]Centers for Disease Control and Prevention (CDC). (2009, December 1). Fraudulent emails referencing CDC-sponsored state vaccination program. Retrieved from http://www.cdc.gov/hoaxes_rumors.html

[57]Chapter 5 explores cyberharassment in further detail.

[58]Altman & Altman. (2008, October 20). Andover, Massachusetts, couple arrested for Internet crimes of harassment, identity fraud and threats. Retrieved from http://www.bostoncriminallawyerblog.com/2008/10/andover_massachusetts_couple_a.html.

[59]McClain, D. L. (2007, October 8). Chess group officials accused of using Internet to hurt rivals. *New York Times* [online]. Retrieved from http://www.nytimes.com/2007/10/08/nyregion/08chess.html

[60]US Department of Justice. (1999, August). Cyberstalking: A new challenge for law enforcement and industry. *Report from the Attorney General to the Vice President.* Retrieved from http://www.justice.gov/criminal/cybercrime/cyberstalking.htm

[61]Dean, E. (2010). 17 year-old arrested for aggravated cyberstalking. Media Release, Marion County Sheriff's Office. Retrieved from http://www.marionso.com/mediarelease.php?id=10029

[62]A cyberstalking case that resulted in the death of the victim (Amy Boyer) is explored in Chapter 5.

[63]Chapter 6 explores cyberbullying in further detail.

[64]Working Group on Unlawful Conduct on the Internet. (1999, August 5). Appendix D: Internet sale of prescription drugs and controlled substances. Retrieved from http://www.justice.gov/criminal/cybercrime/append.htm

[65]US Department of Justice. (2007, September 19). Doctor and pharmacy owner plead guilty in Internet pharmacy conspiracy. Retrieved from http://www.justice.gov/criminal/cybercrime/covinoPlea.htm

[66]DeKeiffer, D. E. (2006). The Internet and the globalization of counterfeit drugs. *Journal of Pharmacy Practice, 19,* 174.

[67]US Department of Justice. (2007, August 2). Eighteen charged with racketeering in Internet drug distribution network. Retrieved from http://www.justice.gov/criminal/cybercrime/affpowerIndict.htm

[68]US Department of Justice. (2007, September 19). Doctor and pharmacy owner plead guilty in Internet pharmacy conspiracy. Retrieved from http://www.justice.gov/criminal/cybercrime/covinoPlea.htm

[69]Tracqui, A., Miras, A., Tabib, A., Raul, J. S., & Ludes B. (2002). Fatal overdosage with sildenafil citrate (Viagra1): First report and review of the literature. *Human & Experimental Toxicology, 21,* 626.

[70]DeKeiffer, D. E. (2006). The Internet and the globalization of counterfeit drugs. *Journal of Pharmacy Practice, 19,* 174.

[71]For instance, Schedule 1 drugs (e.g., heroin) have no medicinal value, are highly addictive, and are extremely harmful to society.

[72]US Department of Justice. (2007, September 24). Operation Phony Pham: Six charged as a result of investigation targeting Internet sale of steroids, human growth hormone. Retrieved from http://www.justice.gov/criminal/cybercrime/porterIndict.htm

[73]DeKeiffer, D. E. (2006). The Internet and the globalization of counterfeit drugs. *Journal of Pharmacy Practice, 19,*174–175.

[74]DeKeiffer, D. E. (2006). Potential liability for counterfeit medications. *Journal of Pharmacy Practice, 19,* 215.

[75]United States Attorney's Office, Northern District of Georgia. (2008, August 15). "Hi-Tech Pharmaceuticals" plead guilty to importing and distributing "knock-off" prescription drugs. US Department of Justice, Computer Crime & Intellectual Property Section. Retrieved from http://www.justice.gov/criminal/cybercrime/wheatPlea.pdf

[76]United States Attorney's Office, District of Arizona. (2010, July 29). Nine indicted in multi-million dollar Internet gambling ring. US Department of Justice, Computer Crime & Intellectual Property Section. Retrieved from http://www.justice.gov/usao/az/press_releases/2010/2010-158(Meisel).pdf

[77]Working Group on Unlawful Conduct on the Internet. (1999, August 5). Appendix F: Internet gambling. Retrieved from http://www.justice.gov/criminal/cybercrime/append.htm

[78]United States Attorney's Office, Southern District of New York. (2010, April 16). Australian man charged in Manhattan federal court with laundering half-billion dollars in Internet gambling proceeds. Federal Bureau of Investigation, New York. Retrieved from http://newyork.fbi.gov/dojpressrel/pressrel10/nyfo041610a.htm

[79]United States Attorney's Office, Southern District of New York. (2009, April 7). Internet gambling company PartyGaming PLC enters non-prosecution agreement with U.S. and will forfeit $105 million. Federal Bureau of Investigation, New York. Retrieved from http://www.justice.gov/usao/nys/pressreleases/April09/partygamingnonprosecutionagreementpr.pdf

[80]United States Attorney's Office, Southern District of New York. (2009, April 7). Internet gambling company PartyGaming PLC enters non-prosecution agreement with U.S. and will forfeit $105 million. Federal Bureau of Investigation, New York. Retrieved from http://www.justice.gov/usao/nys/pressreleases/April09/partygamingnonprosecutionagreementpr.pdf

[81]Ekblom, P. (2005). How to police the future: Scanning for scientific and technological innovations which generate potential threats and opportunities in crime, policing, and crime reduction. In M. J. Smith & N. Tilley (Eds.), *Crime science: New approaches to preventing and detecting crime*. Devon, UK: Willan, p. 32.

Chapter 2

An Introduction to Computer Forensics Investigations and Electronic Evidence

This chapter provides an introduction to computer forensics. It also explores computer forensics investigations, in particular those involving violations of criminal law, civil law, and administrative regulations. It then analyzes the specific processes involved in conducting computer forensics investigations. Additionally, this chapter considers the different types of evidence. Finally, it examines electronic evidence and the methods used to authenticate it.

Computer Forensics: What Is It?

Computer forensics is a branch of forensic science that focuses on criminal procedure law and evidence as applied to computers and related devices. This branch of forensics is not limited to computers, but also includes mobile phone forensics, personal digital assistant (PDA) forensics, and network forensics (which are explored in Chapters 11 and 12 of this book). Computer forensics concerns the process of obtaining, processing, analyzing, and storing digital information for use as evidence in criminal, civil, and administrative cases.[1] This type of information can be obtained from computers and other electronic devices such as printers, scanners, copiers, CDs, DVDs, Blu-ray disks, external hard drives, universal serial bus (USB) flash drives, magnetic tape data storage devices (e.g., linear tape-open [LTO] devices), cameras, mobile phones, fixed telephony, faxes, PDAs, portable media players (e.g., Apple's iPod), and gaming consoles (e.g., Microsoft's Xbox). Data can be retrieved from existing files (even those that have been deleted, encrypted, or damaged) or by monitoring user activity in real time. The information acquired from computers and other electronic devices can be used as evidence of a wide range of traditional crimes, cybercrimes, and computer misuses. It can also assist in the arrest and prosecution of criminals, the prevention of future illicit activity, the investigation of employee misconduct, and the termination of employment. The use of the acquired information depends on the type of investigation being conducted.

Computer Forensics Investigations: The Basics

This section focuses on the role of computer forensics in legal and administrative proceedings and the specific steps required in performing computer forensics investigations. Computer forensics specialists may be involved in both public investigations (conducted by law enforcement and government agencies) and private investigations (conducted by nongovernmental agencies, businesses, and individuals). Today, investigations for criminal, civil, and administrative cases are increasingly involving evidence from computer technologies and related devices.

Investigations Involving Criminal Law

A criminal case is initiated by the state, and the burden of proof is on the state to prove that the defendant (the individual charged with a crime) is guilty beyond a reasonable doubt (the highest standard of proof). **Criminal law** deals with public offenses—that is, actions that are harmful to society as a whole. These actions are prohibited by law. Those who violate criminal law are assigned a punishment, which may consist of a fine paid to the court, community service, probation, a term of incarceration in a penal institution (such as a jail or prison), or capital punishment (i.e., the death penalty).

Government and law enforcement personnel use computer forensics to investigate and prosecute crimes. They do not have the authority to enforce private organizations' policies and procedures. Criminals use computer technology to commit both traditional (e.g., theft, fraud, and extortion) and new crimes (e.g., computer virus distribution and denial of service attacks). As Chapter 1 described, computers can be the target of a crime, can be a tool used to commit the crime, or can be incidental to a crime (i.e., evidence of the crime may be stored on them). Computer forensics investigations can be, and have been, conducted in each type of incident.

In private investigations, computer forensics technologies and methods may be used to collect electronic evidence for crimes. These types of investigations can also be used to identify policy violations, such as those involving employee computer misuse. An offense may be committed or a policy may be violated as follows:

- A company may be the victim of the crime (such as theft of intellectual property, industrial espionage, or destruction of data).

- A computer owned by a company may be used to facilitate the crime (e.g., sending threatening e-mails from a company account).

- A computer owned by a company may be used to store contraband material (e.g., pirated software or child pornography images) or may contain evidence of the crime.

Information that is obtained from workplace computers can be used as evidence in harassment, discrimination, wrongful termination, embezzlement, and other criminal cases.

- Employees may violate company policy by improperly using computer resources, such as surfing the Web for personal reasons during company time.

Some special considerations apply in corporate investigations:[2]

- During a computer forensics investigation, confidential data must be protected. If an incident occurs, company and client data must be secured.

- Computer and information systems must remain available to the company for use in daily operations while the investigation is ongoing.

- The integrity of the data should be maintained and no data should be altered or lost during an investigation.

Investigations Involving Civil Law

Civil law governs the relationships between private parties, including both individuals and organizations (e.g., business, corporations, enterprises, and nongovernmental agencies). The most common form of this law is tort law, which provides remedies for individuals who have been harmed or wronged. Civil law also assigns punishment for harms and wrongs. However, unlike under criminal law, the punishment largely consists of monetary damages or other forms of reparation—incarceration and capital punishment can be imposed only for criminal law violations.

A civil case is initiated by the wronged individual (the plaintiff). The laws of evidence in civil and criminal courts vary, however. Specifically, the burden of proof is lower in civil courts. The standard of proof in civil cases is a preponderance of evidence, which requires that what is being asserted is more likely than not to be true.

Civil law can cover some of the same areas of legal action as criminal law, such as assault. For example, a victim of a criminal act may sue the perpetrator for damages in civil court—even if the perpetrator was found not guilty of the crime in criminal court. A famous case involves former football star O. J. Simpson.[3] Although he was found not guilty of murdering Nicole Brown and Ron Goldman in a criminal trial, the families of the victims successfully sued him for damages in civil court.

Computer forensics investigations can be conducted to provide evidence for both crimes and torts. Moreover, the same cybercrime can be considered as both a crime and a tort. Indeed, a hacker, computer virus distributor, or botherder can be prosecuted in a criminal court and sued in civil court for the same crime. Government agencies also have the option to pursue a legal matter through the criminal justice system or file a tort action.

Investigations Involving Administrative Law

Administrative law focuses on the exercise of government authority by the executive branch and its agencies. It regulates conduct in areas such as safety, welfare, national security, environmental and consumer protection, and taxation. Some U.S. federal administrative agencies are listed here:

- Department of Agriculture
- Department of Commerce
- Department of Defense
- Department of Education
- Department of Energy
- Department of Health and Human Services
- Department of Homeland Security
- Department of Housing and Urban Development
- Department of the Interior
- Department of Justice
- Department of Labor
- Department of State
- Department of Treasury
- Department of Transportation
- Department of Veteran Affairs

Administrative agencies are created through "enabling legislation." For example, the Occupational Safety and Health Administration (OSHA), which regulates private and federal workplace conditions and safety, was created by the Occupational Safety and Health Act of 1970.

Administrative agencies conduct investigations for four major reasons:

1. *To determine if agency rules and regulations have been violated.* Administrative agencies provide specific rules and regulations that stipulate the required duties and responsibilities of individuals and organizations. These agencies have the power to investigate and prosecute violations of these rules and regulations. For instance, OSHA can investigate and prosecute workplace safety violations. Evidence of these violations may exist on workplace computers (e.g., evidence can be stored in a computer or sent via e-mail). An example would be a memo from upper-level management informing mid-level management that workers should disregard a safety hazard that may result in the loss of life.

2. *To investigate crimes committed against them.* The Department of Treasury was the target of hacking that placed malicious software on three of the agency's websites. When visitors tried to access these sites, they were redirected to a website in Ukraine that contained malicious software, which, among other things, launched attacks that exploited flaws in Adobe's reader software. Many federal administrative

departments have been subjected to malicious attacks over the years, including the Department of Defense and the Department of Homeland Security.[4]

3. *To determine if employees of the agency have engaged in misconduct or criminal activity.* The Treasury Inspector General for Tax Administration (TIGTA), which is responsible for enforcing criminal law as it pertains to the operations of the Internal Revenue Service (IRS), has investigated its own employees for accessing taxpayer account information without authorization.[5] For instance, the TIGTA found that one IRS employee, Ericka Duson, had unlawfully inspected approximately 183 tax returns of individuals.[6]

4. *To assist law enforcement agencies in investigations.* IRS Special Agents have assisted the Federal Bureau of Investigation (FBI) in investigations that involved a counterfeit software distribution ring[7] and a mortgage loan fraud scheme.[8] Agents of the IRS have also assisted in investigations involving attempts to defraud the United States. One such case involved Anthony G. Merlo, who promoted, marketed, sold, and administered fraudulent tax shelters, which he called loss-of-income insurance policies, through Security Trust Insurance Company in the U.S. Virgin Islands.[9]

Investigations can occur through the use of inspections, **subpoenas** (either to compel someone to appear for a hearing and testify on a suspected violation or to request that an organization provide the agency with records, documents, books, or papers to assist its personnel in their investigation and prosecution of a case) and searches and seizures. In respect to searches and seizures, government agents must obtain a **search warrant** to enter a premises, search for the objects named in the warrant, and seize them.[10] Although government personnel can conduct searches, these searches are restricted by the **Fourth Amendment** to the **U.S. Constitution**, which states

> The right of the people to be secure in their persons, houses, papers, and effects, against unreasonable searches and seizures, shall not be violated, and no Warrants shall issue, but upon probable cause, supported by Oath or affirmation, and particularly describing the place to be searched, and the persons or things to be seized.

Investigations conducted by these agencies are also restricted by the privilege against self-incrimination of the **Fifth Amendment** to the U.S. Constitution, which states

> No person shall be held to answer for a capital, or otherwise infamous crime, unless on a presentment or indictment of a Grand Jury, except in cases arising in the land or naval forces, or in the Militia, when in actual service in time of War or public danger; nor shall any person be subject for the same offence to be twice put in jeopardy of life or limb; nor shall be compelled in any criminal case to be a witness against himself, nor be deprived of life, liberty, or property, without due process of law; nor shall private property be taken for public use, without just compensation.

Agencies also need to protect privileged data if they are searching areas in which this type of information is stored. The applicable Amendments of the U.S. Constitution and privileged information are explored in further detail in Chapter 4.

Computer Forensics Investigations: A Four-Step Process

There are four distinct steps in computer forensics investigations: acquisition, identification, evaluation, and presentation.

Acquisition

The **acquisition** step involves the process of evidence retrieval in computer forensics investigations—from the search for the evidence to its collection and documentation (see Chapter 8, which explores what investigators should do when called to a scene to investigate a cybercrime). The computer forensics specialist must document all aspects of the computer search, including the following:

- Which evidence was obtained
- Which individual or individuals retrieved the evidence
- Where the evidence was gathered
- When was the evidence collected
- How was the evidence acquired

Computer forensics specialists can acquire computer evidence in several different ways:

- An on-site search of the computer and collection of evidence can be performed.
- A copy can be made of the entire storage device on-site to be later examined off-site.
- The computer and other electronic devices can be seized and their contents reviewed off-site.

On-Site Versus Off-Site Search

You are investigating a cybercrime. One of the objects to be seized pursuant to your search warrant is a computer. After you find the computer during your search of the residence (or company or other area) listed in your warrant, what do you do? Do you search the computer on-site or do your remove the computer from the scene and search it off-site?

If the search can easily occur on-site, then there is no justification to seize a computer (or computers) and search it (them) off-site.[11] In *United States v. Upham*,[12] the prosecution showed that the search performed off-site could not readily have been conducted on-site. Consequently, the court found that the search of the computer off-site was justified.

Conducting on-site searches may not always be possible. The following factors may affect the feasibility of on-site and off-site computer searches:[13]

- *The configuration of the software and hardware.* This factor determines the tools an investigator requires to conduct a successful search. It also determines the specialized knowledge and skills an investigator must have to successfully conduct the search.

- *The overall size and complexity of the computer system.* Off-site searches of large-scale computer systems are often infeasible.

- *The technical demands of the search.* Many different issues are raised here. For example, is the computer password protected? If so, is the password known? Are the tools required for the specific configuration available on-site to search the computer? Will these tools effectively search the computer for evidence?

The most common practice is to seize a computer during a cybercrime investigation and take it off-site—typically to a forensic lab—for a search of its contents for evidence. In fact, numerous court rulings have upheld the validity of seizure of computer systems and subsequent search of them off-site as the only reasonable means to conduct a search. For instance, in *Upham,* a child pornography case, the court held that the seizure of the computer and its subsequent off-site search constituted "the narrowest definable search and seizure reasonably likely to obtain the [child pornography] images." Additionally, in *United States v. Hay,*[14] the court held that the search of the computer off-site was "reasonable" because of the time, expertise, and controlled environment required for its proper analysis. In *United States v. Maali,*[15] the court also pointed out that for some aspects of a computer search, a controlled environment is required. As such, an off-site search of the computer in these situations constitutes the only reasonable means to conduct the search.

The computer forensics process is very time-consuming. When deciding whether to search the computer on-site or remove it from the scene and search its contents off-site, one should also bear in mind that the "Fourth Amendment's mandate of reasonableness does not require the agent to spend days at the site viewing . . . computer screens."[16]

Other factors that limit the wholesale seizure of computer hardware for off-site search include privacy concerns and the presence of privileged information on the computer and/or related devices. Chapter 4 provides further information on these issues.

Identification

In the **identification** step, an investigator explains and documents the origin of the evidence and its significance.[17] Given that evidence can be interpreted from a number of different perspectives, this phase determines the context in which the evidence was found. It looks at both the physical environment and the logical context of the location of electronic evidence. Physically, evidence may reside on a specific medium such as a hard drive. Data can, therefore, be physically extracted from a hard drive using methods such as **keyword searching** (the legality of this and other techniques are explored in Chapter 4). Data can also be extracted in other ways—for instance, by using file carving. **File carving** looks for specific files in a hard drive based on the header, footer, and other identifiers within the file. By searching in this manner, an investigator can recover files or file fragments of damaged or deleted files in corrupt directories or damaged media.

Logically, an investigator looks at where the evidence resides relative to the file system of a computer operating system, which is used to keep track of the names and locations of files that are stored on a storage medium such as a hard disk. This system stores a file on a hard disk wherever it finds free space. The locations of these files are subsequently listed in an index. The logical context thus focuses on the system of storage that is created by the computer, which may, in fact, be different from the actual physical storage of the data. Logical extraction of data can occur through the retrieval of file system information that can reveal file name, location, size, and other attributes. Data can be retrieved from active files, recovered **deleted files**, and **unallocated space**. It can also be extracted from **file slack**, which can be interpreted as follows: Imagine that you have a 10 kilobyte file, named Doc1. Computer operating systems store files in fixed-length blocks of data known as **clusters**. If each cluster holds 4 kilobytes' worth of data, then Doc1 takes up 2.5 clusters and, therefore, needs 3 clusters for its physical size. The leftover space from the end of Doc1 to the end of the cluster is known as file slack or **slack space**.

The identification stage also involves making evidence that was otherwise concealed visible to the forensic analyst for review in the next stage of the computer forensics process—evaluation. An accurate and detailed record of the physical and logical context of the evidence must also be kept by the investigator to maintain the integrity of the process and to ensure that relevant evidence is admissible in a court of law.

Evaluation

In the **evaluation** step of the computer forensics process, the data retrieved during the investigation are analyzed to establish their significance and relevance to the case at hand.

- What does the evidence tell the forensic analyst about the cybercrime, suspect, or victim?
- How did the data end up in the computer or other electronic device?
- Were the data placed there by the suspect or by someone else?

In respect to the last question, on numerous occasions defendants have claimed that they were not responsible for child pornography that was found on their computers. In fact, the defendants in each of the following cases argued that a computer virus had downloaded child pornography onto their computers:

- *United State v. Shiver*[18]
- *United States v. Miller*[19]
- *United States v. Bass*[20]

Investigators try to determine the "who," "what," "when," "where," "why," and "how" of the crime based on the evidence retrieved. Conclusions are drawn in this step as to how the evidence could be used to support employee terminations, proceedings for company policy violations, civil lawsuits, or prosecutions in a criminal court. The validity and **reliability** of the data are also examined. Only valid and reliable information that has been properly and lawfully documented, collected, processed, inventoried, packaged, transported, and analyzed will be used for the next step in the computer forensics process—presentation. Under Rule 401 of the U.S. Federal Rules of Evidence, evidence is considered to be relevant if it has the "tendency to make the existence of any fact that is of consequence to the determination of the action more probable or less probable than it would be without the evidence." Any information that is retrieved from a computer and related electronic devices that is not considered relevant to the case is inadmissible (as per Rule 402, U.S. Federal Rules of Evidence) and will be excluded from presentation in court.

Presentation

The final stage, **presentation**, involves reporting data pertinent to the case that were found during the investigation. The evaluation of evidence by outside parties may be expected in this stage. Accordingly, computer forensics investigators must be prepared to testify in court.[21] During their testimony, they are usually required to defend their personal qualifications, methods, validity of procedures, handling of evidence, and findings. Computer forensics experts must also be able to communicate their findings to a variety of audiences (e.g., juries, judges, lawyers, corporate management, and administrative officials). The chain of custody, which is a chronological record documenting each individual who had the piece of evidence in his or her possession and the points at which the person had it, may also be challenged at this stage.

Electronic Evidence: What Is It?

Evidence is defined as any object or piece of information that is relevant to the crime being investigated and whose collection was lawful. Evidence is sought for the following reasons: [22]

- To prove that an actual crime has taken place ("corpus delicti")
- To link a particular person to the crime
- To disprove or support the testimony of a victim, witness, or suspect
- To identify a suspect
- To provide investigative leads
- To eliminate a suspect from consideration

Electronic evidence consists of any type of information that can be extracted from computer systems or other digital devices and that can be used to prove or disprove an offense or policy violation. Such evidence can illustrate possession and intent. The presence of child pornography images on one's computer can show possession, for example. Intent could be shown if these images were organized and placed in alphabetized files according to the child's screen name. Such data can be used to support a claim or can serve as an alibi. By analyzing the evidence retrieved during computer forensics investigations, investigators try to figure out what happened, when it happened, how it happened, why it happened, and who was involved.

Types of Evidence

Different types of evidence exist that can be used to either prove or disprove a piece of information. Testimonial, documentary, demonstrative, physical, direct, circumstantial, and hearsay evidence are explored in further detail in this section.

Testimonial Evidence

 Testimony is a type of evidence that consists of witnesses speaking under oath, including eyewitness and expert testimony. A witness is considered competent to testify if he or she meets several criteria outlined by the U.S. Federal Rules of Evidence. For instance, the individual who testifies must have personal knowledge of the subject about which he or she is providing information. This can be either first-hand knowledge of the crime—if the individual is an eyewitness—or specialized knowledge directly resulting from education, training, experience, and skills required to authenticate or refute a piece of information that is being presented in the case. This specialized knowledge is required for those individuals who are called to provide expert testimony.

Documentary Evidence

 Documentary evidence consists of any kind of writing, video, or sound recording material whose authenticity needs to be established if it is introduced as evidence in a court of law. Examples include business records, manuals, and computer printouts. The genuineness of each piece of documentary evidence needs to be established. Typically, an expert provides testimony as to the authenticity of the evidence being presented, although certain exceptions to this rule exist. Notably, certain documents do not require extrinsic evidence of their authenticity under Rule 902 of the U.S. Federal Rules of Evidence, such as domestic public documents under seal, official publications, and certified copies of public records.

Demonstrative Evidence

 Demonstrative evidence cannot independently prove a fact. Instead, this type of evidence is used for the following purposes:

- To explain other evidence
- To illustrate, demonstrate, or recreate an event
- To show a situation similar to something being presented in the case

Diagrams, drawings, maps, models, and sketches are used to make evidence more understandable. Photographs and videos are other examples of this type of evidence, which provides a visual depiction of what happened at a crime scene and/or what is being narrated by the individual testifying in court. In particular, according to the court ruling in *Campbell v. Pitt County Memorial Hospital, Inc.*,[23] videos can be used to illustrate the testimony of a witness if their contents are carefully examined to determine their authenticity, relevancy and competency.

Demonstrative evidence is generally admissible in court if it meets the following criteria:[24]

- It relates to other relevant evidence being used in court (Rule 401, Federal Rules of Evidence).
- The individual who is using this type of evidence is familiar with its use (Rules 602 and 703, Federal Rules of Evidence).
- The evidence correctly and fairly reflects other evidence of the case to which it relates (Rule 401, Federal Rules of Evidence).
- The evidence aids the trier of fact[25] in understanding or assessing other related evidence (Rules 401 and 430, Federal Rules of Evidence).

Physical Evidence

Evidence collected from a computer forensics investigation is not limited to that which is extracted from computer hard drives and other electronic devices. **Physical evidence** (sometimes referred to as real evidence) can also be found. This type of evidence can be used to corroborate suspect, witness, and victim statements; to link a suspect to a crime scene or crime; to link a suspect to a victim; or to rule out a person as a possible suspect. This tangible evidence includes objects such as ammunition, firearms, knives, glass, and questioned documents (handwritten or typewritten documents whose authenticity has yet to be established).

Physical evidence found on computers and related electronic devices usually takes the form of trace evidence and impression evidence. Trace evidence includes items that are extremely small (i.e., microscopic), such as soil, hair, fibers, and dust.

Fingerprints are considered impression evidence. Normally, fingerprints are not visible to the human eye (and, therefore, are known as latent fingerprints) and need to be enhanced with some sort of chemical developer. This chemical developer can be either a fingerprint powder or chemical treatment. The developer employed depends on the type of surface to which it is being applied. If it is a hard and nonabsorbent surface, then fingerprint powder will be applied in the following manner:

- Black colored powder is used for white or light surfaces.
- Gray colored powder is used for dark colored surfaces.
- Gray colored powder is applied to mirrors and shiny metallic surfaces. Black colored powders cannot be used on these surfaces, because when they are photographed, they appear black.

If an investigator is dealing with a soft and porous surface, chemical treatments—such as iodine fuming, ninhydrin, silver nitrate, and Super Glue fuming—can be used (typically in the order listed).

Fingerprints may be found on various parts of the computer and related devices. For example, they may be found on the computer screen, keyboard, mouse, printer, scanner, computer tower, external hard drives, CDs, DVDs, and USB sticks.

Direct Evidence

 Direct evidence, such as eyewitness testimony, establishes a fact. An example of direct evidence is as follows: Mike testifies under oath that he was watching the computer screen while Joe wrote and sent a harassing e-mail to his ex-girlfriend, Kate. Another example concerns witness testimony: If the fact at issue is whether "Jake had a gun" and a witness testifies that "Jake had a gun," the witness testimony constitutes direct evidence.

Eyewitness testimony is not the only example of direct evidence. Consider the following scenario: Brandon sits at his company's computer, enters his username and password, and logs on to the system. He hacks into the company's database and deletes certain files. When he is finished, Brandon logs off his computer and leaves the office. The following day, the informational technology (IT) department notices that someone has deleted certain files from the company's database. Its personnel are able to determine that Brandon's computer, username, and password were used in the incident. The direct evidence in this case is that Brandon's computer, username, and password were used to log on to the system and delete the files.

Another form of direct evidence that a particular individual engaged in illicit activity is the confession of the suspect that he or she actually committed the alleged offense.

Circumstantial Evidence

Circumstantial evidence allows someone to infer the truth of a given fact. A prime example of circumstantial evidence is the motive to commit a crime. Other examples abound. For instance, if a woman accused of embezzling money from her company had the money "pop up" in her personal account around the time of the alleged embezzlement, that fact would be circumstantial evidence that she had stolen the money. Another example is as follows: A fingerprint is found on a desktop computer at the crime scene. This fingerprint belongs to the suspect in the crime, Eddie. This fingerprint is both direct evidence that Eddie was at the crime scene and indirect (circumstantial) evidence that Eddie committed the crime.

Consider the difference between direct and circumstantial evidence in a cybercrime: When Hera signed in to her e-mail account last week, she clicked "yes" in response to a message that asked whether she would like her username and password to be saved for the next time she wanted to view her e-mail. Without her knowledge, Nick used her computer and found that Hera's username and password were saved to her e-mail login page. He accessed her e-mail account and sent threatening messages to his ex-girlfriend, Phoebe.

The direct evidence in this case is that Hera's e-mail account was used to send the threatening messages to Phoebe. The circumstantial evidence suggests that Hera was responsible. Someone else may have used Hera's account to send the messages, however—which is what actually happened in this case.

Hearsay Evidence

Rule 801 of the Federal Rules of Evidence defines **hearsay** as a statement that can be "an oral or written assertion or . . . nonverbal conduct of a person, if it is intended by the person as an assertion," "other than one made by the declarant" (the individual who makes the statement) "while testifying at the trial or hearing, offered in evidence to prove the truth of the matter asserted."

The determination as to what constitutes hearsay and whether hearsay is admissible in court has been heavily debated. One ruling dealing with this issue occurred in *United States v. Hamilton*.[26] In this case, Kenneth Hamilton was suspected of downloading child pornography and distributing it on the Internet. Law enforcement officers investigating this incident found a link between Hamilton and the child pornography pictures. Specifically, the computer-generated header information attached to the child pornography pictures revealed the screen name of the person who posted them, the date the images were posted, and the **Internet Protocol (IP) address** of the individual who posted them. An IP address is a unique identifier assigned to a computer by the Internet service provider (ISP) when it connects to the Internet. Whenever a user requests access to a website, connects to another site, or chats online, his or her IP address is sent along with the data. Without this address, the computer on the other end of the communication would not know where to send the reply. IP addresses are used to identify users and can be used by authorities to trace the origin of a message or website—which is exactly what happened in this case. Even though Hamilton's defense attorney tried to have this evidence excluded on the grounds that it was hearsay, the court ruled that the computer-generated header information attached to the child pornography images was not hearsay "as there was neither a 'statement' nor a 'declarant' involved here within the meaning of Rule 801" of the Federal Rules of Evidence.[27]

In another case, the Louisiana Supreme Court recognized that the printout of a computer log of telephone numbers that were the result of the "computer's internal operations is not hearsay evidence. It does not represent the output of statements placed into the computer by out of court declarants."[28]

Courts have made a distinction between computer-generated data and computer-stored data when determining whether evidence constitutes hearsay. One relevant case in this regard is *Louisiana v. Armstead*.[29] In this case, the court affirmed that computer-generated data contained computer printouts of human statements or assertions. The computer-stored data constituted "assertions" because "the output represents only the by-product of a machine operation which uses for its input 'statements' entered into the machine by out of court declarants."[30] Some examples falling under this category are word processing files, spreadsheets, e-mails, instant messaging transcripts, and Internet chat room messages.[31] Courts have also determined that computer printouts that "reflect computer stored human statements are hearsay when introduced for the truth of the matter asserted in statements."[32] The trustworthiness of the statements in computer-stored files must be established for them to be admissible in a court of law.

Hearsay evidence is inadmissible in criminal courts except as provided by the Federal Rules of Evidence, other rules prescribed by the Supreme Court pursuant to law, or an Act of Congress (Rule 802, U.S. Federal Rules of Evidence). Rule 803 of the Federal Rules of Evidence includes several exceptions to the inadmissibility of hearsay evidence, including

The Case of Ali Saleh Kahlah al-Marri

On June 23, 2003, former U.S. President George W. Bush declared Ali Saleh Kahlah al-Marri, a suspected member of al-Qaeda, to be an **enemy combatant**. Al-Marri, a Qatari nationalist, was arrested as a material witness in the FBI's investigation of the attacks on September 11, 2001. He had come to the United States to pursue a degree in computer science at Bradley University.

In July 2004, al-Marri's lawyers filed a request for habeas corpus. The U.S. government opposed it and submitted a declaration, classified as "secret," which was prepared by Jeffery Rapp, the Director of the Joint Intelligence Task Force for Combating Terrorism at the Defense Intelligence Agency. This declaration included accusations that al-Marri was sent to the United States by Osama bin Laden to learn how to hack into U.S. computer systems and to research how bombs were made. Specifically, Rapp stated that al-Marri had plans on his laptop computer for manufacturing bombs by using various cyanide compounds and sulfuric acid—compounds that can be used to create cyanide gas. Rapp alleged that al-Marri was planning to hack into the main computers of U.S. banks to damage the U.S. economy. Rapp further stated that common hacking tools were found on al-Marri's laptop computer. In particular, Rapp claimed that al-Marri's computer was loaded with "numerous computer programs typically utilized by computer hackers; 'proxy' computer software which can be utilized to hide a user's origin or identity when connected to the Internet; and bookmarked lists of favorite websites apparently devoted to computer hacking."[33]

The evidence against al-Marri was largely hearsay evidence. Normally, such evidence would be inadmissible. However, it was deemed admissible in al-Marri's case because of an earlier decision in *Hamdi v. Rumsfeld*,[34] in which the court ruled that hearsay evidence was admissible because Hamdi was an enemy combatant.

Critical Thinking Exercise

Consider the computer programs typically used by hackers that were found on al-Marri's laptop computer. Do they prove that al-Marri was planning to hack computer systems in the United States? If so, explain why. If not, why do you think this is the case? What do these tools tell us about al-Marri? Use the material from this chapter to support your answer.

one concerning public records. According to Rule 803(8)—commonly called the "public record" exception— records excluded from the hearsay rule include

> Records, reports, statements, or data compilations, in any form, of public offices or agencies, setting forth (A) the activities of the office or agency, or (B) matters observed pursuant to duty imposed by law as to which matters there was a duty to report, excluding, however, in criminal cases matters observed by police officers and other law enforcement personnel, or (C) in civil actions and proceedings and against the Government in criminal cases, factual findings resulting from an investigation made pursuant to authority granted by law, unless the sources of information or other circumstances indicate lack of trustworthiness.

For public records to be admissible, they need to be certified. In particular, in *State v. Sherman*,[35] the court held that "transcripts of entries in official records stored within either a data processing device or computer are, if properly certified, admissible as primary evidence of the facts stated." Conversely, the courts have excluded public record evidence that was not certified by the appropriate official. In fact, in *Norton v. Alabama*,[36] the court recognized that "without the proper certification, the printout is merely a written statement made by an individual who was unavailable to testify and, therefore, unavailable for cross-examination."

The hearsay exception rules have been well established in the courts. In fact, in *Robinson v. Commonwealth*,[37] the court asserted "that hearsay evidence is inadmissible unless it falls within one of the recognized exceptions to the hearsay rule and that the party attempting to introduce a hearsay statement has the burden of showing the statement falls within one of the exceptions."

A case that illustrates this point is *United States v. Jackson*.[38] On December 4, 1996, Angela Jackson, a young African American woman, had United Parcel Service (UPS) deliver four packages to her apartment in St. Paul, Minnesota. Jackson reported to UPS that she had received only three packages and that they were all damaged and contained racial statements. Jackson filed a $572,000 claim with UPS. Jackson faxed letters to various African American officials claiming that racist elements within UPS were responsible for damaging her packages. The UPS driver, the building's receptionist, and the concierge who handed the packages directly to Jackson testified that there were four packages and that all were damaged. UPS filed fraud charges against her. Hate mail sent through UPS and addressed to African Americans, including members of Congress, other government officials, newspapers, the National Association for the Advancement of Colored People (NAACP) and the Rainbow Coalition, allegedly from white supremacist groups, such as the Euro-American Student Union and Storm Front, were eventually traced back to Jackson.

In court, Jackson wanted to submit Web postings from the European American Student Union and Storm Front, in which these groups gloated over her situation and took credit for the racist UPS mailings. According to the court, the Web postings that Jackson tried to submit qualified as hearsay evidence because they were being offered to provide the truth of the matter being asserted and did not include statements made by declarants testifying at a trial. Jackson tried to claim that this hearsay evidence was admissible under one of the exceptions listed in Rule 803 of the Federal Rules of Evidence. Specifically, Jackson claimed that the Web postings were admissible under Rule 803(6) (the "business rule" exception), which concerns records of regularly conducted activity. According to the business rule exception, hearsay evidence is admissible if it consists of

> A memorandum, report, record, or data compilation, in any form, of acts, events, conditions, opinions, or diagnoses, made at or near the time by, or from information transmitted by, a person with knowledge, if kept in the course of a regularly conducted business activity, and if it was the regular practice of that business activity to make the memorandum, report, record or data compilation.

In *United States v. Briscoe*,[39] the court recognized that "computer data compilations are admissible as business records . . . if a proper foundation as to the reliability of the records is established." A proper foundation for the reliability of the business records is shown if the records "are kept in the course of regularly conducted business activity, and [that it] was the regular practice of that business activity to make records, as shown by the testimony of the custodian or other qualified witness."[40] The courts, in *United States v. Ramsey*[41] and *Briscoe*, have also recognized that such "records are reliable to the extent they are compiled consistently and conscientiously."

While Jackson argued that the Web postings were business records of the supremacy groups' ISPs and, therefore, admissible under the business rule exception, the court rejected this argument on two grounds. First, it stated that ISPs are mere conduits and, in this case, did not actually post or use the information on this website in the regular course of their business activities. Second, the court found that the Web postings lacked authentication. In particular, Jackson did not prove that the Web postings were actually made by members of the white supremacist groups and not by someone else.

As stated previously, hearsay evidence can consist of both oral and written statements made out of court. Imagine that during a murder investigation, a computer forensics specialist comes across an e-mail of a suspect that states: "You do not have to worry about him. He is gone now. I took care of him." Although the e-mail can be cited as proof that the suspect made that statement, it does not prove the veracity of the statement. Other evidence is required to corroborate this evidence.

Sometimes the information needed to authenticate the e-mail may be included in the message itself. In *United States v. Siddiqui*,[42] Mohamed Siddiqui sent false e-mails, forms (which he signed as well), and letters of recommendation on behalf of individuals without their permission to nominate himself for the Waterman Award from the National Science Foundation. Siddiqui had sent several e-mails to Dr. Hamuri Yamada and Dr. von Gunten requesting their permission to nominate him for the award and telling them to sign the form for the award. Siddiqui argued that the e-mails were hearsay and, therefore, were inadmissible. The court dismissed this argument on the grounds that the e-mails had been authenticated (that is, the trustworthiness of the statements had been established) because Siddiqui used his nickname "Mo" when signing the e-mails—a nickname confirmed by both Yamada and von Gunten. The e-mails were also authenticated because Siddiqui followed up with phone calls reiterating what he stated in the e-mails—requesting Yamada's and von Gunten's nominations for the award. Furthermore, the statements in the e-mail were Siddiqui's own statements. As such, the court held that the content of the e-mails did not constitute hearsay under Rule 801(d)(2)(A) of the Federal Rules of Evidence.

The same could be said about text messages. In the case of *State v. Dickens*,[43] John Dickens sent several text messages to the mobile telephone of his estranged wife before shooting and killing her. The substance of these messages served as sufficient circumstantial evidence for the court to conclude that the messages were sent by Dickens. For instance, the text messages included information concerning the parties' minor child and their wedding vows that only Dickens would know.

Cases that address the authentication of instant messages and chat room postings follow along the same lines as those dealing with e-mails and text messages.[44] In *Bloom v. Commonwealth*,[45] the court recognized that conversations conducted over the Internet were similar to conversations carried out over the phone. As established in *Snead v. Commonwealth*,[46] overheard telephone conversations are admissible in court if direct or circumstantial evidence establishes the identity of the parties involved in the conversation. Likewise, according to *Bloom*, instant messages are admissible as evidence against the sender of the message only if the identity of the sender is established.

In *United States v. Tank*[47] and *United States v. Simpson*,[48] the courts ruled on the admissibility of chat room logs. Chat room logs are admissible if they consist of the actual computer files. In contrast, courts have raised admissibility issues with chat sessions or instant messages logs that have been submitted as evidence but that are not the actual computer files or printouts of the logs. Chat sessions that have been copied and pasted into a word document are one such example. In *United States v. Jackson*,[49] a key issue was evidence consisting of a chat conversation via instant message that occurred between police officer David Margritz, who was posing as a 14-year-old girl, and the defendant, Jackson. Margritz saved the online chats in a Word document and submitted the saved document in court. No original transcripts of the conversations were available because these sessions were not saved or archived on computers. Jackson claimed that the document containing their chat was missing sections, and the missing sections cast doubt as to the trustworthiness of the document.[50] The court explicitly stated that "changes, additions, and deletions have clearly been made to this document, and accordingly, the court finds this document is not authentic as a matter of law."[51] Clearly, authenticating electronic evidence is of the utmost importance in cybercrime cases.

Authentication of Evidence

As indicated earlier, evidence retrieved from a computer system or other electronic device must be authenticated before it can be used in court as evidence in a particular case. To determine that data are genuine and accurate, computer evidence may be authenticated through direct or circumstantial evidence. Evidence must also be collected in such a way as to ensure that it is not altered. **Authentication** is a process whereby the original data are compared to a copy of the data to determine if the two are exactly the same. Key to authentication is the chain of custody because electronic evidence is easy to modify—often without leaving a trace. As such, a detailed log indicating who handled the evidence and any alterations that were made to it must be kept. This chain of custody can indicate whether an unknown individual had an opportunity to access or replace the data.

The authenticity of electronic evidence depends on the computer program that generated or processed the data in question. The **validity**[52] of the chosen program used in the computer forensics process must be proven. That is, the tools and supplies used to search for, collect, and preserve electronic evidence for computer forensics investigations must be validated. Standards have been developed to determine the validity of computer forensics tools and, therefore, their admissibility in a court of law. The most notable of these standards are the Frye, Coppolino, and Daubert standards.

Frye Standard

In *Frye v. United States*,[53] the defendant tried to submit evidence of expert testimony on the systolic blood pressure deception test, which was an early version of the polygraph test. Frye wanted to show his innocence by presenting evidence that he passed the test. To do so, he sought to admit expert testimony to explain the test and the significance of its result. The court rejected his request on the grounds that the systolic blood pressure deception test had not received general acceptance in the scientific community. Forensic tools, techniques, procedures, and evidence are admissible in court if they have a "general acceptance" within the scientific community. Accordingly, to introduce them in court, one must provide scientific papers (reviewed by other known and respected scientists) and books that have been written about the testing procedures.

The **Frye standard** does not leave much room for admitting novel but valid tests that have not yet won general acceptance within the scientific community. Significant debate has been generated about the inability of the Frye standard to do so. Some courts have rejected the Frye standard, instead using the Federal Rules of Evidence to determine the admissibility of evidence and to admit novel but relevant evidence in court.[54]

Coppolino Standard

The **Coppolino standard** did not follow a similar path as the Frye standard. In fact, the Coppolino standard rejected the "general acceptance" test, which was a prominent feature of the Frye standard. Instead, in *Coppolino v. State*,[55] the court held that a novel test may be admitted if its validity could be proven, even if the general scientific community was unfamiliar with it.

Daubert Standard

Much like the Coppolino standard, the **Daubert standard** allows novel tests to be admitted in court, albeit with different criteria. According to the ruling in *Daubert v.*

Merrell Dow Pharmaceuticals Inc.,[56] four criteria are used to determine the reliability of a particular scientific theory or technique:[57]

- Testing: Has the method in question undergone empirical testing?
- Peer review: Has the method been peer reviewed?
- The potential rate of error: What is the error rate in the results produced by the method?
- Acceptability: Has the method received general acceptance in the relevant scientific community?

The Daubert standard requires an independent, judicial assessment of the reliability[58] of the scientific test or evidence. A "reliability assessment does not require, although it does permit, explicit identification of a relevant scientific community and an express determination of a particular degree of acceptance within that community."[59] Moreover, in *Daubert v. Merrell Dow Pharmaceuticals Inc.,*[60] the court ruled that the fact that a theory or technique has not been subjected to peer review or has not been published does not automatically render the method in question inadmissible. In *Daubert,* the court thus recognized that the inquiry must be flexible and must focus on the specific principles and methodology in question.

Standards of Evidence

According to the **National Institute of Standards and Technology** (NIST), computer forensics test results must be repeatable and reproducible to be admissible as electronic evidence.[61] Computer forensics test results are repeatable when the same results are obtained when using the same methods in the same testing environment (same hard drive, disk, and so on). Test results are reproducible when the same test results are obtained using the same method in a different testing environment (different hard drive, disk, and so on). More specifically, the NIST (2001) defines these terms as follows:

- **Repeatability** means that the same results must be obtained when using "the same method on identical test items in the same laboratory by the same operator using the same equipment within short intervals of time."
- **Reproducibility** means that the same results are obtained when using "the same method on identical test items in different laboratories with different operators using different equipment."

Chapter Summary

Public and private computer forensics investigations differ in terms of which types of crimes are involved, who conducts the investigation and how the investigation is conducted. Investigations in the criminal, civil, and administrative arenas and the rules of evidence within these areas of law differ, for example.

The computer forensics process consists of four steps: acquisition, identification, evaluation, and presentation. Electronic evidence is provided in court to prove the veracity—if any—of a given piece of information. Categories of evidence include testimonial, documentary, demonstrative, physical, direct, and circumstantial. Hearsay evidence can also be admitted in court as evidence if it falls under one of the exceptions under the Federal Rules of Evidence or if it is substantiated with direct or circumstantial evidence. Several court rulings have sought to clarify whether certain types of electronic evidence are considered to be hearsay evidence and, therefore, whether they are admissible in court. To be admissible in court, electronic evidence must be authenticated either by circumstantial and direct evidence and/or by use of specific standards to validate the program that was used to collect, search, analyze, and preserve the data. Computer forensics tests must be repeatable and reproducible to be admissible in court.

Key Terms

Acquisition	Fourth Amendment
Administrative law	Frye standard
Authentication	Hearsay
Circumstantial evidence	Identification
Civil law	Internet Protocol (IP) address
Clusters	Keyword searching
Computer forensics	National Institute of Standards and
Coppolino standard	Technology
Criminal law	Physical evidence
Daubert standard	Presentation
Deleted files	Reliability
Demonstrative evidence	Repeatability
Direct evidence	Reproducibility
Documentary evidence	Search warrant
Electronic evidence	Slack space
Enemy combatant	Subpoena
Evaluation	Testimony
File carving	Unallocated space
File slack	U.S. Constitution
Fifth Amendment	Validity

Practical Exercise

For this project, you must search for and interpret information in a cybercrime case that dealt with admitting hearsay evidence in criminal or civil court. Provide a brief description of the case and discuss the role of hearsay evidence in it. You should also identify how or why the evidence was admissible (or inadmissible). Within your analysis, do not forget to identify and include other types of evidence presented in the case (e.g., direct and circumstantial). Finally, discuss the impact of the challenges to the authenticity of the evidence on the outcome of the case.

Review Questions

1. What is computer forensics?

2. What are the major differences between public and private investigations?

3. What are the similarities and differences between criminal and civil law?

4. Why do administrative agencies conduct investigations?

5. Describe the computer forensics process.

6. When should a search be conducted on-site?

7. When should a search be conducted off-site?

8. What is slack space?

9. Which different types of evidence exist?

10. What is the difference between circumstantial and direct evidence?

11. When is hearsay evidence admissible in court?

12. How can electronic evidence be authenticated?

13. What are the standards of evidence?

Footnotes

[1]There is no single universally accepted definition of computer forensics.

[2]Corporate investigations are further explored in Chapter 9.

[3]The chain of custody issues associated with the O. J. Simpson trial are explored in Chapter 8.

[4]See Chapter 5 for examples of such occurrences.

[5]The IRS Restructuring and Reform Act of 1998 transferred all powers and responsibilities of the IRS Inspection Service to the Treasury Inspector General for Tax Administration, except one: The IRS Inspection Service is still responsible for conducting background checks and providing physical security.

[6]Treasury Inspector General for Tax Administration (TIGTA). (2007, December 31). IRS employee pleads guilty to unauthorized inspection of tax return information. *Highlights—2007 Archive*. Retrieved from http://www.ustreas.gov/tigta/oi_highlights_2007.shtml

[7]United States Attorney, Eastern District of New York. (2003, February 26). Six charged with trafficking in counterfeit computer software—$9 million worth of counterfeit software seized. US Department of Justice, Computer Crime & Intellectual Property Section. Retrieved from http://www.justice.gov/criminal/cybercrime/maArrest.htm

[8]United States Attorney, Southern District of Mississippi. (2010, March 23). Three found guilty in mortgage loan fraud scheme. Federal Bureau of Investigation, Jackson. Retrieved from http://jackson.fbi.gov/dojpressrel/pressrel10/ja032310.htm

[9]US Department of Justice. (2009, May 22). Tax shelter promoter pleads guilty to conspiring to impede and impair the IRS. Retrieved from http://www.justice.gov/tax/txdv09508.htm

[10]There are certain exceptions to a search warrant requirement, which are explored in detail in Chapter 4.

[11]*United States v. Upham,* 168 F 3d 532, 535 (1st Cir. 1999). See also Baron-Evans, A., & Murphy, M. F. (2003, May/June). The Fourth Amendment in the digital age: Some basics on computer searches. *Boston Bar Journal, 47,* 12.

[12]168 F 3d 532, 535 (1st Cir. 1999).

[13]Brenner, S. W., & Frederiksen, B. A. (2002). Computer searches and seizures: Some unresolved issues. *Michigan Telecommunications and Technology Law Review, 8,* 62–63.

[14]231 F.3d 630 (9th Cir. 2000).

[15]346 F. Supp. 2d 1226, 1264. (M.D. Fla., 2004).

[16]*United States v. Gawrysiak,* 972 F. Supp. 853, 866 (D.N.J. 1997).

[17]National Institute of Justice. (2001, July). Electronic crime scene investigation: A guide for first responders. Retrieved from http://www.ncjrs.org/pdffiles1/nij/187736.pdf

[18]305 F. App'x 640, 643 (11th Cir. 2008).

[19]527 F.3d 54, 65 (3d Cir. 2008).

[20]411 F.3d 1198, 1200 (10th Cir. 2005).

[21]Chapter 13 explores the role of a computer forensics investigator in providing testimony in pretrial hearings and the courtroom.

[22]Girard, J. E. (2008). *Criminalistics: Forensic science and crime.* Boston: Jones and Bartlett, pp. 36–37.

[23]352 S.E.2d 902, 905-06 (N.C. App. 1987).

[24]Butera, K. D. (1998). Seeing is believing: A practitioner's guide to the admissibility of demonstrative computer evidence. *Cleveland State Law Review, 46,* 511–532.

[25]The trier of fact is an individual (judge) or group of individuals (juries) that determines the facts of a case in legal proceedings.

[26]413 F.3d 1138 (10th Cir. 2005).

[27]*United States v. Hamilton,* 413 F.3d 1138, 1142-43 (10th Cir. 2005).

[28]*State v. Armstead,* 432 So. 2d 837 (La. 1983).

[29]432 So. 2d 837, 839-840 (1983).

[30]*Louisiana v. Armstead,* 432 So. 2d 837, 839 (1983).

[31]Kerr, O. S. (2001, March). Computer records and the Federal Rules of Evidence *USA Bulletin* [online]. Retrieved from http://www.cybercrime.gov/usamarch2001_4.htm

[32]*United States v. Ruffin,* 575 Fed. 2d 346, 356 (2d Cir. 1978); *Louisiana v. Armstead,* 432 So. 2d 837, 839 (1983).

[33]Jeffery Rapp's report about al-Marri may be obtained at http://www.washingtonpost.com/wp-srv/nation/documents/jeffreyrapp_document.pdf

[34]542 U.S. 507 (2004).

[35]48 Or App 881, 884 (1980).

[36]502 So. 2d 393 (1987).

[37]258 Va. 3, 6, 516 S.E.2d 475, 476-77 (1999).

[38]208 F.3d 633 (7th Cir. 2000).

[39]896 F.2d 1476, 1494 (7th Cir. 1990); see also *United States v. Croft,* 750 F.2d 1354 (7th Cir. 1984).

[40]*United States v. Chappell,* 698 F.2d 308, 311 (7th Cir. 1983); see also *United States v. Briscoe,* 896 F.2d 1476, 1494 (7th Cir. 1990).

[41]785 F.2d 184, 192 (7th Cir. 1986).

[42]235 F.3d 1318 (11th Cir. 2000).

[43]175 Md. App. 231 (July 2, 2007).

[44]Goode, S. (2009). The admissibility of electronic evidence. *Review of Litigation, 29,* 31; Morrison, C. P. (2004). Instant messaging for business: Legal complications in communication. *Journal of Law and Commerce, 24,* 155.

[45]34 Va. App. 364, 369 (2001).

[46]4 Va. App. 493, 496, 358 S.E.2d 750, 752 (1987); see also *Bloom v.* Commonwealth, 34 Va. App. 364, 369–370 (2001).

[47]200 F.3d 627 (9th Cir. 2000).

[48]152 F.3d 1241, 1249-50 (10th Cir. 1998).

[49]488 F. Supp. 2d 866, 869 (D. Neb. 2007).

[50]*United States v. Webster,* 84 F.3d 1056, 1064 (1996); *United States v. Jackson,* 488 F. Supp. 2d 866, 871 (D. Neb. 2007).

[51]*United States v. Jackson,* 488 F. Supp. 2d 866, 871 (D. Neb. 2007).

[52]Validity is defined as "the extent to which measures indicate what they are intended to measure." See Bachman, R., & Schutt, R. K. (2003). *The practice of research in criminology and criminal justice* (2nd ed.). London: Sage/Pine Forge Press, p. 68.

[53]293 F. 1013 (1923).

[54]E.g. *United States v. Downing,* 753 F.2d 1224 (1985).

[55]223 So. 2d 68 (Fla. Dist. Ct. App. 1968).

[56]509 U.S. 579 (1993).

[57]*Daubert v. Merrell Dow Pharmaceuticals Inc.,* 509 U.S. 579, 593-595 (1993); see also *Kumho Tire Company, Ltd. v. Carmichael,* 526 U.S. 137, 145 (1999).

[58]Reliability is established when a measure "yields consistent scores or observations of a given phenomena on different occasions." See Bachman, R., & Schutt, R. K. (2003). *The practice of research in criminology and criminal justice* (2nd ed.). London: Sage/Pine Forge Press, p. 72.

[59]*United States v. Downing,* 753 F.2d 1224, 1238, quoted in *Daubert v. Merrell Dow Pharmaceuticals Inc.,* 509 U.S. 579 (1993).

[60]509 U.S. 579 (1993).

[61]National Institute of Standards and Technology. (2001). *General test methodology for computer forensic tools.* US Department of Commerce.

Chapter 3

Laws Regulating Access to Electronic Evidence

This chapter explores the various laws that govern investigators' access to electronic evidence. More specifically, it examines existing legal privacy protections of telecommunications and electronic communications data by reviewing the Electronic Communications Privacy Act of 1986 (ECPA), Communications Assistance for Law Enforcement Act of 1994 (CALEA), the Cable Communications Privacy Act of 1984 (CCPA), the Uniting and Strengthening America by Providing Appropriate Tools Required to Intercept and Obstruct Terrorism Act of 2001 (USA Patriot Act), and the Telecommunications Act of 1996. This chapter also reviews, among other laws, the (Federal) Privacy Act of 1974, the e-Government Act of 2002, the Computer Security Act of 1987, the Privacy Protection Act of 1980 (PPA, the Children's Online Privacy Protection Act of 1998 (COPPA)1998, and the Sarbanes-Oxley Act of 2002 (Sarbox).

In considering these laws, this chapter focuses on the following questions:

- What do these laws cover?
- What do they prohibit?
- When is the disclosure of subscriber (or customer) records and the contents of telecommunications and electronic communications data to government agencies allowed?

Telecommunications and Electronic Communications Data

Several different types of telecommunications and electronic communications data are regulated by law. Telecommunications data include traffic data, location data, and other

51

related data that is necessary to identify a customer from fixed network telephony and mobile telephony. Electronic communications data also include traffic and location data about Internet access, e-mail, and telephony.

Traffic and Location Data

Traffic data consists of data about a communication. Among other things, it includes data that permit investigators to do the following:

- Determine the type of communication used (e.g., telephone or Internet)
- Trace and identify the source of a communication (e.g., the telephone number, name, and home address of the customer)
- Identify the destination of a communication (e.g., the numbers dialed)
- Determine the date, time, and duration of a communication (e.g., the start and end times of a telephone call or the log-on and log-off times of the Internet e-mail service)

Location data from mobile telephony, Internet telephony, access, and e-mail include data about the locations from which a particular communication was made, such as the geographic location from which a phone call was made.

Content and Non-content Data

Under 18 U.S.C. § 2510(8), the content of communication is defined as "any information concerning the substance, purport, or meaning of [a] communication." **Content data** are the spoken words in a conversation or the words written in a message (through either texting or e-mail). The content of telephone calls is protected under the Fourth Amendment to the United States Constitution.[1]

Non-content data include, but are not limited to, telephone numbers dialed, customer information (name and address), and e-mail addresses of the message sender and recipient. The U.S. courts, however, have held that such data are not protected under the Fourth Amendment. For instance, in *United States v. Forrester*,[2] the court ruled that the sender (to) and receiver (from) e-mail messages and the IP addresses of websites are not protected by the Fourth Amendment.

Stored non-content telecommunications and electronic communications data can reveal private information about the citizens using these forms of communications by revealing their contacts (who they phone or e-mail), movements (from mobile phone location data), and interests, political views, sexual preferences, and religious affiliations (through the websites they browse). When viewed together, these types of data can provide a detailed account of an individual's private life.

Despite claims to the contrary, non-content data are not retained because they are less intrusive than the content. As Walden argues, "The volume and value of [non-content] communications data has expanded considerably."[3] The value of non-content communications data has increased because, as mentioned previously, such data can reveal an individual's contacts, activities, locations, habits, and sensitive personal data (e.g., race, ethnic origin, political opinions, religious beliefs, sexual preferences). The value of non-content communications data has also expanded as a result of technological advances in communications. For example, third-generation mobile phones can pinpoint a user's location to a few meters, requiring only that the phone is switched on to find the user. Indeed, mobile phones that are used in conjunction with other technology such as the global positioning system (GPS) provide the means with which to "monitor, observe and trace their users continuously in real time."[4] This ability has been realized in the United States with the implementation of the **Wireless Communications and Public Safety Act of 1999**, which required all mobile phones created after the year 2000 to have GPS tracking capabilities.[5]

Finally, the value of non-content data has increased because traffic and location data are easier to obtain than content data as a result of new technologies that inhibit access to the content of communications data. For example, Pretty Good Privacy (PGP) software (a technique for encrypting messages that is often used to protect messages on the Internet) provides cryptographic privacy and authentication for users, thereby hampering law enforcement agencies' ability to gain access to the content of the communication. Consequently, "law enforcement agencies are increasingly reliant on evidence derived from communications data rather than content."[6]

The Statutory Background of Privacy Protection and Government Access to Data

In the United States, privacy is protected by numerous statutes that cover the collection, storage, access, use and disclosure of different types of personal data such as communications information.

Electronic Communications Privacy Act of 1986

Sections 2510 through 2522 and Sections 2701 through 2712 of Title 18 of the United States Code (which concerns crimes and criminal procedures) encompass **the Electronic Communications Privacy Act of 1986** (ECPA). The ECPA governs the privacy and collection, access, disclosure, and interception of content and traffic data related to electronic communications.

ECPA amended Title III of the **Omnibus Crime Control and Safe Streets Act of 1968** (the Wiretap Statute), which protected the privacy of only the content of telecommunications. This protection did not extend to electronic communications, however. Accordingly, the ECPA was created to specify the situations in which the government could and could not intercept citizens' electronic communications. Under Section 2510(12) of Chapter 119 of Title 18 of the U.S. Code, electronic communications are defined as "any transfer of signs, signals, writing, images, sounds, data, or intelligence of any nature transmitted in whole or in part by a wire, radio, electromagnetic, photoelectronic or photo-optical system that affects interstate or foreign commerce."

The ECPA places restrictions on government and law enforcement agencies' access to telecommunications and electronic communications data by requiring them to specify:

- The individual (or individuals) who is (are) the target of the investigation
- The crimes for which the individual (or individuals) is (are) being investigated
- The system from which the desired communications are to be accessed

Who Seeks Information Pursuant to This Act?
The ECPA applies to government agencies.

Who Must Provide Information Under This Act?
Providers of the following services are subject to this act:

- **Electronic communications services** (ECS): Defined by 18 U.S.C. § 2510 as "any service which provides to users thereof the ability to send or receive wire or electronic communications."
- **Remote computing services** (RCS): Defined by 18 U.S.C. § 2711(2) as "the provision to the public of computer storage or processing services by means of an electronic communications system."

In *United States v. Kennedy*,[7] the court pointed out that an Internet service provider (ISP) was considered as both an ECS and RCS provider. A private provider of electronic communications can also be considered as a provider of ECS. For example, in *United States v. Mullins*,[8] the court stated that an airline was an ECS because it provided an electronic travel reservation system that could be accessed through separate computer terminals. Similarly, in *Bohach v. City of Reno*,[9] the court ruled that the city was an ECS because it provided pager services to its police officers.

Types of Information Disclosed
A government agency can require the disclosure of subscriber information, electronic transmission (or transactional information), and the content of wire or electronic communications. The ECPA limits the telecommunications and electronic communications data that may or may not be voluntarily disclosed to others (including government agencies).

Content of Communications
Sections 2510 through 2522 of Chapter 119 (which concerns wire and electronic communications interception and interception of oral communications) of Title 18 of the U.S. Code govern the real-time surveillance of the contents of communications by government agencies. In particular, these regulations establish the procedures that govern-

ment agencies must follow to legally intercept the content of communications. In addition, 18 U.S.C. § 2702(b) specifies exceptions to when service providers can disclose the contents of communications. For instance, the contents of communications may be disclosed to government agencies with the consent of any party of the communication. Only one party needs to consent (see 18 U.S.C. § 2511). The content of communications may also be disclosed by a service provider if the provider is required to do so by law. For example, the **Child Protection and Sexual Predator Punishment Act of 1998** requires service providers that become aware of child pornography to report it.

In *Warshak v. United States*,[10] the court ruled that government investigators could not obtain e-mails from an ISP unless they had a search warrant. Government access to e-mail depends on whether an e-mail is unopened or has been accessed by the user. Unopened communications, including both e-mail and voice mail, that have been in electronic storage for 180 days or less can be obtained by government entities only pursuant to a search warrant. Unopened e-mails and voice mail that have been stored for more than 180 days can be obtained with a subpoena, a court order that meets the requirements of 18 U.S.C. § 2703(d),[11] or a search warrant. Opened e-mails can also be obtained by subpoena, a court order pursuant to 18 U.S.C. § 2703(d), or a search warrant.

According to the Department of Justice, "providers of services not available 'to the public' may freely disclose both contents and other records relating to stored communications. ECPA imposes restrictions on voluntary disclosures by providers of services to the public, but it also includes exceptions to those restrictions."[12] Public service providers cannot voluntarily provide content data to authorities unless one of the exceptions listed in § 2702(b) is met.

Non-content Data of Communications

Sections 2701 through 2712 of Chapter 121 (which concerns stored wire and electronic communications and transactional records access) of Title 18 of the U.S. Code prohibit unauthorized access to stored electronic communications and require specific procedures to be followed by government agencies before stored electronic communications can be accessed.

18 U.S.C. § 2702(c) contains exceptions regarding when service providers can disclose the non-content data of communications. Under 18 U.S.C. § 2703(c)(2), subscriber (or customer) records, which include, but are not limited to, information such as the name, address, and telephone number of the subscriber, can be obtained by government agencies. In *United States v. Allen*,[13] the court pointed out that a log identifying the "date, time, user, and detailed internet address of sites accessed" by a subscriber constituted subscriber information under the ECPA. Criminal investigators can use subpoenas to obtain basic subscriber information.

In *McVeigh v. Cohen*,[14] the court stated that the ECPA prohibits the government from obtaining the private information of a subscriber from an electronic communications service provider without a warrant, subpoena, or court order. Authorities cannot access customer records that belong to electronic communications service providers without a search warrant or court order, unless they have received the consent of the customer. Furthermore, public service providers cannot voluntarily provide this type of information unless it meets one of the exceptions listed in § 2702(c).

The ECPA in a Nutshell

- *Covers*: The content of electronic communications in ECS and RCS.
- *Prohibits*: Service providers from disclosing customer records, transactional data, and content to government agencies except as provided by the provisions of the ECPA.
- *Limits*: Government access to records held by third-party service providers. It does not limit access to stored electronic communication records on the suspect's computer.

Data Preservation Versus Data Retention

Data preservation—that is, the ad hoc "freezing" of communications data—is based on targeted surveillance, where "it affects only a limited number of individuals during specific periods rather than the entire population all of the time."[15] This practice provides "authorities with the power to order the logging and disclosure of traffic data in regards to . . . communications" on a case-by-case basis.[16] 18 U.S.C. § 2703 requires the preservation and disclosure of customer communications records to government entities.[17] Under 18 U.S.C. § 2703(f)(1), a provider of wire or electronic communication services, upon request of a governmental entity, is required "to preserve records and other evidence in its possession pending the issuance of a court order or other process." As per § 2703(f)(2), the data should be "retained for a period of 90 days, which shall be extended for an additional 90-day period upon a renewed request by the governmental entity."

Today, the United States does have certain communications providers retain data, but only for a limited time. Specifically, toll telephone service providers must retain records required for business purposes (such as billing) for a period of 18 months.[18] Any record that is not required for business purposes may be discarded.

Europe has not only followed suit in mandating some data preservation, but has also taken this requirement a step further. The Council of Europe's **Convention on Cybercrime** contains provisions for the collection of communications data. Among its various provisions, this Convention includes procedures for the preservation of telecommunications and electronic communications traffic data. The Cybercrime Convention allows law enforcement agencies to request data to be preserved upon notice for certain periods of time (the "fast freeze–quick thaw" model). Under this model, service providers are to store the data quickly, upon request by law enforcement authorities. This model provides for only individual secure storage of data; those data must subsequently be quickly released to the authorities upon receipt, for example, of a court order.

The Data Protection Directive (**Directive 95/46/EC**), which was created to protect individuals with regard to the processing of their personal data and the free movement of such data, in accordance with European Community law, required communications service providers to erase data or make the data anonymous after they were no longer needed for business purposes.[19] The Telecommunications Directive (Directive 97/66/EC),[20] which supplemented Directive 95/46/EC, translated the principles set out in the Data Protection Directive into specific rules for the protection of privacy and personal data in telecommunications sectors.

Due to the rise of the Internet and the ever more frequent use of electronic communications, this directive was already outdated by the time it was finally implemented in 1997. For that reason, Directive 97/66/EC was repealed and replaced by the Electronic Communications Directive (**Directive 2002/58/EC**). Directive 2002/58/EC, which translated the principles set out in Directive 95/46/EC into specific rules for the protection of privacy and personal data in all forms of communications, including electronic communications (e.g., e-mail), also required the deletion of all traffic data no longer required for specific and limited services (Article 5 and 6). Yet, under Article 15 of Directive 2002/58/EC, member states of the European Union were permitted to restrict the scope of the rights in Articles 5 and 6, albeit only if the interference was proportionate and necessary within a democratic society to safeguard national security, to provide for defense, to ensure public security, and for the prevention, investigation, detection, and prosecution of criminal offenses. If Article 15 was not invoked, then Article 6 required the mandatory destruction of data. Nevertheless, the Counsellor for Justice Affairs asserted that the failure of a single country to enact an exception to the default data destruction requirements would significantly hinder authorities' efforts to prevent and investigate serious crime.[21] Therefore, it was argued that Directive 95/46/EC and Directive 2002/58/EC could not ensure the availability of data beyond business purposes.

To remedy this situation in Europe, the Data Retention Directive (**Directive 2006/24/EC**)[22] was implemented. This measure makes compulsory the a priori retention of traffic, location, and other related data of individuals to make sure that such information is available to authorities.[23] Data are retained to ensure their availability for both domestic and international criminal investigations and prosecutions of serious crimes, such as terrorism.

Since the Data Retention Directive's implementation in Europe, U.S. government officials have seriously considered implementing mass data retention. In fact, in 2009, two bills with similar titles were presented that contained provisions mandating mass data retention: S. 436, Internet Stopping Adults Facilitating the Exploitation of Today's Youth (SAFETY) Act of 2009, in the Senate and H.R. 1076, Internet Stopping Adults Facilitating the Exploitation of Today's Youth (SAFETY) Act of 2009, in the House of Representatives. These bills seek to amend 18 U.S.C. § 2703 by requiring ECS and RCS providers to retain "for a period of at least two years all records or other information pertaining to the identity of a user of a temporarily assigned network address the service assigns to that user."

(continues)

Critical Thinking Questions *(Continued)*

Given that data retention may be implemented in the United States, think about the following questions:

1. Should such laws be implemented in the United States?

2. What impact do you think these laws might have on Internet users? What about criminal investigations and prosecutions?

3. What are the implications of this legislation for the privacy of Internet users?

4. Will these laws help or hinder authorities in their investigations and prosecutions of serious crime?

Communications Assistance for Law Enforcement Act of 1994

The **Communications Assistance for Law Enforcement Act of 1994** (CALEA) was originally designed for telecommunications. With the advent of the Internet, it was later extended to include electronic communications. Nowadays, interconnected Voice over Internet Protocol (VoIP) and broadband Internet access service providers are also covered by CALEA.

Under CALEA (codified at 47 U.S.C. § 1001–1021), telecommunications and electronic communications service providers are required to ensure that government entities, pursuant to a lawful authorization (e.g., a search warrant), have access to all wire and electronic communications and "call identifying information."[24] As such, service providers are required to place the necessary equipment on the network to enable and facilitate wiretapping by law enforcement agencies. This law is intended to make sure that new technologies will not hamper law enforcement agencies' ability to conduct surveillance when they have the legal authority to do so. Pursuant to 47 U.S.C. § 1002(a)(4)(A), it is the service provider's responsibility to ensure the privacy of the wire and electronic communications and call-identifying information that has not been authorized for interception by government entities.

Cable Communications Privacy Act of 1984

47 U.S.C. § 551 encompasses the **Cable Communications Privacy Act of 1984** (CCPA). The CCPA both restricts cable operators from using their systems to collect personally identifiable information and from disclosing such data under § 551(b) and § 551(c), respectively. More specifically, according to § 551(b), cable operators are prohibited from using the cable system to collect personal data concerning any customer without his or her prior consent, unless the information is needed for business purposes (e.g., billing) or to detect unauthorized reception. The CCPA also prohibits cable operators from dis-

closing data to third parties without consent, unless the information is necessary for business purposes or is provided to government agencies pursuant to a court order as provided by § 551(h).

When it was passed in 1984, the CCPA contained provisions that severely hindered criminal investigations. This was particular true for cable companies that offered Internet access and telephone service along with their cable services, as the disclosure of Internet and telephone data of these cable services to law enforcement agencies was still governed by the CCPA. Section 211 of the Uniting and Strengthening America by Providing Appropriate Tools Required to Intercept and Obstruct Terrorism Act of 2001 (USA Patriot Act) sought to remedy this problem by stating that where a cable company provides telephone and/or Internet services, it must comply with the laws that govern the interception and disclosure of wire, oral. and electronic communications of telecommunications and Internet service providers (18 U.S.C. § 2510–2522). They must also comply with the laws regulating pen registers and trap-and-trace devices (18 U.S.C. § 3121–3127) and those governing access to stored wire and electronic communications and transactional records access (18 U.S.C. § 2701–2712).

Basically, the CCPA requires the cable service companies to inform customers about the collection of their personal data and when (and if) their information is disclosed to third parties.

USA Patriot Act of 2001

On October 26, 2001 (just six weeks after the terrorist attacks of September 11), Congress passed the broadest anti-crime act in American history, the USA Patriot Act of 2001 (also known more simply as the Patriot Act). This law sought, among other things, to deter and punish terrorist acts in the United States and worldwide and to enhance law enforcement surveillance powers. In support of these goals, Section 210 of the Act expanded the types of data that law enforcement agencies could access by meeting only the lowest ECPA standards. For instance, the Patriot Act made subscriber and customer records, such as transaction history and IP logs, available by subpoena. This information was previously available only by search warrant or court order. Essentially, this legislation was implemented to ease the restrictions on government and law enforcement access to communications data.

The Patriot Act amended existing legislation to remove any unnecessary burdens in criminal investigations. Of particular note were the following provisions:[25]

- *The Patriot Act made it easier to obtain the personal information of Internet users.* Prior to this Act, court orders were required to obtain limited user information. Section 210 of the Patriot Act allows the name, address, telephone numbers, Internet session times and duration, length and type of service used, IP address, and means of payment for services (bank account or credit card number) of the subscriber to be obtained with a subpoena.

- Law enforcement authorities have used Section 210 to obtain critical subscriber data in child pornography and child molestation cases.[26] In one case, these data helped investigators locate the creator and distributor of child pornography images, who turned out to be the father of the child in the photographs. Such data also provided critical investigative clues in Operation Artus, in which investigators used this information to dismantle a child pornography ring. Section 210 was also used during Operation Hamlet to track down child molesters who would videotape themselves sexually assaulting children and post these videos on the Internet—often using live feeds. This operation led to the rescue of more than 100 child victims of sexual assault.

- *The Patriot Act expanded the list of offenses for which wiretap orders can be obtained.* Section 202 of the Patriot Act provided authorities with the ability to intercept wire, oral, and electronic communications relating to computer fraud and abuse offenses (e.g., hacking) covered in the Computer Fraud and Abuse Act of 1984 (codified in 18 U.S.C. § 1030) and its subsequent amendments. Prior to the Patriot Act, these offenses were not listed under 18 U.S.C. § 2516(1). Accordingly, an investigator could not obtain a wiretap order to intercept the wire communications of those persons violating the provisions of 18 U.S.C. § 1030.

- *The Patriot Act allowed investigators to obtain voice mails with a warrant.* Prior to the Patriot Act, voice mails were treated like telephone conversations. As such, investigators needed a wiretap order to obtain these messages because the ECPA did not cover voice mail in its provisions. Under the Patriot Act, they are considered to be equivalent to e-mails and, therefore, can be obtained with a warrant.

- *The Patriot Act extended the use of pen registers and trap-and-trace devices to electronic communications.* Prior to the Patriot Act, pen registers and trap-and-trace devices were associated with telecommunications. Basically, pen registers would record the phone numbers that a person of interest called, while trap-and-trace devices would record the phone numbers of individuals who called the person of interest. The Patriot Act authorized the use of these technologies for electronic communications (see Section 216). To do so, it amended the definition of both pen registers and trap-and-trace devices in 18 U.S.C. § 3127(3) and 18 U.S.C. § 3127(4), respectively, as follows:

  - A pen register is "a device or process which records or decodes dialing, routing, addressing, or signaling information transmitted by an instrument or facility from which a wire or electronic communication is transmitted."

  - A trap-and-trace device "captures the incoming electronic or other impulses which identify the originating number or other dialing, routing, addressing, and signaling information reasonably likely to identify the source of a wire or electronic communication."

Section 216 has been successfully used to track down the perpetrators of cybercrime. For example, in California, pen/traps were used to find a hacker group that was responsible for committing distributed denial of service (DDoS) attacks against military networks.[27]

- *The Patriot Act authorized nationwide search warrants for electronic evidence.* Section 220 of the Patriot Act eliminated the jurisdictional restrictions placed on authorities regarding their access to electronic evidence under a search warrant. Pursuant to this Act, the court with jurisdiction over the crime being investigated executes the warrant. This practice is quite beneficial in time-sensitive investigations, which in the past were often hindered by unnecessary delays in the retrieval of critical telecommunications and electronic communications data.

  Section 220 proved quite useful in the investigation of the "Portland Seven," for example. This group of terrorists, whose members resided in Portland, Oregon, was planning to attack U.S. soldiers fighting in Afghanistan. According to Christopher Wray, the former Assistant Attorney General of the Criminal Division of the U.S. Department of Justice, Section 220 of the Patriot Act assisted in the investigation of these terrorists by allowing the "judge who was most familiar with the case . . . to issue the search warrants for the defendant[s'] e-mail accounts from providers in other districts, which dramatically sped up the investigation and reduced . . . unnecessary burdens on . . . prosecutors, agents and courts."[28]

- *The Patriot Act authorized the nationwide execution of court orders for pen registers, trap-and-trace devices, and access to stored e-mail or communications records.* Prior to the Patriot Act, usually several court orders for pen registers and trap-and-trace devices (or, more simply, pen/trap orders) were placed for a communication, because a communication is normally routed through various jurisdictions. Consider the following scenario, which depicts how pen/trap orders might be placed for each jurisdiction where the communication was routed: A pen/trap order is placed with a phone company in New York. The New York phone company replies that it does not have the information on the call that the law enforcement agency (LEA) requested. The New York phone company refers the LEA to a phone company in Florida. The LEA gets a pen/trap order for the Florida phone company, contacts the company, and places a pen/trap order. The Florida phone company informs the LEA that it does not have the data on the call requested and refers the LEA to a phone company in California. Subsequently, the LEA gets a pen/trap order for the California phone company, contacts the company, and places a trap order. The California phone company has the call data for which LEA was looking.

  With the Patriot Act, a single, nationwide court order suffices, which in turn eliminates the unnecessary burden of placing separate pen/trap orders in each jurisdiction. When the nationwide pen/trap order is obtained, the LEA does not have to name the specific telecommunications or electronic communications service providers that will be involved. Instead, the names of these service providers may be certified at a later date.

Telecommunications Act of 1996

The **Telecommunications Act of 1996**, which amended the **Communications Act of 1934**, was implemented in response to widespread concerns over telecommunications companies' misuse of customers' personal information records. This Act requires telecommunications companies to obtain the consent of the customer before using customer proprietary network information (CPNI). Section 222(h)(1)(A) and Section 222(h)(1)(B) of this Act define CPNI as "information that relates to the quantity, technical configuration, type, destination, location, and amount of use of a telecommunications service subscribed to by any customer of a telecommunications carrier, and that is made available to the carrier by the customer solely by virtue of the carrier-customer relationship" and "information contained in the bills pertaining to telephone exchange service or telephone toll service received by a customer of a carrier," respectively. Principally, CPNI covers data about usage and calling patterns of customers.

47 U.S.C. § 222 concerns the privacy of customer information and requires telecommunications providers to keep CPNI confidential. Specifically, under 47 U.S.C. § 222(c)(1), these service providers may only use, disclose, or permit access to this information in the course of providing their services and for other authorized business purposes. Certain exceptions to this rule are included in 47 U.S.C. § 222(d). For example, one exception deals with the provision of the location data of a call placed by a mobile phone to law enforcement personnel so that they may respond to a call for emergency services.[29]

Privacy Act of 1974

The (Federal) **Privacy Act of 1974** (codified in 5 U.S.C. § 552a) was implemented to "promote accountability, responsibility, legislative oversight, and open government with respect to the use of computer technology in the personal information systems and databanks of the Federal Government."[30] To protect individuals' privacy, this Act regulates the collection, maintenance, use, and dissemination of information by government agencies. It also seeks to limit the collection of personal data by government agencies.

The Privacy Act protects U.S. citizens and permanent residents[31] from intrusions by the federal government into their private life. However, it protects only those records of personal information that are held and handled by government agencies—not those maintained by private agencies. Under 5 U.S.C. § 552a(a)(4), these government records include

> Any item, collection, or grouping of information about an individual that is maintained by an agency, including, but not limited to, his education, financial transactions, medical history, and criminal or employment history and that con-

tains his name, or the identifying number, symbol, or other identifying [data] particular . . . to the individual, such as a finger or voice print or a photograph.[32]

The private information in these records cannot be disclosed without the written consent of the individual. Certain exceptions to this rule are listed in 5 U.S.C. § 552a(k). For instance, a federal agency does not need to receive an individual's consent when the agency is seeking to protect the health or safety of the individual.

The Privacy Act prohibits government agencies from having or maintaining secret record systems on individuals. It also requires that government agencies inform individuals of these record systems by listing them in the *Federal Register.*

The Privacy Act was amended by the **Computer Matching and Privacy Protection Act of 1988**. Computer matching can be defined as a computer-supported process in which personal data from records about individuals held by government agencies are compared for the purpose of determining eligibility for federal benefits programs. A major criticism of computer matching relates to the possibility of errors in the information held in the databases. Inaccurate records can result from erroneously reported or incorrectly entered information into databases. Given that this information is used to determine the eligibility of the individual for federal benefit programs, an individual with errors in his or her record may be wrongly denied access to government services (e.g., welfare). The Computer Matching and Privacy Protection Act does not provide individuals with the right to access or correct the information in their records, although the Privacy Act does. In fact, the Privacy Act allows individuals to review the personal information in their records to ensure that it is timely, accurate, relevant, and complete. If one or more errors are found in a person's file, the individual can subsequently correct them.

Under the provisions of the Privacy Act, the requesting individual can see only his or her own record. If other information is sought, such as government records on specific matters (e.g., a document created by a federal agency), it can be obtained through the **Freedom of Information Act of 1966** (FOIA; codified in 5 U.S.C. § 552) and the **Electronic Freedom of Information Act of 1996** (e-FOIA; see Public Law No. 104–231, 110 Stat. 3048). The FOIA and e-FOIA do not apply to private companies and organizations, nor do they apply to state and local governments. For their part, state and local governments have implemented similar laws to FOIA to allow individuals to access their records. If an individual is requesting information under the FOIA, e-FOIA, or Privacy Act about a third party, written consent must be obtained from the third party (i.e., the individual whose record is sought).

e-Government Act of 2002

Section 208 of the **e-Government Act of 2002** contains specific privacy provisions for personal information. Namely, this legislation requires government agencies to perform privacy impact assessments on computer and information technology systems that are designed to collect, maintain, and disseminate information about individuals in "identifiable form."[33]

According to Section 208(b)(2)(B)(ii), privacy impact assessments should address the following questions:

- Which information is being collected?

- Why is this information being collected?

- How is this information being used by the agency?

- Is this information being shared with other agencies? If so, with whom?

- Does the agency notify individuals of the information that is collected about them and shared with other agencies?

- Which opportunities, if any, does the agency provide for individuals to consent to the collection and dissemination of their data?

- How are individuals' personal information secured?

- Is the agency creating a system of records as defined by the Privacy Act?[34]

The e-Government Act also requires certain privacy protections to be adopted on federal government websites. In particular, it requires government agencies to post privacy notices on their sites. These notices must include information similar to that covered in the privacy impact assessments, including which information is being collected, for what purposes, how the information is being used, who is the information being shared with, and so on.[35]

Computer Security Act of 1987

The **Computer Security Act of 1987** amended the **National Bureau of Standards Act of 1901**[36] and the **Federal Property and Administrative Services Act of 1949** (the Brooks Act).[37] The main reason for implementing this legislation was to improve privacy protections and establish minimum acceptable security practices for federal computer systems that contain sensitive information. As defined in this Act, sensitive information includes "any information, the loss, misuse, or unauthorized access to or modification of which could adversely affect the national interest or the conduct of Federal programs or the privacy to which individuals are entitled under" the Privacy Act of 1974. The Computer Security Act, however, excludes information that has been "specifically authorized under criteria established by an Executive Order or an Act of Congress to be kept secret in the interest of national defense or foreign policy."

Section 2(b)(1) of the Computer Security Act assigns the responsibility for developing standards and guidelines for government computer systems to the National Bureau of Standards (which has been since renamed as the National Institute for Standards and Technology). These standards and guidelines are required to secure government computer and information systems and protect the privacy of sensitive information within these systems. According to Section 3 of the Act, the primary purposes of these standards and guidelines are to control and prevent the loss, unauthorized modification, or disclosure of sensitive information in these systems.

Section 2(b)(3) requires the development of security plans for these systems by all operators that handle systems containing sensitive information. As such, it requires government agencies to identify and secure the computer systems that contain sensitive information.

Section 2(b)(4) requires the "mandatory periodic training for all persons involved in management, use, or operation of Federal computer systems that contain sensitive information." This training is designed to both increase security awareness and educate employees on the proper handling of sensitive data.

Since its passage, the Computer Security Act has been superseded by the **Federal Information Security Management Act of 2002** (FISMA), which is Title III of the e-Government Act of 2002.[38] FISMA is currently the primary legislation that governs the information security of government systems. Under 44 U.S.C. § 3542(b)(1), information security focuses on the protection of "information and information systems from unauthorized access, use, disclosure, disruption, modification, or destruction" so as to maintain the integrity, confidentiality, and availability of data.[39]

Privacy Protection Act of 1980

42 U.S.C. § 2000aa encompasses the **Privacy Protection Act of 1980**. The Privacy Protection Act was implemented to reduce the **chilling effect** that searches and seizures had on publishers. A chilling effect occurs "when individuals otherwise interested in engaging in a lawful activity are deterred from doing so in light of perceived or actual government regulation of that activity."[40] These searches and seizures, therefore, may deter socially desirable conduct—namely, the dissemination of information and freedom of speech—and may ultimately confine individuals' freedom to pursue their own interests and activities. Such searches and seizures may further interfere with expression of political dissent and views, "some of which may be unpopular with those exercising state power."[41]

Consider *Zurcher v. Stanford Daily*.[42] In this case, the police went to the *Stanford Daily* newsroom and searched for unpublished photographs of violent demonstrators after the newspaper ran a story with photographs of the demonstration. Recognizing the need for journalists to be able to collect and distribute the news without fear of government interference, Congress passed the Privacy Protection Act. This legislation protects journalists, publishers, and other such individuals by not requiring them to turn over work that falls under the protection of the **First Amendment**[43] to law enforcement agencies. These materials can be seized only if the law enforcement agencies have probable cause to believe that the publisher of these materials is involved in a criminal offense. 42 U.S.C. § 2000aa(a)(2) also allows these materials to be seized if "there is reason to believe that the immediate seizure of such materials is necessary to prevent the death of, or serious bodily injury to, a human being."

42 U.S.C. § 2000aa also protects "documentary materials, other than work product materials, possessed by a person in connection with a purpose to disseminate to the public a newspaper, book, broadcast, or other similar form of public communication, in or affecting interstate or foreign commerce," from searches and seizures. There are, however, certain exceptions to this rule.[44] One such exception exists if law enforcement agencies have "reason to believe that the giving of notice pursuant to a **subpoena** *duces tecum* [emphasis added] would result in the destruction, alteration, or concealment of such materials."[45]

Children's Online Privacy Protection Act of 1998

The **Children's Online Privacy Protection Act of 1998** (COPPA) applies to children younger than 13 years of age and protects them from the collection and misuse of their personal information by commercial websites. Under 15 U.S.C. § 1302(8), the type of personal information that is collected about children and included in the purview of this legislation includes their name, home address, e-mail address, telephone number, and Social Security number. 15 U.S.C. § 1303(b)(1)(A) requires websites to provide notices of their information collection and use practices with respect to children. 15 U.S.C. § 1303(b)(1)(B) affords parents with the opportunity to review the information collected about their children.

The COPPA does not control the dissemination of information to children. The **Child Online Protection Act of 1998** (COPA), which was declared unconstitutional in 2007, tried to do so by prohibiting websites from knowingly making available harmful material to minors. In contrast, the COPPA seeks to give parents control over the collection, use, and dissemination of the personal information of their children. The main purpose of this legislation is to protect the privacy of children when they are using electronic communications.

The Communications Decency Act of 1996

The **Communications Decency Act of 1996** (47 U.S.C. § 223) was designed to protect children from exposure to indecent material. A violation of this Act occurs when an individual knowingly transmits indecent material to a person younger than 18 years old. This violation can occur even if an individual younger than 18 years old is the one who accesses a legal website. The website may be held liable if it has not taken adequate precautions to prevent children from accessing this type of material. The sections of this Act[46] that attempted to criminalize the content of sites that contained indecent and patently offensive materials was struck down by the Supreme Court in *Reno v. American Civil Liberties Union*.[47]

Sarbanes-Oxley Act of 2002

The **Sarbanes-Oxley Act of 2002**[48] was created as part of Congress' attempts to improve the accuracy and reliability of corporate disclosures so as to protect investors from fraudulent business practices. In fact, this law was implemented as a response to corporate scandals such as those involving Enron, Worldcom, and Tyco, to name a few.

The purpose of the Sarbanes-Oxley Act is twofold. First, this law seeks to make business practices more transparent. Second, it seeks to make corporate entities accountable for errors, fraudulent business practices, and obstruction of justice.

Of particular importance and relevance are *Title VIII* and *Title XI* of this Act.

Title VIII: Corporate and Criminal Fraud Accountability

Title VIII of the Sarbanes-Oxley Act is also known as the **Corporate and Criminal Fraud Accountability Act of 2002**. Section 802 of this Act provides for criminal penalties for

---

The Events That Triggered the Creation of the Sarbanes-Oxley Act

Enron Scandal

In 2001, terrorism was not the only thing that plagued the United States—uncontrolled corporate greed was also running rampant. That year an energy trading company known as **Enron** engaged in one of the most infamous and complex fraudulent accounting schemes of all time. In the Enron scandal, the firm's personnel showed shareholders that the company was making a profit when, in fact, it was billions of dollars in debt. In the end, shareholders and employees lost billions of dollars in stock prices and pensions.

Worldcom Scandal

A year later, the scandals continued with the fall of a telecommunications company, **Worldcom**. Worldcom was also involved in fraudulent accounting practices—albeit not as intricate as those of Enron. By falsifying business records (i.e., reclassifying operating expenses into corporate investments), company employees were able to show billions in profits, even though the company was actually in debt.

Tyco International Scandal

That same year another scandal erupted in the United States. The chief executive officer (CEO) of **Tyco International**, Dennis Kozlowski, and the chief financial officer (CFO), Mark Swartz, siphoned millions of dollars from the company through fraudulent stock sales and unauthorized business loans. To cover up their actions, they disguised the money as executive bonuses and benefits. Investors in this company lost millions of dollars as a result of Kozlowski's and Swartz's stock manipulations and false business reports.

altering documents. Specifically, this section amended Chapter 73 of Title 18 of the United States Code by adding two sections, § 1519 and § 1520. 18 U.S.C. § 1519 covers the "[d]estruction, alteration, or falsification of records in Federal investigations and bankruptcy." In particular, this section calls for the payment of fines or imprisonment (for a period no greater than 20 years) for

> Whoever knowingly alters, destroys, mutilates, conceals, covers up, falsifies, or makes a false entry in any record, document, or tangible object with the intent to impede, obstruct, or influence the investigation or proper administration of any matter within the jurisdiction of any department or agency of the United States or any case filed under title 11 [of the U.S.C.].

Moreover, 18 U.S.C. § 1520 concerns the "destruction of corporate audit records." Anyone who violates this section is liable for a fine or a term of incarceration of no more than 10 years. The Corporate and Criminal Fraud Accountability Act also includes a section that administers criminal penalties to companies that defraud their shareholders (see Section 807).

Title IX: Corporate Fraud Accountability

Title IX of the Sarbanes-Oxley Act is also known as the **Corporate Fraud Accountability Act of 2002**. Section 1102 of the Corporate Fraud Accountability Act concerns the actions of "tampering with a record or otherwise impeding an official proceeding." Section 1102 of the Act amended 18 U.S.C. § 1512 (tampering with a witness, victim, or informant). According to this section, a person may be fined or imprisoned (for a period no greater than 20 years) if he or she "alters, destroys, mutilates, or conceals a record, document, or other object, or attempts to do so, with the intent to impair the object's integrity or availability for use in an official proceeding" or "otherwise obstructs, influences, or impedes any official proceeding, or attempts to do so."[49]

The Sarbanes-Oxley Act also contains the following provisions:

- Defines the types of records that must be recorded and for how long
- Deals with falsification and destruction of data
- Holds executives directly responsible for problems
- Defines e-mails as corporate documents
- Requires all business records, including electronic records and electronic messages, to be saved for no less than 5 years

Chapter Summary

This chapter concerned the regulation of telecommunications and electronic communications data. First, it explained which types of data are included in both types of communication. The differences between content and non-content data were then explored. Next, the focus shifted to the various laws that govern the privacy of personal data. Government agencies must follow specific rules when they seek to retrieve data that may assist in their investigations and may serve as evidence in a court of law. Issues concerning the voluntary and compelled disclosure of certain types of data to government agencies by communications service providers were also addressed in this chapter.

Key Terms

Cable Communications Privacy Act of 1984
Carnivore
Child Online Protection Act of 1998
Child Protection and Sexual Predator Punishment Act of 1998
Children's Online Privacy Protection Act of 1998
Chilling effect
Communications Act of 1934
Communications Assistance for Law Enforcement Act of 1994
Communications Decency Act of 1996
Computer Matching and Privacy Protection Act of 1988
Computer Security Act of 1987
Content data
Convention on Cybercrime
Corporate and Criminal Fraud Accountability Act of 2002
Corporate Fraud Accountability Act of 2002
Directive 95/46/EC
Directive 2002/58/EC
Directive 2006/24/EC
e-Government Act of 2002
Electronic Communications Privacy Act of 1986
Electronic communications services
Electronic Freedom of Information Act Amendments of 1996
Enron scandal
Executive Order 13181
Family Education Rights and Privacy Act of 1974

Federal Information Security Management Act of 2002
Federal Property and Administrative Services Act of 1949
First Amendment
Freedom of Information Act of 1966
Gramm-Leach-Bliley Act
Health Insurance Portability and Accountability Act of 1996
Location data
National Bureau of Standards Act of 1901
Non-content data
Omnibus Crime Control and Safe Streets Act of 1968
Privacy Act of 1974
Privacy Protection Act of 1980
Regulation of Investigatory Powers Act of 2000
Remote computing services
Right to Financial Privacy Act of 1978
Sarbanes-Oxley Act of 2002
Stored Wire and Electronic Communications and Transactional Records Act of 1996
Subpoena *duces tecum*
Telecommunications Act of 1996
Traffic data
Tyco International scandal
USA Patriot Act of 2001
Wireless Communications and Public Safety Act of 1999
Worldcom scandal

Critical Thinking Questions

The **Regulation of Investigatory Powers Act of 2000** (RIPA) requires all Internet service providers in the United Kingdom to "install black boxes on their network to monitor all data as it passes and subsequently feed it to a central processing location controlled by the United Kingdom's security service MI-5."[50] U.K. authorities do not need warrants "to monitor patterns, such as websites visited, to and from whom email was sent, which pages are downloaded, of which discussion groups a user is a member, and which chat rooms an individual visits."[51]

A similar program, known as **Carnivore**, exists in the United States. When Carnivore is installed at facilities of ISPs, it can monitor all the traffic moving through them. In fact, before the USA Patriot Act was implemented, if the FBI suspected someone of a criminal activity, it would request authorization from the court to use Carnivore. After a court order was obtained, the FBI would install Carnivore at the suspect's ISP. While Carnivore could collect specific data packets included and authorized by the court order, it could not collect any other individual packets in the process. Specifically, according to a statement made by the former Deputy Assistant Attorney General for the U.S. Department of Justice:

> Carnivore is, in essence, a special filtering tool that can gather the information authorized by court order, and only that information. It permits law enforcement, for example, to gather only the email addresses of those persons with whom the drug dealer is communicating, without allowing any human being, either from law enforcement or the service provider, to view private information outside of the scope of the court's order. [52]

By contrast, the Associate Director of the American Civil Liberties Union has claimed that

> Carnivore permits access to the e-mail of every customer of an ISP and the [e-mail] of every person who communicates with them. Carnivore is roughly equivalent to a wiretap capable of accessing the contents of the conversations of all of the phone company's customers, with the "assurance" that the FBI will record only conversations of the specified target . . . In essence, Carnivore is a black box into which flows all of the service provider's communications traffic. The service provider knows what goes in, but it has no way of knowing what the FBI takes out.[53]

Additionally, it is feared that the USA Patriot Act also permits government and law enforcement agencies to go on fishing expeditions.

1. What do you believe? And why?
2. How does Carnivore fit with existing laws?

Review Questions

1. What is traffic data?

2. What is location data?

3. What are the differences between content and non-content telecommunications and electronic communications data?

4. What does the ECPA permit law enforcement agencies to do?

5. How can U.S. government agencies obtain subscriber records from telecommunications and electronic communications service providers?

6. When can a U.S. government agency obtain a suspect's e-mail?

7. When can providers disclose e-mails and records to the U.S. government voluntarily?

8. How do the ECPA and the USA Patriot Act regulate the interception of electronic communications, government access to those communications, and government access to ISP records?

9. Which laws regulate personal information stored in government databases?

10. Which sections of the Sarbanes-Oxley Act of 2002 were direct results of the financial scandals that occurred in 2001 and 2002?

Footnotes

[1]*Katz v. United States*, 389 U.S. 347, 351-53 (1967).

[2]495 F.3d 1041, 1048-49 (9th Cir. 2007).

[3]Walden, I. (2004, March 19–April 1). Addressing the data problem: The legal framework governing forensics in an online environment. In C. Jensen, S. Poslad, & T. Dimitrakos (Eds.), *Lecture notes in computer science: Trust management.* Online Book Series, 1995, p. 11. Retrieved from http://www.springerlink.com/content/gdw4gpplx9k23ctl/ fulltext.pdf

[4]Lyon, D. (2006). Why where you are matters: Mundane mobilities, transparent technologies, and digital discrimination. In T. Monahan (Ed.), *Surveillance and security: Technological politics and power in everyday life.* New York: Routledge, p. 211.

[5]This feature enables 9-1-1 operators to trace the location of callers who are in distress. Under this Act, telecommunications companies are required to provide their customers with the chance to opt out of this service in non-emergency situations.

[6]Walden, I. (2001). Balancing rights: Surveillance and data protection. *Electronic Business Law, 3*(11), 11.

[7]81 F. Supp. 2d 1103 (D. Kan. 2000).

[8]992 F. 2d 1472 (9th Cir. 1993).

[9]932 F. Supp. 1232 (D. Nev. 1996).

[10]No. 06-4092 (6th Cir. 2007).

[11]§ 2703(d) provides court orders if a less rigorous standard than probable cause is met. That is, if the government has "reasonable grounds to believe" that the communications sought are pertinent to an ongoing criminal investigation, then a court order will be issued.

[12]US Department of Justice. (2002). Searching and seizing computers and obtaining electronic evidence in criminal investigations: Search and seizure manual. Retrieved from www.justice.gov/criminal/cybercrime/ssmanual/ssmanual2009.pdf

[13]53 M.J. 402, 409 (C.A.A.F. 2000).

[14]983 F. Supp. 215 (D.D.C. 1998).

[15]Rauhofer, J. (2006). Just because you're paranoid, doesn't mean they're not after you: Legislative developments in relation to the mandatory retention of communications data in the European Union. *SCRIPT-ed, 3*(4), 340. Retrieved from http://www.law.ed.ac.uk/ahrc/script-ed/vol3-4/rauhofer.pdf

[16]Breyer, P. (2005). Telecommunications data retention and human rights: The compatibility of blanket traffic data retention with the ECHR. *European Law Journal, 11*(3), 373.

[17]A section of the **Stored Wire and Electronic Communications and Transactional Records Act of 1996** (18 USC § 2701–2712).

[18]47 C.F.R. § 42.6.

[19]See Article 6, Council Directive (EC) 95/46 of 24 October 1995, on the protection of individuals with regard to the processing of personal data and on the free movement of such data [1995] OJ L281/31.

[20]See Council Directive (EC) 97/66 of 15 December 1997, concerning the processing of personal data and the protection of privacy in the telecommunications sector [1998] OJ L24/1.

[21]Richard, M. M. (2005, April 14). Prepared statement of Counselor for Justice Affairs, U.S. Mission to the European Union. Presented at Meeting of EU's Article 29 Working Group, p. 2. Retrieved from http://www.usdoj.gov/ /cybercrime/mmrArt29DRstmt041405.pdf

[22]See Council Directive (EC) 2006/24 of 15 March 2006, on the retention of data generated or processed in connection with the provision of publicly available electronic communications services or of public communications networks and amending Directive 2002/58/EC [2006] OJ L105/54.

[23]See Articles 1–5, Council Directive (EC) 2006/24 of 15 March 2006, on the retention of data generated or processed in connection with the provision of publicly available electronic communications services or of public communications networks and amending Directive 2002/58/EC [2006] OJ L105/54; Rowland, D. (2003, April). Privacy, data retention and terrorism. *8th BILETA Conference: Controlling information in the online environment.* London: Author, p. 2. Retrieved from http://www.bileta.ac.uk/Document%20Library/1/Privacy,%20Data%20Retention%20and%20Terrorism.pdf

[24]47 U.S.C. § 1001(2) defines "call-identifying information" as "dialing or signaling information that identifies the origin, direction, destination, or termination of each communication generated or received by a subscriber by means of any equipment, facility, or service of a telecommunications carrier."

[25]Doyle, C. (2002, April 15). The USA Patriot Act: A legal analysis. *US Congressional Research Report RL31377*, p. 60. Retrieved from http://www.fas.org/irp/crs/RL31377.pdf; US Department of Justice. (2001, November 5). Field guidance on new authorities that relate to computer crime and electronic evidence enacted in the USA Patriot Act 2001. Computer Crime and Intellectual Property Section. Retrieved from http://www.cybercrime.gov/PatriotAct.htm; Plesser, R., Halpert, J., & Cividanes, M. (2001, October 31). Summary and analysis of key sections of the USA Patriot Act 2001. Center for Democracy and Technology. Retrieved from http://optout.cdt.org/security/011031summary.shtml

[26]US Department of Justice. (2004, July). Report from the field: The USA Patriot Act at work, pp. 19–20. Retrieved from http://www.justice.gov/olp/pdf/patriot_report_from_the_field0704.pdf

[27]US Department of Justice. (2004, July). Report from the field: The USA Patriot Act at work, p. 25. Retrieved from http://www.justice.gov/olp/pdf/patriot_report_from_the_field0704.pdf

[28]Parsky, L. H. (2005, April 21). Reauthorization of the USA Patriot Act. Statement of the Deputy Assistant Attorney General before the US House of Representatives, Committee on the Judiciary, Subcommittee on Crime, Terrorism and Homeland Security. Retrieved from http://www.justice.gov/criminal/cybercrime/parskyTestimony042105.htm

[29]47 U.S.C. § 222(d)(4)(a).

[30]S. Rep. No. 1183, 93rd Cong., 2d Sess. 1 (1974).

[31]Keep in mind that only U.S. citizens or permanent residents can sue if their rights under this Act have been violated.

[32]The privacy of medical, financial, and education information is also protected by other acts. Examples of such laws include the **Health Insurance Portability and Accountability Act of 1996**, which governs the privacy of health information in electronic form; **Executive Order 13181** (December 2000), which protects consumers' health information by preventing the disclosure to and use of this information by government agencies (there are certain exceptions to this rule); the Financial Modernization Act of 1999 (otherwise known as the **Gramm-Leach-Bliley Act**), whose primary purpose is to protect the financial information of individuals that is held by financial institutions; the **Right to Financial Privacy Act of 1978** (12 U.S.C. § 3401), which requires the notification of a subject if his or her financial records are sought and provides individuals with the ability to challenge the disclosure in court; and the **Family Education Rights and Privacy Act of 1974**, which protects the confidentiality of student records by requiring consent by the student or parent before disclosing personally identifiable information about a student to a third party.

[33]Section 208(d) of the e-Government Act (2002) defines "identifiable form" as "any representation of information that permits the identity of an individual to whom the information applies to be reasonably inferred by either direct or indirect means."

[34]For further information on this issue, see 5 U.S.C. § 552a(a)(4).

[35]Section 208(c)(1)(B).

[36]15 U.S.C. §§271–278h.

[37]40 U.S.C. § 759 (d).

[38]44 U.S.C. § 3541, et seq.

[39]44 U.S.C. § 3542(b)(1)(A) defines the integrity of data as the "guarding against improper information modification or destruction, and includes ensuring information nonrepudiation and authenticity"; 44 U.S.C. § 3542(b)(1)(B) defines confidentiality as the means of "preserving authorized restrictions on access and disclosure, including means for protecting personal privacy and proprietary information"; and 44 U.S.C. § 3542(b)(1)(B) defines availability as the means with which to ensure the "timely and reliable access to and use of information."

[40]Horn, G. (2005). Online searches and offline challenges: The chilling effect, anonymity, and the new FBI guidelines. *New York University Annual Survey of American Law*, 60, 49.

[41]Davis, H. (2003). *Human rights and civil liberties*. Devon, UK: Willan, p. 127.

[42]436 U.S. 547 (1978).

[43]The First Amendment states that "Congress shall make no law respecting an establishment of religion, or prohibiting the free exercise thereof; or abridging the freedom of speech, or of the press; or the right of the people peaceably to assemble, and to petition the government for a redress of grievances."

[44]42 U.S.C. § 2000aa(b).

[45]42 U.S.C. § 2000aa(b)(3). Subpoena *duces tecum* is a subpoena used to command the production of evidence.

[46]47 U.S.C. § 223(a) & (d) (1996).

[47]521 U.S. 844 (1997).

[48]Public Law 107–204.

[49]18 U.S.C. § 1512(c).

[50]Nabbali, T., & Perry, M. (2004). Going for the throat: Carnivore in an ECHELON world—Part II. *Computer Law and Security Report, 20*(2), 86.

[51]Donohue, L. K. (2006). Criminal law: Anglo-American privacy and surveillance. *Journal of Criminal Law and Criminology, 96*(3), 1179–1180.

[52]Di Gregory, K. V. (2000, June 24). Carnivore and the Fourth Amendment. Statement made before the Subcommittee on the Constitution of the House Committee on the Judiciary. Retrieved from http://www.justice.gov/criminal/cybercrime/carnivore.htm

[53]Steinhardt, B. (2000, July 24). The Fourth Amendment and Carnivore. Statement of Associate Director of the American Civil Liberties Union before the House Judiciary Committee, Subcommittee on the Constitution. Retrieved from http://judiciary.house.gov/legacy/stei0724.htm

Chapter 4

Searches and Seizures of Computers and Electronic Evidence

This chapter examines the notion of privacy and its importance. In addition, it explores what constitutes a search and seizure in computer forensics investigations, looking in particular at the Fourth Amendment to the U.S. Constitution and the restrictions it places on the **searches and seizures** of computers. Moreover, it considers the various exceptions to the Fourth Amendment related to the search and seizure of computer and electronic evidence. Finally, the chapter analyzes the Fourth Amendment's implications for searching a seized computer for electronic evidence and searching computers that contain privileged material.

What Is Privacy and Why Is It Important?

Privacy is a multivalent social and legal concept, where the definitions of privacy have ranged from the right to be let alone,[1] to the capacity to keep certain things secret,[2] to the right to control other individuals' access to oneself and information about oneself.[3] Despite its complexity and ambiguity, "privacy is closely connected with the emergence of a modern sense of self."[4] The emergence of the modern self, however, depends on individuals' ability to achieve self-determination and develop their personality free from coercion. That is, they must be free to make choices about who they are, who they associate with, and with whom they want to share information.

Choice is a prerequisite for leading a successful, fulfilling, and authentic existence, which requires that an individual must be able to "deliberate about and choose projects he or she will take up in life from an adequate range of options accommodating the diversity of human aptitudes, abilities, interests and tastes."[5] As an autonomous being, an individual is morally entitled to act in ways, or under conditions, of his or her own choosing, so long as there is no compelling moral reason to override his or her choice.[6] It is the right for an individual to choose for himself or herself—with only extraordinary exceptions being made in the interest of society—when and on which terms his or her actions should be revealed to the general public.[7]

The requirement that individuals should have control over information about themselves is an important aspect of privacy. The deprivation of control over what individuals do and who they are is considered as the "ultimate assault on liberty, personality, and the self."[8] Indeed, self-disclosure is one of the major mechanisms individuals use to regulate their privacy.[9] Privacy, therefore, functions as a means to control "access to information about, or to the intimate aspects, of oneself" by limiting access to information about the individual unless he or she chooses to reveal those details.[10] This notion of privacy reveals its connection to human dignity, to the extent that dignity requires nonexposure.[11] Specifically, certain types of information should be exposed only under conditions of trust, such as intimate details of an individual's private life. An individual's sexual preference is one such example. When intimate details of an individual's private life are collected, stored, and disclosed to others without that person's consent, it is damaging to the individual. The disclosure of this information may trigger emotions such as anxiety, fear, and humiliation. Here, the understanding of privacy is based in the intimate sphere, where invaded privacy can lead to dignitary harms such as exposure and shame.[12] Accordingly, courts have ruled that reasonable suspicion that a crime was committed is required before conducting searches and seizures because they affect the dignity and privacy interests of persons who are subjected to these intrusions.[13]

The right to privacy is the principle that "protects personal writings and any other production of the intellect or of the emotions."[14] Computer and other electronic storage devices are capable of amassing vast amounts of personal information. As Kerr observed, in 2005 alone, the computer hard drives that were sold had approximately 80 gigabytes of storage capacity, which is "roughly the equivalent to forty million pages of text—about the amount of information contained in the books on one floor of a typical academic library."[15] The storage capabilities of computers and related electronic devices have significantly increased since 2005, of course.

As one court ruling noted, "a laptop and its storage devices have the potential to contain vast amounts of information. People keep all types of personal information on computers, including diaries, personal letters, medical information, photos and financial records."[16] Furthermore, "opening and viewing confidential computer files implicates dignity and privacy interests. Indeed, some may value the sanctity of private thoughts memorialized on a data storage device above physical privacy."[17] As such, the privacy of these files should be afforded protection, because they can provide an enormously detailed account of an individual's private life.

Constitutional Source of Privacy Protection: The Fourth Amendment

The Fourth Amendment of the United States Constitution encompasses an individual's right to privacy. The Fourth Amendment to the U.S. Constitution provides that

> The right of the people to be secure in their persons, houses, papers, and effects, against unreasonable searches and seizures, shall not be violated, and no Warrants shall issue, but upon probable cause, supported by Oath or affirmation, and particularly describing the place to be searched, and the persons or things to be seized.

What You Should Know About the Fourth Amendment

- The Fourth Amendment provides everyone in the United States (e.g., citizens, noncitizens, illegal immigrants, foreign nationals visiting the United States for work or education) with the right to be free from unreasonable searches and seizures.

- The Fourth Amendment does not apply to searches conducted by private individuals, businesses, and nongovernmental agencies; in other words, it applies only if government action is involved. Specifically, the Fourth Amendment is not applicable "to a search or seizure, even an unreasonable one, effected by a private individual not acting as an agent of the Government or with the participation or knowledge of any governmental official."[18] Conversely, private searches may have Fourth Amendment implications if the individual conducting the search is acting as an "instrument" or agent of the government.[19] Private individuals who act to further their own ends and who have not been directed by the government to act in a certain way are not considered to be acting as agents of the government. For instance, a court ruled that a hacker who had provided police with child pornography images that he had found on a suspect's computer was not acting as an agent of the government.[20] Likewise, an individual who lives in a house with a suspect and turns over disks containing child pornography images to the government is not acting as an instrument of the government.[21]

 but not all

- The Fourth Amendment is enforced with the **exclusionary rule**, which makes evidence that was obtained in violation of the Fourth Amendment generally inadmissible in court.[22] The exclusionary rule seeks to:

  - Deter law enforcement agencies from conducting unreasonable searches and seizures
  - Motivate law enforcement agencies to comply with warrant requirements to ensure the admissibility of this evidence in a court of law
  - Protect individuals from being convicted based on illegally seized evidence

Prior to the decision in *Weeks v. United States*,[23] all evidence was admitted in court, no matter how it was obtained by law enforcement agencies.

Certain exceptions to the exclusionary rule do exist, however. For example, the **inevitable discovery exception** allows evidence that has been illegally obtained to be introduced in court if it would have inevitably been discovered through lawful means. The **good faith exception** holds that illegally seized evidence is admissible in court if a law enforcement agent acted in good faith belief that he or she was acting according to a valid search warrant that is later found defective.[24]

A Reasonable Expectation of Privacy in Communications

In *Katz v. United States*,[25] the court stated that the Fourth Amendment protection applies only to situations where an individual has a subjective expectation of privacy that society willingly recognizes as reasonable.[26] The "reasonable expectation of privacy" test provided by the decision in *Katz* has been used to define what constitutes a "search" under the Fourth Amendment.

In the case of *Katz v. United States*,[27] government agencies monitored the conversations of Katz in a public phone booth. The information retrieved from the communications was then introduced in court as evidence against him. Katz argued that such evidence violated his Fourth Amendment rights. The court agreed. Specifically, it stated that "one who occupies [the phone booth], shuts the door behind him, and pays the toll . . . is surely entitled to assume that the words he utters into the mouthpiece will not be broadcast to the world."[28] Accordingly, "what a person knowingly exposes to the public, even in his own home or office, is not a subject of Fourth Amendment protection. But what he seeks to preserve as private, even in an area accessible to the public, may be constitutionally protected."[29] Based on this rationale, Katz had a reasonable expectation of privacy in his communications in the public phone booth.

In contrast, exposure of communications to outsiders provides an indication that this information was not considered private by the individual. The courts have ruled, for example, that individuals do not have a reasonable expectation of privacy in respect to information they post on an online public bulletin board (discussion forum)[30] and in chat room conversations.[31] A case in point is *United States v. Charbonneau*,[32] where the court affirmed that individuals who sent messages to the "public at large" are at risk of having their information read by law enforcement agencies. In particular, the court affirmed that in *Charbonneau*, "when [the defendant] engaged in chat room conversations, he ran the risk of speaking with an undercover agent."[33]

An individual does not have a reasonable expectation of privacy in regard to the messages that he or she transmits to a third party. This applies to e-mails,[34] text messages, and messages sent to pagers. For example, in *United States v. Meriwether*,[35] the court held that an individual had no expectation of privacy for a text message the individual transmitted to a third party's pager.

Expectations of Privacy in the Workplace

Fourth Amendment principles apply to physical searches conducted by government employers in areas in which employees have a reasonable expectation of privacy.[36] Similar considerations espoused in cases involving physical searches have been adapted to employees' expectations of privacy in computer, Internet, e-mail, and text use monitoring by employers.

Consider *K-Mart Corporation Store No. 7441 v. Trotti*.[37] In this case, Trotti, who was an employee of K-Mart, claimed she had a reasonable expectation of privacy in respect to the locker K-Mart had provided so that she could store her personal items. While K-Mart owned the locker, it afforded Trotti (and other employees) with the opportunity to provide their own locks and did not require Trotti (or any other employee for that matter) to provide the combination or a duplicate key to management. The court ruled that Trotti did, indeed, have a reasonable expectation of privacy based on the provision of the lock. Had Trotti purchased her own lock but provided management with the combination or duplicate key to the lock, then she would not have a reasonable expectation of privacy. The same argument could be made about the lockers at schools, gyms, or other establishments.

In *Trotti*, the court also found that the policy on workplace searches was not clear to employees. Many cases have turned to favor the plaintiff who complains of privacy violations when employers have failed to take all necessary steps to inform employees of the search policy of their spaces and computers. That is, clear policies need to be posted in areas where employees can clearly see them; companies also protect themselves when they have employees sign a statement that says that they are aware that their spaces and computers are subject to searches (which is one of the best methods of protecting employers from invasion of privacy lawsuits).

Generally, the courts have determined the expectation of privacy that an individual has based on several factors, including whether an individual reasonably expected that his or her computer files would remain private and whether he or she took steps to conceal them from disclosure.[38] Recall the *Trotti* case. A personal lock she provided on her locker indicated that Trotti had a reasonable expectation of privacy; she was also the only person who had access to her locker. By analogy, a "lock" on computers and computer files should indicate an individual's expectation of privacy in respect to those objects. Indeed, the courts have held that an individual who uses passwords to protect his or her files from disclosure to a third party has a reasonable expectation of privacy in regard to those files.[39] It is important to note that an individual cannot be compelled by authorities to reveal passwords to computers or files.[40] By doing so, the individual may incriminate himself or herself—and the Fifth Amendment to the U.S. Constitution protects individuals against such self-incrimination.[41]

In *United States v. Arnold*,[42] the court found that an individual has a reasonable expectation of privacy for the data and files within a computer or other electronic storage devices where the person has taken steps to conceal the data files from view. For instance, an individual has a reasonable expectation of privacy relative to files that have been deleted. In fact, in *United States v. Upham*,[43] the court rejected the government's argument that "by deleting the images, [the defendant] 'abandoned' them and surrendered his right of privacy. Analogy is a hallowed tool of legal reasoning; but to compare deletion to putting one's trash on the street where it can be searched by every passer-by . . . is to

---

Attorney–Client Privilege in E-mails

The **attorney–client privilege** protects the communications between a client and an attorney concerning a legal matter. To ensure that the attorney–client privilege is protected, the communication must be made confidentially.[52] In *United States v. Schwimmer*,[53] the court held that the attorney–client privilege "requires a showing that the communication in question was given in confidence and that the client reasonably understood it to be so given." For the privilege to be protected, it must be made "under circumstances from which it may reasonably be presumed that it will remain in confidence."[54] Disclosure should occur only to those parties involved in the legal action that the communication concerns.[55] If the communication is disclosed either directly or inadvertently to a third party, then the privilege is waived. Information may be disclosed inadvertently if a communication is monitored, for example. Accordingly, if an employee has been notified that computer use and communications will be monitored, the attorney–client privilege is waived.

One case dealing with this issue concerns a former physician of a hospital who sued his employer, Beth Israel Medical Center, for a breach of contract. The individual sought the return of the communications he had with his attorney via the hospital's e-mail server, arguing that these communications were protected by attorney–client privilege. To determine whether these communications were privileged, the court looked at Beth Israel's e-mail policy, which stated the following:[56]

1. All Medical Center computer systems, telephone systems, voice mail systems, facsimile equipment, electronic mail systems, Internet access systems, related technology systems, and the wired or wireless networks that connect them are the property of the Medical Center and should be used for business purposes only.

2. All information and documents created, received, saved, or sent on the Medical Center's computer or communications systems are the property of the Medical Center. Employees have no personal privacy right in any material created, received, saved, or sent using Medical Center communication or computer systems. The Medical Center reserves the right to access and disclose such material at any time without prior notice.

Critical Thinking Questions

1. What do you think the court held in this case?

2. Was the physician's e-mail communications on the hospital's computer confidential?

3. Did attorney–client privilege apply? Why or why not?

reason by false analogy." Essentially, if a person does not take any steps to limit other employees' access to a workplace computer, the individual cannot have a reasonable expectation of privacy. Computers can be hidden from view by merely locking them in a private office.[44] In *Leventhal v. Knapek*,[45] for example, the court held that Leventhal had a reasonable expectation of privacy in respect of the computer in his private office because he had exclusive use of the office and computer. Leventhal's employer also did not have computer use and monitoring policies.

In *United States v. Slanina*,[46] the court held that the defendant had an expectation of privacy because his employer did not distribute a policy to employees stating that the storage of personal information was prohibited. In contrast, when an employer has policies that inform employees that the computers they use are subject to monitoring and that the personal or other objectionable use of workplace computers and networks is strictly prohibited, the courts have held that employees do not have a reasonable expectation of privacy.[47] For instance, in *United States v. Simmons*,[48] the court found that a government employee did not have a reasonable expectation of privacy in his Internet use because his employer had a policy stating that the Internet could be used only for business purposes.[49] The court has applied the same reasoning in cases involving mobile devices. In *Quon v. Arch Wireless Operating Company*,[50] for example, the court held that employers must have a policy that informs employees that their communications via these devices may be monitored and should have a history of monitoring such communications. If not, employees have a reasonable expectation of privacy in their communications via mobile phones.[51]

Search Warrants

Search warrants provide law enforcement agencies with the authority to enter a premises, search for the objects named in the warrant, and seize them. To be valid, the warrant must specifically state the crime (or crimes) being investigated, the location to be searched, and the items to be seized. Exceeding the scope of the warrant and making any errors or omissions in the search warrant may result in evidence being deemed inadmissible in court.

Probable Cause and Particularity

Under the Fourth Amendment, search warrants must be supported by probable cause and the objects of the search must be described with sufficient particularity. **Probable cause** "exist[s] where the known facts and circumstances are sufficient to warrant a man of reasonable prudence in the belief that contraband or evidence of a crime will be found."[57] When applying for a search warrant, an investigator must demonstrate probable cause both that a crime has been committed and that the evidence of the crime will be found in the location specified in the warrant.

Consider the crime of sending or receiving child pornography. Some courts have held that an individual's membership for a child pornography site shows probable cause of the receipt or distribution of child pornography.[58] In *United States v. Bailey*,[59] the court stated that when an individual

Knowingly becom[es] a computer subscriber to a specialized Internet site that frequently, obviously, unquestionably and sometimes automatically distributes electronic images of child pornography to other computer subscribers [that action] alone establishes probable cause for a search of the target subscriber's computer even though it is conceivable that the person subscribing to the child pornography site did so for innocent purposes and even though there is no direct evidence that the target subscriber actually received child pornography on his or her computer.

Other courts, however, have not agreed that membership in a child pornography website shows probable cause of the sending or receiving of child pornography.[60] According to these courts, additional information is required to demonstrate probable cause that a crime is being committed. This supplementary information may include, among other things, prior convictions on child pornography or sex offenses involving children and evidence of downloaded child pornography on the suspect's computer.[61]

In regards to **particularity**, the search warrant needs to specify the place that will be searched and the items that will be seized. "Mere reference to 'evidence' of a violation of a broad criminal statute or general criminal activity provides no readily ascertainable guidelines for the executing officers as to what items to seize."[62] The courts have found warrants that seize all documents and computer files without specifying how the items relate to a suspected criminal activity invalid and contrary to the Fourth Amendment.[63] By contrast, a warrant is considered sufficiently particular if it includes a description of the crime under investigation and describes which types of evidence are sought in relation to that crime.[64] For instance, if the crime being investigated is child pornography, a warrant authorizing the search and seizure of computers and electronic storage devices containing images of minors engaging in sexual activity as defined by child pornography statutes is sufficiently particular.[65] Basically, the search and seizure of items should be limited to the electronic devices that are connected to the criminal activity being investigated.

In *United States v. Upham*,[66] the court pointed out that the particularity requirement of the Fourth Amendment was implemented to protect citizens from general warrants authorizing "the wholesale rummaging through a person's property in search of contraband or evidence." A general warrant either provides too much discretion to law enforcement agencies to decide what to seize or permits law enforcement officers to indiscriminately seize both incriminating evidence and innocent items.

When Search Warrants Are Not Required for Searches and Seizures

Certain exceptions that justify a search without a warrant, include stop-and-frisk procedures, open fields, automobile exceptions, search incident to arrest, exigent circumstances, plain view, consent searches, and border searches.[68] Searches may occur during a stop-and-frisk procedure if the law enforcement agent believes that the individual being

No-Knock Warrants

When executing a search warrant, law enforcement agencies must knock before entering the premises, announce their presence, and give the resident (if present) a reasonable amount of time to comply with the requirements of the warrant. Failure to do so can result in the application of the exclusionary rule. However, law enforcement authorities could obtain a **no-knock warrant**, which would exempt them from having to knock and announce their presence before executing a search.[67] These warrants can prove quite beneficial, especially when dealing with situations where authorities believe that announcing their presence may lead to vital evidence being destroyed by the offender. This is highly likely with cybercrime, as computer evidence can be damaged, modified, concealed, or deleted in a matter of seconds.

"patted down" is armed and dangerous.[69] Searches may also occur in open fields without a warrant because these areas are exposed and accessible to the public.[70] A **warrantless search** of an automobile may occur if a law enforcement agent has probable cause to believe that the vehicle holds evidence of a crime.[71] The remaining four exceptions are explored in further detail in the following subsections, as they are particularly applicable to computers and related electronic devices.

Search Incident to Arrest

Searches that occur upon the arrest of individuals (**search incident to arrest**) encompass searches of the arrestees and the areas under their immediate control. As the court stated in *United States v. Robinson*,[72] "in the case of a lawful custodial arrest a full search of the person is not only an exception to the warrant requirement of the Fourth Amendment, but is also a 'reasonable' search under that Amendment." These searches are authorized to protect the arresting officer, to ensure that the evidence in the arrestee's possession is not destroyed, and to prevent an individual from escaping arrest with the items on his or her person. Law enforcement agents may also search the arrestee for evidence of the crime he or she is being suspected of committing.[73] Indeed, in *United States v. Ortiz*,[74] the court affirmed that a law enforcement agent's "need to preserve evidence is an important law enforcement component of the rationale for permitting a search of a suspect incident to a valid arrest."[75]

The primary reason for searching for such evidence is to prevent its destruction. For instance, in respect of pagers, the court stated that due to "the finite nature of a pager's electronic memory, incoming pages may destroy currently stored telephone numbers in a pager's memory. The contents of some pagers also can be destroyed merely by turning off the power or touching a button."[76] Essentially, the courts have upheld the searches of portable electronic devices, such as pagers, cell phones, and personal digital assistant (PDAs), incident to arrest.[77] The applicability of this doctrine to electronic devices such as laptops, however, has yet to be addressed by the courts.

Exigent Circumstances

Another exception to the warrant requirement occurs in **exigent circumstances**. In *United States v. David*,[78] the court stated that evidence can be seized without a warrant if "the destruction of [this] evidence is imminent" and "there is probable cause to believe that the item seized constitutes evidence of criminal activity." Generally, emergency circumstances arise with computers if a suspect seeks to damage the computer or damage or delete its files. The suspect can accomplish this by either physically damaging the computer (e.g., by breaking it) or using computer commands or program designed to destroy evidence (e.g., by deleting files or formatting entire disks).

Usually, this exception allows a law enforcement agent to seize a computer or other electronic device. The search of the device, however, usually requires another search warrant. For example, in *United States v. David*,[79] the law enforcement agent seized the defendant's computer memo book because the defendant was deleting files in it. After the memo book was seized from the suspect's possession, the law enforcement agent was required to obtain a warrant to search it.

Sometimes emergency access to electronic devices is required because evidence may be destroyed independent of any action by the suspect. In *United States v. Parada*,[80] the defendant's cell phone records were accessed due to exigent circumstances. Specifically, swift access to these records was required because incoming calls had the potential of overwriting call memory, thereby possibly destroying vital evidence in the case.

Consent

One of the most relevant exceptions to the requirement for a warrant to conduct the search and seizure of computers is the **consent search**. According to the ruling in *Schneckloth v. Bustamante*,[81] "consent searches are part of the standard investigatory techniques of law enforcement agencies. They normally occur on the highway, or in a person's home or office, and under informal and unstructured conditions." Searches can occur without a warrant and without probable cause if an individual who has authority over the place or items to be searched has consented to the search.[82] For the consent to be legal, the individual must not have been tricked or coerced into consenting to the search. If an individual consents to a search when a law enforcement agent falsely claims to have a warrant to search the premises, the court has found the consent is invalid and has deemed the search unconstitutional.[83] The individual must also have voluntarily consented to the search. If the "consent was not given voluntarily—that it was coerced by threats or force, or granted only in submission to a claim of lawful authority—then [the courts] have found the consent invalid and the search unreasonable."[84] As such, it is imperative that investigators advise subjects that the search is voluntary and that the subject may withdraw his or her consent to the search at any time.

If a computer is searched with the consent of a third party, its legitimacy may be challenged. In addition, a search can be contested if the search of a subject's property or computer exceeds the scope of the consent given. The scope of consent defines the area that the individual is allowing to be searched. For instance, if a police officer asked a subject,

"Can I search your car for drugs?" and the subject consents, it is quite clear that the scope of the consent includes the inside of the car, underneath the seats, and so on.

Principally, a warrant is not required in the following circumstances:

1. The suspect himself or herself consents to the search.

2. A third party with authority over the property consents to the search.

   - *Employers/Employees.* They can consent to searches of areas not exclusively set aside for the employee suspected of the crime. However, the court has also found that an employee can consent to the search of an employer. Specifically, in *United States v. Longo*,[85] an employee—the secretary—was found to have authority to consent to a search of her employer's computer. Additionally, coworkers can consent to searches of computers that are shared with the suspect of a crime.

   - *Parents.* They can give consent to search their child's computer even if the parents do not use or have access to the computer and the computer is located in the child's room or another private space of the child. As long as the child is dependent on the parents and not paying his or her parents rent to live in the room, the parents have the authority to consent to the search.[86]

   - *Spouses.* Generally, they have the authority to consent to a search of the computer of the other spouse as long as the computer of the nonconsenting spouse is not used exclusively by him or her and is not kept in a separate room (where only the nonconsenting spouse enters or has access to). For instance, in *Walsh v. State*,[87] the court ruled that a wife could consent to the search of a computer used by the defendant because she had bought the computer and it was used by the entire family.

   - *Relatives.* The court has held that a relative can consent to a search of a defendant's property if the relative has access and control over the place or object that is the target of the search. For example, a court ruling stated that a son-in-law had the authority to consent to a search over an area he had access to and control over.[88]

   - *Roommates/Housemates.* Similarly to the requirements with spouses and relatives, roommates have the authority to consent to searches of spaces and objects that they share with the defendant. For example, in the case of *United States v. Smith*,[89] a roommate was able to consent to the search of the defendant's computer because she had joint access to it and it was part of the house that she shared with the defendant.[90]

If a third party does not have a key to a locked item of a suspect, he or she cannot consent to its search. For instance, even though parents have the authority to consent to the search of their child's room, they do not have the authority to consent to a search of their child's locked property (e.g., locked toolbox, locked closet).[91] The court reached a similar conclusion in a case involving the defendant's girlfriend. Specifically, although the defendant's girlfriend consented to the search of the defendant's locked safe, the court found that she did not have the authority to consent to the search.[92] In general, the courts have ruled that boyfriends, girlfriends, roommates (or housemates), parents, spouses, and other relatives are unable to consent to a search of locked items they do not have access to in homes shared with the defendants.

What happens when a third party has partial access to a shared computer? Can a third party consent to the part of the computer that he or she does not have access to? Consider a computer that is shared by two users, A and B. User A has password-protected certain files. User B does not know the password to those files. Can user B consent to a search of user A's password-protected files? In *United States v. Buckner*,[93] the court compared password-protected files to "locked boxes" in common areas. As previously noted, the court has ruled that third parties cannot consent to the search of locked objects in common areas if they do not have the combination or key to open those locks. Consequently, if user A password-protected certain files and user B does not know the password to access these files (because user A has not shared it with user B), then user B cannot consent to have those files searched by law enforcement officers.

The best practice is to get consent in writing because it can serve as evidence that consent was given voluntarily. Written consent can also show the scope of the consent—that is, which locations, property, computers, or electronic devices the subject consented to have searched by law enforcement officers.[94]

Border Searches

The U.S. Congress has authorized customs searches at borders. Each sovereign nation has the right to regulate the entry and exit of individuals at its borders and under what conditions this may occur—the so-called **border search doctrine**. In most cases, U.S. courts have found warrantless border searches reasonable primarily due to the belief that the sovereign nation has the right to "protect itself by stopping and examining persons and property crossing into this country."[95] The interests of the sovereign state to exclude undesirable persons and prohibited goods have been cited to justify warrantless searches and searches conducted without reasonable suspicion or probable cause.[96]

While routine border searches do not require reasonable suspicion, probable cause, or a warrant,[97] the same cannot be said about nonroutine searches. The courts have ruled that "reasonable suspicion is required for the detention of a traveler at the border 'beyond the scope of a routine customs search and inspection.'"[98] Thus any nonroutine search must be preceded by reasonable suspicion of criminal activity and the search must not exceed that which is necessary to find the evidence of the crime.[99] The suspicious behavior of the traveler may also trigger a nonroutine search (e.g., if an individual seems extremely nervous during the routine search or questioning or if an individual purchased a one-way ticket in cash the day of the flight).[100]

Consider two landmark cases concerning border searches as they pertain to computers, *United States v. Romm*[101] and *United States v. Arnold*.[102] In *Romm*, the defendant flew from the

United States to Canada. Upon arriving in Canada, he was questioned by agents from Canada's Border Services Agency concerning his criminal history.[103] The agents also checked his computer and found several child pornography websites listed in his Internet history. Romm was denied entry to Canada, detained, and subsequently sent back to the United States. Upon his arrival in the United States, his computer was searched again and the evidence retrieved from it was used to charge Romm with crimes related to child pornography. Romm sought to have the evidence suppressed by claiming that the search and seizure of his computer violated his rights under the Fourth Amendment. The court disagreed, holding that international airports are the functional equivalent of a border and, therefore, the search of Romm's computer fell within the scope of the border search exception.[104]

In *Arnold*, the defendant was selected for secondary questioning at the Los Angeles International airport after arriving from the Philippines.[105] Upon inspecting his luggage, customs officials found a laptop computer, a hard drive, a computer memory stick, and six CDs. A customs official asked Arnold to turn the computer on, which he did. The customs official then accessed two folders on Arnold's computer and found something that the official believed warranted the attention of special agents from U.S. Immigration and Customs Enforcement (ICE). Subsequently, the ICE special agents searched Arnold's laptop and found what they believed to be child pornography. The special agents then seized Arnold's laptop and storage devices. According to Arnold, the warrantless search of his laptop violated his rights under the Fourth Amendment. As such, he insisted that the evidence retrieved from his laptop and storage devices should be suppressed. Although the District Court of the Central District of California agreed with him by finding that the search and seizure of Arnold's laptop by customs agents violated the Fourth Amendment, this decision was later overturned by the Ninth Circuit Court of Appeals.[106] The courts have rendered similar judgments, upholding warrantless searches under the border exception, in other cases involving the transport of child pornography.[107]

Plain View

Another relevant exception to a warrant for the search and seizures of computers is the "plain view" exception. The **plain view doctrine** allows law enforcement agencies conducting a search to seize evidence (not specified in a search warrant) that is in plain view, whose incriminating nature is immediately apparent to the officers.[108] Some courts have held that the plain view doctrine is applicable to computers and electronic evidence;[109] others have claimed that it should not be applied to such devices.[110]

The plain view doctrine has been applied to cybercrime cases where the search of a computer pursuant to a valid warrant subsequently led to the discovery of incriminating information not specified in the warrant. For example, in *United States v. Carey*,[111] the warrant specified that law enforcement personnel could search the suspect's computer for "documentary evidence pertaining to the sale and distribution of controlled substances."[112] While searching the suspect's computer for evidence of the sale and distribution of drugs, the investigator found images of child pornography. The investigator then abandoned the initial search for drug files and started searching for more child pornography images. However, every image after the initial discovery of an image of child pornography was ruled inadmissible in court. Why? Because if incriminating evidence other than that specified in the warrant is found during a search, the investigator should stop the search and get a new warrant based on the plain view evidence discovered.

Changing the search based on the new incriminating evidence found on a computer will likely result in the evidence being deemed inadmissible in court, as was the case in *Carey*.[113] As a general rule, if a warrant authorizes a search of the files pertaining to a specific crime, it will not authorize the search of files for other crimes not specified in the warrant.[114]

Searching the Computer for Evidence

In *United States v. Barth*,[115] the court held that "the Fourth Amendment protection of closed computer files and hard drives is similar to the protection it affords a person's closed containers and closed personal effects." Because intimate information may be stored on computers, the courts have ruled that these devices should be placed "into the same category as suitcases, footlockers, or other personal items that command a high degree of privacy."[116] Additionally, given that government agents need to obtain search warrants to access closed containers, the court held that the same requirement should apply to closed computer files and other electronic storage devices.[117] The question that follows is this: How should the search of the computer for electronic evidence of the crime be conducted?

Search Protocols

The process of searching a computer for electronic evidence can easily turn into a sweeping examination of a wide array of information. Computer forensics technology, however, provides investigators with a range of methods by which they can more narrowly target their search of computers for electronic evidence, such as by "limiting the search by date range, doing key word searches [and phrases], limiting the search to text files or graphics files, and focusing on certain software programs."[118] The mere existence of these search tools demonstrates that investigators have the ability to conduct more targeted searches of computers. In one child pornography case, *United States v. Carey*,[119] the court stated that investigators can and should limit searches by searching the computer using filenames, directories, and a sector-by-sector search of the hard drive.

Some courts have stated that a search protocol must be formulated before the search. A **search protocol** is a document that describes what is being searched for in a computer and which methodology will be used. By formulating a search protocol before the search, it is believed that the search will be narrowed and less privacy invasive. This is what Ralph Winick proposed. Specifically, he suggested that, "before a wide-ranging exploratory search is conducted, the magistrate should require the investigators to provide an outline of the methods that they will use to sort through the information"[120] (the **Carey-Winick approach**). *In re Search of 3817 W. West End*,[121] the magistrate of the court did just that by requiring the investigating officers in the case to specify which search protocol they planned to use to search the computer for electronic evidence.

Kerr argues that Winick and the *Carey* court have failed to realize how difficult it is to specify which search strategy is required without first having looked at the types of files that are present on the hard drive of the computer.[122] Indeed, some courts have argued that predetermined search protocols are impractical. For example, in *United States v. Scarfo*,[123] the court stated that when searching computers for information whose nature cannot be known in advance, "law enforcement officers must be afforded the leeway to wade through a potential morass of information in the target location."

Incriminating evidence can be stored on a computer in numerous ways. From the perspective of law enforcement officers, computer searches must be broad because suspects may encrypt, hide, or mislabel incriminating files to evade detection. In fact, in *United States v. Gray*,[124] the court noted that suspects may "intentionally mislabel files, or attempt to bury incriminating files within innocuously named directories"; for this reason, the agents searching computer files should not be "required to accept as accurate any file name or suffix" and to limit their search accordingly. Put simply, the seizing agents are not required to accept the labels of objects as indicative of their contents.[125]

Consider the practice of changing the extension at the end of the filename. This widely used and popular method for hiding a file is quite simple to use. To see how it works, follow these steps: First, create an Excel document and save it. Change the .xls extension on the document to .doc. The icon attached to the filename in the directory should change from Excel to Word, but when you try to open the file by clicking on it, the attempt should fail. Now launch Excel and then open the .doc file you created; it should open correctly now.

In the child pornography case known as *United States v. Hill*,[126] the defendant claimed that the search of his computer should have been solely "limited to certain files more likely to be associated with child pornography, such as those with a '.jpg' suffix . . . or those containing the word 'sex' or other key words." The court held that this search methodology was unreasonable:

> Criminals will do all they can to conceal contraband, including the simple expedient of changing the names and extensions of files to disguise their content from the casual observer . . . Forcing police to limit their searches to files that the suspect has labeled in a particular way would be much like saying police may not seize a plastic bag containing a powdery white substance if it is labeled "flour" or "talcum powder."[127]

As the court recognized in *Hill*, "images can be hidden in all manner of files, even word processing documents and spreadsheets."[128] Indeed, a vast number of text file extensions can be used to hide files:

- .doc (Word document)
- .docx (Word Open XML document)
- .pages (Pages document)
- .rtf (rich text file)
- .txt (plain text file)
- .wpd (WordPerfect document)
- .wps (Microsoft Works word processor document)

Of course, an experienced computer user will probably not use common text files to hide evidence of his or her crimes, nor is such a criminal likely to use common image file extensions, such as the following:

- .bmp (bitmap image file)
- .gif (Graphical Interchange Format file)
- .jpg (JPEG image file)
- .png (Portable Network Graphic)
- .psd (Photoshop document)
- .tiff (Tagged Image File Format)

Instead, the criminal will likely employ file extensions that are rarely used, and with which investigators may be unfamiliar. For instance, someone seeking to hide child pornography photos may save the images under the following image file extensions:[129]

- .411 (Mavica thumbnail image)
- .fbm (Fuzzy bitmap image)
- .imj (JFIF bitmap image)
- .kfx (Kofax image file)
- .mip (Multiple Image Print file)
- .mrb (Multiple Resolution bitmap file)

Perhaps a criminal will save the file as an attachment in an e-mail. Given the many different operating systems available today and the many different filename extensions that they use, it can be extremely difficult for an investigator to narrow down which information is included in each file type (even before a suspect seeks to alter the contents or change the file extensions to evade detection by authorities). Accordingly, if a warrant specifies the search of specific file formats, such as text files, it would exclude from the search image and other files that may potentially hold evidence of the crime.

In summary, most computer forensics investigations cannot follow a predetermined search protocol. To overcome this difficulty, the courts, in cases such as *United States v. Triumph Capital Group, Inc.*,[130] have endorsed the practice of investigators keeping detailed notes of their search of the computer for electronic evidence and explaining the rationale for each part of their search.[131]

Dealing with Privileged Data During a Search of a Computer

What should an investigator do if the computer that he or she is about to search contains **privileged information**? Which steps should be taken to protect such information during a search? The courts have recognized that computers may contain personal, sensitive, confidential, and proprietary information. Some cases may require the collection of evidence from the computer of a lawyer, a medical professional, a member of the clergy, a journalist, or a business person. Special care must be exercised when planning to search such a computer because of the privileged—and, therefore, protected—information that it may contain (e.g., medical data and attorney-client communications). As the court stated in *United States v. Arnold*,[132] "attorneys' computers may contain confidential client information. Reporters' computers may contain information about confidential sources or story leads. Inventors' and corporate executives' computers may contain trade secrets." The search of computers that contain privileged documents requires different rules because these computers may contain hundreds (maybe even thousands) of files and records of disinterested third parties unrelated to the investigation being conducted.

Three main strategies are used for screening computers that may contain privileged data. The first option is for the investigator to review the files while being recorded on camera. The court has stated that the in-camera review of potentially privileged documents "is a relatively costless and eminently worthwhile method to [e]nsure that the balance between petitioners' claims of irrelevance and privilege and plaintiffs' asserted need for the documents is correctly struck."[133] This method has been viewed by courts as "a highly appropriate and useful means of dealing with claims of governmental privilege."[134]

The second and third options involve third parties in the sorting and searching of these files. For the files to be given to the court, one or more third parties must review the files to see which are considered privileged and which can be given over to the prosecution team.

The second method involves **taint teams**, which are made up of prosecutors and agents—not in any way related to the case at hand—whose task is to view the electronic records seized and screen out any privileged information. In *Khadar v. Bush*,[135] the court approved the government's use of a "filter team" to review potentially privileged documents seized during searches of prisoners' cells at Guantanamo Bay. The search of the prisoners' cells in Guantanamo Bay was triggered by the coordinated suicides of three prisoners there. Such use of filter or taint teams has also been favored by other courts. For example, in *United States v. Triumph Capital Group, Inc.*,[136] taint teams were used to review privileged information. In *Triumph*, the court stated that "the use of a taint team is a proper, fair, and acceptable method of protecting privileged communications when a search involves property of an attorney."

Other courts,[137] however, have expressed discomfort with the use of taint teams. For instance, in the case known as *In re Search Warrant for Law Offices Executed on March 19, 1992 and Grand Jury Subpoena Duces Tecum Dated March 17, 1992*,[138] the court stated that "reliance on the implementation of a [taint team], especially in the context of a criminal prosecution, is highly questionable, and should be discouraged." It is believed that the prosecutors and agents in taint teams may not be able to ignore other crimes they may potentially find in the electronic records they are reviewing, which creates an appearance of unfairness. One way to try to ensure that these teams remain neutral is to perform regular audits of their practices and to have team members submit a detailed report of the actions they perform during their review.

To alleviate the concerns raised by the use of filter or taint teams, a third method may be used in cases potentially involving privileged information. With this method, a presiding judge appoints a neutral third party known as a "special master" to review potentially privileged files. The use of a special master to determine whether certain seized electronic documents contained privileged information was approved by the court in *United States v. Abbell*.[139] Another case where a court preferred the use of a special master to sort through potentially privileged electronic documents was *United States v. Hunter*.[140] In *Hunter*, law enforcement agents obtained a warrant authorizing the search and seizure of an attorney's computer systems. A special master, who was not employed by the law enforcement agency or the prosecutor's office, was used to screen the privileged material. For the search of the computer systems, a detailed search protocol was also provided to explain how the analyst would access relevant documents while avoiding the observation of privileged files.

Chapter Summary

This chapter focused on the right to privacy, including how it applies to computers. It also considered the areas where individuals have a reasonable expectation of privacy. Specifically, case law indicates that individuals have a reasonable expectation of privacy in the contents of their computers and related electronic devices (with certain exceptions).

Investigators are bound by the Fourth Amendment to the U.S. Constitution when they conduct searches and seizures. In particular, searches conducted by police officers (and anyone acting as an agent of the government) are limited to items and areas described in a warrant. Warrantless searches may occur under certain exceptions.

Special issues arise in searching computers for evidence. Some believe that a search protocol should be used during this process to minimize the intrusion into an individual's privacy. Others argue that by doing so, vital evidence may be overlooked because an investigator may not know how evidence of the crime was stored in the computer. Computers may also contain private and confidential information that should not be accessed by investigators. To conduct searches in such cases, special rules and procedures need to be followed to deal with and protect privileged information in computers.

Key Terms

Attorney–client privilege
Border search doctrine
Carey-Winick approach
Consent search
Exclusionary rule
Exigent circumstances
Good faith exception
Inevitable discovery exception
No-knock warrant

Particularity
Plain view doctrine
Privacy
Privileged information
Probable cause
Search and seizure
Search incident to arrest
Search protocol
Warrantless search

Critical Thinking Questions

1. What are your thoughts on the Carey-Winick approach? Is it beneficial or bad news?

2. In your opinion, what is the best strategy for reviewing privileged information on computers and why?

Review Questions

1. Why is privacy important?

2. Is all evidence that is illegally searched and seized inadmissible in court? Why do you think this is the case?

3. How is the "reasonable expectation of privacy" test applied to computers?

4. Does an employee have a reasonable expectation of privacy in the workplace?

5. When does the government need a search warrant to search and seize a suspect's computer?

6. What are some examples of warrantless searches, and under which circumstances may they be conducted?

7. Under which circumstances can a portable electronic device be seized and searched after a suspect is arrested?

8. Which type of exigent circumstances may arise in respect to computers?

9. When can a third party consent to a search?

10. Should search protocols be used in investigations? Why or why not?

11. What should investigators do if a computer that is being searched might contain privileged information?

Footnotes

[1]Cooley, T. (1907). *A treatise on the law of torts.* Chicago: Callaghan; Wong, R. (2005). Privacy: Charting its developments and prospects. In M. Klang & A. Murray (Eds.), *Human rights in the digital age.* London: GlassHouse, p. 148.

[2]Janis, M., Kay, R., & Bradley, A. (2000). *European human rights law: Text and materials* (2nd ed.). Oxford, UK: Oxford University Press, p. 300.

[3]Fried. C. (1970). *An anatomy of values.* Cambridge, MA: Harvard University Press; Wong, R. (2005). Privacy: Charting its developments and prospects. In M. Klang & A. Murray (Eds.), *Human rights in the digital age.* London: GlassHouse, p. 141.

[4]Galison, P., & Minow, M. (2005). Our privacy, ourselves in the age of technological intrusions. In R. A. Wilson (Ed.), *Human rights in the "War on Terror."* Cambridge, UK: Cambridge University Press, p. 258.

[5]Roberts, P. (2001). Privacy, autonomy, and criminal justice rights. In P. Alldridge & C. Brants (Eds.), *Personal autonomy, the private sphere and the criminal law: A comparative study.* Oxford, UK: Hart, p. 59.

[6]Schoeman, F. D. (1984). Privacy: Philosophical dimensions. In F. D. Schoeman (Ed.), *Philosophical dimensions of privacy: An anthology.* Cambridge, UK: Cambridge University Press, p. 20.

[7]Westin, A. F. (1966, June). Science, privacy and freedom: Issues and proposals for the 1970s: Part I—The current impact of surveillance on privacy. *Columbia Law Review, 66*(6), 1031.

[8]Fried, C. (1984). Privacy [a moral analysis]. In F. D. Schoeman (Ed.), *Philosophical dimensions of privacy: An anthology.* Cambridge, UK: Cambridge University Press, p. 212.

[9]Archer, R. L. (1980). Self-disclosure. In D. M. Wegner & R. R. Vallacher (Eds.), *The self in social psychology.* Oxford, UK: Oxford University Press, p. 199.

[10]Schoeman, F. D. (1984). Privacy: Philosophical dimensions. In F. D. Schoeman (Ed.), *Philosophical dimensions of privacy: An anthology.* Cambridge, UK: Cambridge University Press, p. 3.

[11]See footnote 27 in Gavison, R. (1980, January). Privacy and the limits of the law. *Yale Law Journal, 89*(3),469.

[12]Ehrenreich, R. (2001, June). Privacy and power. *Georgetown Law Journal, 89*(6), 2051.

[13]For example, see *United States v. Couch,* 688 F.2d 599, 604 (9th Cir. 1982).

[14]Warren, S., & Brandeis, L. (1890). The right to privacy. *Harvard Law Review,* 4, 213.

[15]Kerr, O. S. (2005). Searches and seizures in a digital world. *Harvard Law Review, 119,* 542.

[16]*United States v. Arnold,* 454 F. Supp. 2d 999, 1003–1004 (C.D. Cal. 2006).

[17]*United States v. Arnold,* 454 F. Supp. 2d 999, 1003 (C.D. Cal. 2006); *United States v. Molina-Tarazon,* 279 F.3d 709, 716 (9th Cir. 2002).

[18]*Walter v. United States,* 447 U. S. 649, 662 (1980) (Justice Blackmun, dissenting); *United States v. Jacobsen,* 466 U. S. 109, 113 (1984).

[19]*Coolidge v. New Hampshire,* 403 U.S. 443, 487 (1971).

[20]See *United States v. Jarrett,* 338 F. 3d 339 (4th Cir. 2003); *United States v. Steiger,* 318 F. 3d 1039, 1042–1046 (11th Cir. 2003).

[21]*United States v. Ellyson,* 326 F. 3d 522 (4th Cir. 2003).

[22]Although there are a few exceptions to this, such as the good faith exception, which holds that if a law enforcement agent acted in good faith belief that he or she was acting according to valid search warrant that is later found defective, the illegally seized evidence is admissible in court.

[23]232 U.S. 383 (1914).

[24]See *United States v. Leon,* 468 U.S. 897 (1984).

[25]389 U.S. 347 (1967).

[26]*California v. Ciraolo,* 476 US 207, 211 (1986).

[27]389 U.S. 347, 351 (1967).

[28]389 U.S. 347, 352 (1967).

[29]389 U.S. 347, 351 (1967).

[30]*Guest v. Leis*, 255 F.3d 325, 333 (6th Cir. 2001).

[31]*Commonwealth v. Proetto*, 771 A.2d 823, 831 (Pa. Super. 2001).

[32]979 F. Supp. 1177 (S.D. Ohio 1997).

[33]*United States v. Charbonneau*, 979 F. Supp. 1177, 1185 (S.D. Ohio 1997).

[34]See, for example, *State v. Evers*, 815 A.2d 432, 439-40 (N.J. 2003); *United States v. Bach*, 310 F.3d 1063, 1066 (8th Cir. 2002); *United States v. Maxwell*, 45 M.J. 406, 418-19 (C.A.A.F. 1996).

[35]917 F.2d 955, 958-59 (6th Cir. 1990).

[36]*O'Connor v. Ortega*, 480 U.S. 709, 721 (1987).

[37]677 S.W.2d 632, 637-38 (Tex. App. 1984).

[38]*United States v. Long*, 61 M.J. 539, 543 (N-M. Ct. Crim. App. 2005); *United States v. Mendoza*, 281 F.3d 712, 715 (8th Cir. 2002).

[39]*United States v. Buckner*, 473 F.3d 551, 554 (4th Cir. 2007).

[40]*In re Boucher*, No. 2:06-mj-91, 2007 WL 4246473 at 5 (D.Vt. 2007).

[41]The Fifth Amendment states: "No person shall be held to answer for a capital, or otherwise infamous crime, unless on a presentment or indictment of a Grand Jury, except in cases arising in the land or naval forces, or in the Militia, when in actual service in time of War or public danger; nor shall any person be subject for the same offence to be twice put in jeopardy of life or limb; nor shall be compelled in any criminal case to be a witness against himself, nor be deprived of life, liberty, or property, without due process of law; nor shall private property be taken for public use, without just compensation."

[42]523 F.3d 941 (9th Cir. 2008).

[43]168 F.3d 532, 537 (1st Cir. 1999).

[44]*United States v. Ziegler*, 474 F.3d 1184, 1189-90 (9th Cir. 2007); *State v. Young*, 974 So. 2d 601, 609 (Fla. Dist. Ct. App. 2008).

[45]266 F.3d 64, 73–74 (2d Cir. 2001).

[46]283 F.3d 670 (5th Cir. 2002).

[47]*Biby v. Board of Regents*, 419 F.3d 845, 850-51 (8th Cir. 2005); *United States v. Thorn*, 375 F.3d 679, 683 (8th Cir. 2004); *United States v. Angevine*, 281 F.3d 1130, 1134 (10th Cir. 2002); *Muick v. Glenayre Electronics*, 280 F.3d 741, 743 (7th Cir. 2002); *United States v. Bailey*, 272 F. Supp. 2d 822, 824 (D. Neb. 2003); *Wasson v. Sonoma County Junior College District*, 4 F. Supp. 2d 893, 905-06 (N.D. Cal. 1997).

[48]206 F. 3d 392 (4th Cir. 2000).

[49]In this case, Simmons had downloaded pornographic material from the Internet to his work computer.

[50]529 F.3d 892 (9th Cir. 2008).

[51]*Quon v. Arch Wireless Operating Company*, 529 F.3d 892, 906–907 (9th Cir. 2008).

[52]*United States v. Melvin*, 650 F.2d 641, 645 (5th Cir. 1981).

[53]892 F.2d 237, 244 (2d Cir. 1989).

[54]*Wilcoxon v. United States*, 231 F.3d 384, 386 (10th Cir. 1956).

[55]*United States v. Evans*, 954 F. Supp. 165, 170 (N.D. Ill. 1997); *United States v. Ryans*, 903 F.2d 731, 741 n.13 (10th Cir. 1990); *United States v. Schwimmer*, 892 F.2d 237, 237 (2d Cir. 1989); *State v. Colton*, 384 A.2d 343, 345-46 (Conn. 1977); *United States v. Tellier*, 255 F.2d 441, 447 (2d Cir. 1958).

[56]*Scott v. Beth Israel Medical Center, Inc.*, 847 N.Y.S.2d 436, 439 (N.Y. Sup. Ct. 2007).

[57]*Ornelas v. United States*, 517 U.S. 690, 696 (1996); *Illinois v. Gates*, 462 U.S. 213, 238 (1983).

[58]See, for example, *United States v. Martin*, 426 F.3d 3 (2d Cir. 2005); *United States v. Wagers*, 339 F. Supp. 2d 934 (E.D. Ky. 2004).

[59]272 F. Supp. 2d 822, 824-25 (D. Neb. 2003).

[60]For example, in *United States v. Corcas*, 419 F.3d 151 (2d Cir. 2005), the court stated that the precedent that membership to child pornography websites provides probable cause of the receipt or distribution of child pornography was unsound. See also *United States v. Gourde*, 382 F.3d 1003, 1006, 1011-13 (9th Cir. 2004); *United States v. Perez*, 247 F. Supp. 2d 459, 483-84 (S.D.N.Y. 2003).

[61]For prior convictions as additional evidence see, for example, *United States v. Wagers*, 339 F. Supp. 2d 934, 941 (E.D. Ky. 2004), and *United States v. Fisk*, 255 F. Supp. 2d 694, 706 (E.D. Mich. 2003). For evidence of downloading as supplemental evidence, see, for example, *United States v. Perez*, 247 F. Supp. 2d 459, 483-84 (S.D.N.Y. 2003), and *United States v. Zimmerman*, 277 F.3d 426, 435 (3d Cir. 2002).

[62]*United States v. George*, 975 F.2d 72, 76 (2d Cir. 1992); *United States v. Maxwell*, 920 F.2d 1028, 1033 (D.C. Cir. 1990); *United States v. Holzman*, 871 F.2d 1496, 1509 (9th Cir. 1989); *Voss v. Bergsgaard*, 774 F.2d 402, 405 (10th Cir. 1985); *United States v. Cardwell*, 680 F.2d 75, 77 (9th Cir. 1982).

[63]*United States v. Kow*, 58 F.3d 423, 427 (9th Cir. 1995); *In re Search Warrant for K-Sports Imports, Inc.*, 163 F.R.D. 594, 597-98 (C.D. Cal. 1995); *Lafayette Academy, Inc. v. United States*, 610 F.2d 1, 5-6 (5th Cir. 1979).

[64]*State v. Askham*, 86 P.3d 1224, 1227 (Wash. App. 2004).

[65]*United States v. Thorn*, 375 F.3d 679, 684-85 (8th Cir. 2004); *United States v. Gleich*, 293 F. Supp. 2d 1082, 1088 (D.N.D. 2003); *State v. Wible*, 51 P.3d 830, 837 (Wash App. 2002); *United States v. Hay*, 231 F.3d 630, 637 (9th Cir. 2000); *United States v. Campos*, 221 F.3d 1143, 1147-48 (10th Cir. 2000); *Davis v. Gracey*, 111 F.3d 1472, 1479 (10th Cir. 1997); *State v. One Pioneer CD-ROM Changer*, 891 P.2d 600, 604 (Okla. Ct. App. 1995).

[66]*United States v. Upham*, 168 F. 3d 532, 535 (1st Cir. 1999).

[67]*Wilson v. Arkansas*, 514 U.S. 927 (1995).

[68]Harr, J. S., & Hess, K. M. (2005). *Constitutional law and the criminal justice system* (3rd ed.). New York: Thomson-Wadsworth, p. 219.

[69]*Terry v. Ohio*, 392 U.S. 1 (1968).

[70]See *Hester v. United States*, 265 U.S. 57 (1924), where the court stated that "the special protection accorded by the Fourth Amendment to the people in their 'persons, houses, papers, and effects,' is not extended to the open fields."

[71]*Wyoming v. Houghton*, 526 U.S. 295 (1999); *United States v. Ross* 456 U.S. 798 (1982).

[72]414 U.S. 218, 235 (1973); See also *New York v. Belton*, 453 U.S. 454, 459 (1981).

[73]*United States v. Robinson*, 414 U.S. 218, 233–234 (1973); *Abel v. United States*, 362 U.S. 217 (1960); *Agnello v. United States*, 269 U.S. 20 (1925).

[74]84 F.3d 977 (7th Cir. 1996).

[75]See also *United States v. Robinson*, 414 U.S. 218, 226 (1973).

[76]*United States v. Robinson*, 414 U.S. 218, 226 (1973); *United States v. Meriwether*, 917 F.2d 955, 957 (6th Cir. 1990).

[77]*United States v. Finley*, 477 F.3d 250, 259-60 (5th Cir. 2007); *United States v. Mercado-Nava*, 486 F. Supp. 2d 1271, 1278-79 (D. Kan. 2007); *United States v. Romero-Garcia*, 991 F. Supp. 1223 (D. Or. 1997); *United States v. Thomas*, 114 F.3d 403, 404 n.2 (3d Cir. 1997); *United States v. Reyes*, 922 F. Supp. 818, 833 (S.D.N.Y. 1996).

[78]756 F Supp. 1385 1392 (D. Nev. 1991).

[79]756 F. Supp. 1385, 1392 (D. Nev. 1991).

[80]289 F. Supp. 2d 1291, 1304 (D. Kan. 2003).

[81]412 U.S. 218, 231– 232 (1973).

[82]See *Schneckloth v. Bustamonte*, 412 U.S. 218, 219 (1973); *United States v. Matlock*, 415 U.S. 164, 171 (1974); *Stoner v. California*, 376 U.S. 483 (1964).

[83]*Bumper v. North Carolina*, 391 U.S. 543, 550 (1968).

[84]*Schneckloth v. Bustamonte*, 412 U.S. 218, 233 (1973); See also *Bumper v. North Carolina*, 391 U.S. 543, 548–549 (1968); *Johnson v. United States*, 333 U.S. 10 (1948); *Amos v. United States*, 255 U.S. 313 (1921).

[85]70 F Supp. 2d 255, 256 (WDNY 1999).

[86]See *United States v. Rith*, 164 F 3d 1323 (10th Cir.), cert. denied, 528 US 827 (1999).

[87]512 S. E. 2d 408, 411–412 (Ga. Ct. App. 1999).

[88]*State v. Guthrie*, 627 N.W. 2d 401 (S.D. 2001).

[89]27 F. Supp. 2d 1111 (C.D. Ill. 1998).

[90]*United States v. Smith*, 27 F. Supp. 2d 1111, 1115–1116 (C.D. Ill. 1998).

[91]*State v. Harris*, 642 A 2d. 1242 (Del. 1993); *People v. Snipe*, 841 N.Y.S. 2d 763 (N.Y. Sup. Ct. 2007).

[92]*State v. Smith*, 966 S. W. 2d 1 (Mo. Ct. App. 1997).

[93]473 F.3d 551, 554 (4th Cir. 2007); See also *Trulock v. Freeh*, 275 F.3d 391 (4th Cir. 2001).

[94]*United States v. Block*, 590 F. 2d 535, 537 (4th Cir. 178).

[95]*United States v. Flores-Montano*, 541 U.S. 149, 152-53 (2004); *United States v. Ramsey*, 431 U.S. 606, 616 (1977).

[96]*United States v. Flores-Montano*, 541 U.S. 149, 152-53 (2004); *United States v. Montoya de Hernandez*, 473 U.S. 531, 538 (1985).

[97]*United States v. Montoya de Hernandez*, 473 U.S. 531, 538 (1985). See also *United States v. Ramsey*, 431 U.S. 606, 620 (1977).

[98]See, for example, *United States v. Sandoval Vargas*, 854 F.2d 1132, 1134 (9th Cir. 1988).

[99]*United States v. Couch*, 688 F.2d 599, 604 (9th Cir. 1982); *United States v. Summerfield*, 421 F.2d 684, 685 (9th Cir. 1970).

[100]See, for example, *United States v. Sokolow*, 490 U.S. 1, 4-5 (1989); *United States v. Carter*, 590 F.2d 138, 139 (5th Cir. 1979); *United States v. Olcott*, 568 F.2d 1173, 1174-75 (5th Cir. 1978).

[101]455 F.3d 990 (9th Cir 2006).

[102]454 F. Supp. 2d 999 (C.D. Cal. 2006).

[103]In an unrelated case, Romm had pleaded nolo contendere (no contest) to crimes involving children (including child exploitation by means of a computer).

[104]*United States v. Romm*, 455 F.3d 990, 997 (9th Cir 2006).

[105]*United States v. Arnold*, 454 F. Supp. 2d 999, 1004 (C.D. Cal. 2006).

[106]*United States v. Arnold*, 533 F.3d 1003 (9th Cir. 2008).

[107]See, for example, *United States v. Ickes*, 393 F.3d 501, 503 (4th Cir. 2005).

[108]*Horton v. California*, 496 US 128, 134 (1990).

[109]*United States v. Mann*, No. 08-3041, 210 U.S. App. LEXIS 1264 (7th Cir. Decided January 20, 2010); *United States v. Farlow*, 2009 U.S. Dist. LEXIS 94778 (D. Maine September 29, 2009).

[110]*United States v. Comprehensive Drug Testing, Inc.*, 513 F.3d 1085 (9th Cir. 2008).

[111]172 F. 3d 1268 (10th Cir. 1999).

[112]*Ibid.* at 1271.

[113]See *Carey* for more information on this issue.

[114]*People v. Carratu*, 755 N.Y.S. 2d 800, 807 (N.Y. Sup. Ct. 2003).

[115]26 F. Supp. 2d 929, 936–937 (W.D. Tex. 1998).

[116]*United States v. Andrus*, 483 F.3d 711, 718 (10th Cir. 2007).

[117]*United States v. Roberts*, 86 F. Supp. 2d 678, 688–689 (S.D. Tex. 2000); *United States v. Barth*, 26 F. Supp. 2d 929, 936–937 (W.D. Tex. 1998); *United States v. David*, 756 F. Supp. 1385, 1390 (D. Nev. 1991).

[118]*In re Search of 3817 W. West End*, 321 F. Supp. 2d 953 (N.D. Ill. 2004).

[119]172 F. 3d at 172.

[120]Winick, R. (1994). Searches and seizures of computers and computer data. *Harvard Journal of Law and Technology, 8,* 108; Kerr, O. S. (2005). Searches and seizures in a digital world. *Harvard Law Review, 119,* 572–573.

[121]321 F. Supp. 2d 953 (N.D. Ill. 2004).

[122]Kerr, O. S. (2005). Searches and seizures in a digital world. *Harvard Law Review, 119,* 575.

[123]180 F. Supp. 2d 572, 578 (D.N.J. 2001).

[124]Kerr, O. S. (2005, January). Digital evidence and the new criminal procedure. *Columbia Law Review, 105,* 302.

[125]*United States v. Abbell,* 963 F. Supp. 1178, 1201 (S.D. Fla. 1997).

[126]459 F.3d 966, at 978 (9th Cir. 2006).

[127]459 F.3d 966, at 978 (9th Cir. 2006).

[128]459 F.3d 966, at 978 (9th Cir. 2006).

[129]This list is by no means exhaustive. There are hundreds of file extensions that could be used.

[130]211 F.R.D. 31 (D. Conn. 2002).

[131]Orton, I. (2006). The investigation and prosecution of a cybercrime. In R. D. Clifford (Ed.), *Cybercrime: The investigation, prosecution and defense of a computer-related crime* (2nd ed.). North Carolina: Carolina Academic Press, p. 162.

[132]454 F. Supp. 2d 999, 1004 (C.D. Cal. 2006), rev'd, 533 F. 3d 1003 (9th Cir. 2008).

[133]*Kerr v. United States,* 426 U.S. 394, 405 (1976).

[134]*Kerr v. United States,* 426 U.S. 394, 406 (1976); *United States v. Nixon,* 418 U.S. 683 (1974); *United States v. Reynolds,* 345 U.S. 1 (1953).

[135]No. 04-1136, 2006 U.S. Dist. LEXIS 65973 (D.D.C. Sept. 15, 2006).

[136]211 F.R.D. 31 (D. Conn. 2002).

[137]See, for example, *United States v. Neill,* 952 F. Supp. 834, 840-841 (D.D.C. 1997).

[138]153 F.R.D. 55, 59 (S.D.N.Y. 1994).

[139]914 F. Supp. 519 (S.D. Fla. 1995).

[140]13 F. Supp. 2d 574, 578 (D. Vt. 1998).

Chapter 5

Cybercrime Laws: Which Statute for Which Crime?

This chapter explores the types of crimes and incidents involved in computer forensics investigations. It further considers the various laws that cover these crimes. More specifically, it examines the different crime categories—computer threats and intrusions, financial crimes and fraud, intellectual property crimes and economic espionage, and personal crimes—and their relevant computer crime statutes.

Computer Threats and Intrusions

The earliest law on cybercrime was the Electronic Communications Privacy Act of 1986 (ECPA). One notable predecessor of this legislation was the Computer Fraud and Abuse Act of 1984 (18 U.S.C. § 1030) and its subsequent amendments (in 1986, 1994, and 1996); collectively, these laws constitute the main federal statute that covers computer threats and intrusions. 18 U.S.C. § 1030 is violated if an individual accesses a protected computer without authorization to obtain data.[1] Under 18 U.S.C. § 1030(e)(2)(A), a protected computer is one that is reserved exclusively for

> [T]he use of a financial institution or the United States Government, or, in the case of a computer not exclusively for such use, used by or for a financial institution or the United States Government and the conduct constituting the offense affects that use by or for the financial institution or the Government.

Moreover, pursuant to 18 U.S.C. § 1030(e)(2)(B), a protected computer is one that "is used in or affecting interstate or foreign commerce or communication, including a computer located outside the United States that is used in a manner that affects interstate or foreign commerce or communication of the United States." Computers connected to the Internet can be considered as "protected computers." In addition, the Computer Fraud and Abuse Act protects against international access to computers that contain restricted (or classified) data.[2]

This statute can also be violated if individuals access a computer in excess of their level of authorization. Indeed, numerous government employees have run afoul of this aspect of the law. While they may have had authority to access the computer, such employees exceeded their authorization when they accessed or altered certain information. Just viewing information from a computer has been considered to qualify as "access" by the courts. However, the information that is viewed must be subsequently used in some manner for the individual to be considered to have violated 18 U.S.C. § 1030(a)(4).[3] A case in point occurred in *United States v. Rice*.[4] The defendant in this case, an Internal Revenue Service (IRS) employee, exceeded his authorized access to an IRS computer by accessing information from a division he was not a member of and disclosing the information he accessed to a third party. The **Taxpayer Browsing Protection Act of 1997** was passed to criminalize federal and state employees' unauthorized browsing of taxpayer information.

A few crimes included in the category of computer threats and intrusions are profiled here.[5]

Hacking and Website Defacement

In addition to 18 U.S.C. § 1030, 18 U.S.C. §1361–1362 covers crimes such as computer hacking and website defacement, and prohibits malicious mischief.

Hacking

Most hackers are charged with violating 18 U.S.C. §1030 when they gain unauthorized access to a protected computer. If a hacker alters files (e.g., by editing files in a database or creating a backdoor to the system to enable access at a later date) or deletes files in a protected computer system, he or she has caused damage according to 18 U.S.C. § 1030(a)(5)(C). This section is violated if someone "intentionally accesses a protected computer without authorization and as a result of such conduct, causes damage and loss." One example of such a violation involved a prior employee of the Airline Coach Service and Sky Limousine Company, Alan Giang Tran. Tran hacked into the company's computer system and wiped out critical data by deleting the customer database and other important records.[6]

At times, hackers have changed the data in a domain name server to redirect traffic to a hacker's website or any other website of the hacker's choice. Anyone altering, substituting, or redirecting a website without authorization is considered to have violated 18 U.S.C. § 1030(a)(5)(A). For example, John William Racine II met these criteria when he hacked the website of al-Jazeera (an Arabic-language media organization) during the Iraq War and redirected traffic to a website that he had created that had an American flag on it and a message stating "Let Freedom Ring."[7]

Individuals may also hack websites to gain unauthorized access to databases. The most common way an individual hacks into a website is with a Structured Query Language (SQL) injection. Using this method, a hacker directly inserts SQL code into a vulnerable Web application[8] to trick the website into submitting the hacker's SQL code into the targeted database. The hacker's code is then executed, allowing the intruder to gain access to the desired database.

The most commonly employed hacking technique is password cracking. With this method, a hacker attempts to crack the password of a legitimate user. To crack a password, the individual can use either dictionary or brute-force attack. A dictionary attack is one of the easiest ways in which a hacker can crack a password. Such an attack loads text files containing words in the dictionary or passwords that specifically relate to the hacker's target (e.g., if the target is a New York Rangers fan, then a text file containing words that fans would most probably use as passwords—such as the players' names— would be used) into a cracking application (e.g., L0phtCrack). The cracking application is then run against the target's accounts. If a user has set up simple passwords, a dictionary attack will suffice to crack his or her passwords. If not, then a brute-force attack will be used. With this method, a hacker uses a software tool that attempts to crack the password by using all possible combinations of letters, numbers, and symbols.

Website Defacement

Website defacement occurs when one or more individuals alter or replace the content of a website without authorization. Such an attack seeks to render the data targeted inaccessible to the public. The majority of the time, activists—who are often politically motivated— are responsible for website defacement. This type of behavior has been termed as "hacktivism." For example, hackers from Pakistan (otherwise known as GForce Pakistan) hacked into several U.S. government websites. In one particular incident, they defaced a server that belonged to the U.S. National Oceanographic and Atmospheric Agency and posted a message glorifying al-Qaeda's jihad (holy war) against the United States.[9] Moreover, during the Iraq War in 2003, hackers placed anti-war, anti-United States and anti-Iraq slogans on several government websites.[10] In a different sort of activism, in 2009, hackers replaced the front page of the International Federation of the Phonographic Industry's (IFPI) Swedish website with a message declaring war on antipiracy entertainment groups.[11]

On other occasions, individuals have defaced websites merely for amusement purposes. For instance, Brazilian hackers defaced several Canadian websites by posting messages with cartoon photos of deranged-looking cows to protest Canada's ban on Brazilian beef.[12] A more recent example involved an unidentified hacker who altered the content of Spain's European Union Presidency website by replacing the photo of Spain's Prime Minister Jose Luis Rodriguez Zapatero with the photo of a comic English actor, Rowan Atkinson, who plays a character known as Mr. Bean.[13]

Malware Creation and Distribution

This category of offenses covers the writing and distributing of malicious code (malware).

Worms and Viruses

Worms automatically self-propagate and are delivered to their target via a network of physical media. In contrast, viruses are not self-executing, although they are designed to self-replicate.

18 U.S.C. § 1030(a)(5)(A) was specifically designed to deal with individuals who use a virus, worm, or other program with the intent to cause damage to a protected computer. One important point to consider is that the statute a person violates can vary depending on whether an individual intended to cause damage. If the person intended to cause damage, then 18 U.S.C. § 1030(a)(5)(A) is applicable. An individual can be prosecuted under this statute if he or she "knowingly causes the transmission of a program, information, code, or command, and as a result of such conduct, intentionally causes damage without authorization, to a protected computer." For example, the creator of the Melissa virus (discussed in Chapter 1) pleaded guilty to intentionally causing damage to a computer under 18 U.S.C. § 1030(a)(5)(A).

By contrast, if a person did not intend to cause damage intentionally (i.e., the damage caused by the virus code was accidental), 18 U.S.C. § 1030(a)(5)(B) is applicable. An individual can be prosecuted under this statute if he or she "intentionally accesses a protected computer without authorization, and as a result of such conduct, recklessly causes damage."

Trojan Horses

A Trojan horse imitates legitimate files and can be either a joke program or a software delivery mechanism for viruses. The statute a person violates by launching a Trojan horse may vary depending on whether the individual caused damage to a computer as defined under 18 U.S.C. § 1030(e)(8). This section considers damage to be "any impairment to the integrity or availability of data, a program, a system, or information." If the Trojan horse is harmless, then it may be difficult to meet the "damage" requirement.

Computer Spying and Intrusions

This section covers both adware and spyware. **Adware** is used to track users' online activities and deliver targeted pop-up ads to users. In particular, it has been used to collect information about users' online habits (e.g., websites browsed) and target advertisements based on users' activities. For instance, if a user has conducted several searches for bachelor's degrees offered online, the advertisements shown to that user might include those for online universities. The main difference between spyware and adware is that spyware is often used for malicious purposes, whereas adware is often used for benign marketing purposes (though it can be used for malicious purposes as well).

One example of spyware is the Loverspy program, which enabled users to intercept the electronic communications of others. This spyware was designed to conceal itself in an electronic greeting card. When the card was opened by the target, the program surreptitiously installed itself on the victim's computer. The Loverspy program was designed to collect and send all of the activities that the victim performed on his or her computer (e.g., user names and passwords typed, websites browsed, e-mails sent and received) to the individual who bought the program and/or to the program's creator.[14]

In one notable case, Craig Matthew Feigin developed, installed, and used spyware that secretly created and downloaded images of other individuals.[15] Feigin pleaded guilty to a violation of 18 U.S.C. § 1030(c)(2)(B) because he marketed the spyware software he created, which he called Web Cam Spy Hacker, on eBay and other websites.

Denial of Service Attacks and Distributed Denial of Service Attacks

Denial of service (DoS) and distributed denial of service (DDoS) attacks are other crimes in this category. DoS attacks flood a network server with phony authentication requests to such an extent that individuals with legitimate requests are prevented from accessing it and ultimately the server shuts down. A prime example of this kind of attack is a **SYN flood attack**. To understand this type of attack, knowledge of how a **Transmission Control Protocol** (TCP) connection is established is required.[16] A TCP connection occurs as follows:

Step 1: Computer X sends a TCP request to start a session, consisting of a SYN packet to Computer Y.

Step 2: Computer Y receives X's SYN (synchronization request) packet.

Step 3: Computer Y sends a SYN-ACK (synchronization acknowledgment) packet.

Step 4: Computer X receives Y's SYN-ACK packet.

Step 5: Computer X sends an ACK (acknowledgment) packet.

Step 6: Computer Y receives the ACK packet.

This process, which is also known as a **TCP handshake**, is how a TCP connection is established. A SYN flood attack attempts to take advantage of the TCP handshake as follows:

Step 1: Computer X sends a SYN packet to Computer Y.

Step 2: Computer Y receives X's SYN packet.

Step 3: Computer Y sends a SYN-ACK packet.

Step 4: Computer X receives Y's SYN-ACK packet.

Step 5: Computer X does not respond to Y's SYN-ACK packet but rather sends another SYN packet.

This process is repeated several times. Consequently, legitimate users are denied connection requests because Y is overwhelmed with multiple open connections waiting for X's ACK. As a result, Y can no longer respond to legitimate users' requests for connection.

18 U.S.C. § 1030(a)(5)(A) criminalizes DoS and DDoS attacks. For example, Anthony Scott Clark plead guilty to launching a DDoS attack against eBay and other entities in

violation of this statute.[17] Another example involved a juvenile who developed the RPCSDBOT computer worm, which attacked the same vulnerability in Microsoft software as the Blaster worm (discussed in Chapter 1). Specifically, the juvenile directed infected computers to launch a DDoS attack against Microsoft's main website, causing it to be unavailable to the public.[18] If an individual targets government computers with this type of attack, such conduct may also be charged under 18 U.S.C. § 1362, which deals with interference with government communications systems.

Sometimes a DoS or DDoS attack is used in conjunction with extortion. For instance, an individual may try to extort money from an individual, business, or organization by threatening to damage a protected computer if a specific sum of money is not paid. In so doing, the individual violates 18 U.S.C. §1030(a)(7).

At other times, individuals may conduct DoS or DDoS attacks in conjunction with other crimes covered in 18 U.S.C. § 1030. A case in point concerned Kenneth Patterson. Patterson not only launched DoS attacks against an American Eagle Outfitters computer network, but also posted the usernames and passwords of American Eagle Outfitter users on a Yahoo hacker group posting board.[19] As a result of the latter action, he was also charged under 18 U.S.C. § 1030(a)(6) for trafficking in computer passwords.

Cyberterrorism

Certain aspects of cyberterrorism are dealt with in Title VIII of the USA Patriot Act of 2001. The most relevant section of the USA Patriot Act to this crime is Section 814, which deals with the deterrence and prevention of cyberterrorism, including the penalties to be meted out for certain cyberterrorism offenses. For example, an individual violates this section if he or she causes physical injury, causes loss of life, or in any way threatens public health or safety as a result of damaging or gaining unauthorized access to a protected computer. Section 814 also covers damage to a computer system that the government uses in the administration of justice, national defense, or national security.

Terrorist groups have been known to target government websites, flooding them with attacks that take days and (sometimes weeks) for the sites to become operational again. The Liberation Tigers of Tamil Eelam (otherwise known as the Tamil Tigers), a terrorist group in Sri Lanka, engaged in such attacks against Israeli government websites during the riots on the West Bank in 2000.[20] These attacks, however, cannot be classified as true cyberterrorism. Cyberterrorists seek to attack the critical infrastructure systems (e.g., water, energy, communications) so as to intimidate or coerce a government for ideological, religious, or political reasons. Cyberterrorists may seek unauthorized access to critical infrastructure computer systems with the following aims:[21]

- Block communications systems.
- Disrupt air traffic control, public surface transport (e.g., trains), and emergency systems.
- Shut down electric utilities and damage gas utilities.

- Contaminate water, food, or medicine supplies. One worst-case scenario posited for this type of an attack is a hacker gaining access to a pharmaceutical manufacturing process and altering medication formulas, resulting in significant loss of life.

According to a report by the Working Group on Unlawful Conduct on the Internet:[22]

[T]he Department of Justice has encountered several instances where intruders have attempted to damage critical systems used in furtherance of the administration of justice, national defense, or national security, as well as systems (whether publicly or privately owned) that are used in the provision of "critical infrastructure" services such as telecommunications, transportation, or various financial services.

An example of this kind of crime occurred in 1997, when a teenager from Worcester, Massachusetts, hacked into the public telephone switching network, disabling telephone service for a number of residents, the fire department, and the airport control tower in the area.[23] U.S. air traffic control systems have been targeted many times since 1997. One of these incidents occurred in 2009, when individuals hacked into the air traffic control mission-support systems of the U.S. Federal Aviation Administration (FAA) and stole the personal information (e.g., names and Social Security numbers) of current and former FAA employees.[24] Authorities believe that such threats can soon spread—through network connections—from the support systems to operational systems that process real-time flight information and communications.[25] The consequences of terrorists disabling operational air traffic control systems could be devastating.

Critical infrastructures have also been targeted in countries other than the United States. To date, there has been only one known case where an individual used a computer system to cause significant environmental harm. In 2000, Vitek Boden hacked into a computer system of an Australian sewage treatment plant and leaked hundreds of thousands of gallons of sewage in rivers and parks near the facility.[26] That same year, an individual hacked into a Russian computer system and gained control of the natural gas that flowed through pipelines.[27] Hypothetically, this hacker could have easily increased the gas pressure until the valves broke, causing an explosion to occur. Luckily, this devastating outcome did not occur in this case. Even so, this case illustrates the ease with which access could be gained to such a system and what could be accomplished with this access.

Supervisory Control and Data Acquisition (SCADA) systems are the computer systems of choice for most of the critical infrastructure industries in the United States. According to a report to Congress, SCADA systems "are often placed in remote locations, are frequently unmanned, and are accessed only periodically by engineers or technical staff via telecommunications links."[28] This report further stated that to increase the efficiency of SCADA systems, many have been connected to administrative networks and/or the Internet.[29] This practice has made the systems susceptible to an attack from a cyberspace. In 1998, a 12-year-old boy hacked into a SCADA computer system that controlled the Roosevelt Dam in Arizona. According to authorities, this young hacker gained complete control over the dam. It was estimated that he could have flooded two towns with a total population of 1 million people if he had opened the floodgates.[30]

Credible intelligence exists that al-Qaeda has a significant interest, capability, and intent to engage in cyberterrorism. Computers seized from al-Qaeda operatives in Afghanistan have been found to contain information on U.S. computer systems controlling critical infrastructure, for example.[31] Al-Qaeda and its allies also have expressed interest in learning about the U.S. water supply. Specifically, a seized computer of an al-Qaeda affiliate had numerous items concerning the structural engineering of dams and other water-retaining structures in the United States.[32]

When investigating crimes in the computer threats and intrusions category, an investigator looks for evidence such as source code, Internet data, computer files, and any suspect programs. He or she also looks for any manuals, reference guides. and "how to" books on computer intrusions and attacks, in both hard copy and electronic form.

Financial Crimes and Fraud

An individual engages in fraudulent behavior when he or she "falsely represent[s] a fact, either by conduct or by words or writing, in order to induce a person to rely on the misrepresentation and surrender something of value."[33] Different types of **fraud** are explored in this section, including auction fraud, investment fraud, economic fraud, property theft, and identity theft. Scams involving these crimes are more fully explored in Chapter 6.

Auction Fraud

The federal statute that has been used to prosecute **auction fraud** is 18 U.S.C. § 1343, which covers fraud by wire, radio, or television. A person violates this statute if he or she has

> devised or intend[s] to devise any scheme or artifice to defraud, or for obtaining money or property by means of false or fraudulent pretenses, representations, or

---

Interesting Facts About IP Addresses

Even though IP addresses can be used to identify users, it is common practice for criminals to use fake identification when registering with Internet service providers. Applications that allows users to conceal themselves and their transactions are also readily available on the Internet. In addition, a user can register for a website without identifying himself or herself by first visiting a site called an anonymizer, which replaces the user's IP address with another IP address that cannot be traced back to the user.[37] While investigators can determine that a message was sent from (or that a website was registered from) an anonymizer, they will be unable to identify the individual who visited the anonymizer.[38] Accordingly, IP addresses may not always be traced back to the user.

promises, transmits or causes to be transmitted by means of wire, radio, or television communication in interstate or foreign commerce, any writings, signs, signals, pictures, or sounds for the purpose of executing such scheme or artifice.

In auction fraud, normally a consumer is informed that he or she has won a bid for a particular item being auctioned online. The consumer, in turn, sends money for the item. The item, however, is never delivered to the consumer.

The Internet Crime Complaint Center cautions users from engaging in online transactions with sellers from a foreign country or those who claim to reside in the United States but are currently in a foreign country for business or a family emergency.[34] It further warns users against engaging in any transaction where payment is sought through a money wire transfer service, such as Western Union or MoneyGram. The importance of heeding such warnings is illustrated in the following case.

An individual bid for a television on UBID.com. An e-mail was sent from the seller telling the individual that he had won the bid for the television; if he wanted the television, the message stated, he needed to send the payment to the seller. The buyer sent the money via Western Union, but the television was never sent. After several unsuccessful attempts to contact the seller, the buyer reported the fraud to Western Union and UBID.com. In doing his own investigative work to locate the seller, the buyer expanded the original e-mail headers and was able to identify four distinct IP addresses.[35] He subsequently attempted to trace them back to the seller. While he was able to trace the IP addresses to the Internet service providers (ISPs), it was impossible to trace the IP addresses back to the actual seller.[36]

Examples of such auction frauds abound. The tactics used are more or less the same. Consider the actions of David Deane Kopp, who claimed to sell computers and computers parts on Internet auction sites (e.g., eBay).[39] After receiving payment from more than 100,000 buyers for these goods, Kopp failed to deliver the items.

Jay Nelson is another individual who was convicted of engaging in numerous fraudulent online auction activities.[40] To avoid detection, he took the following steps:

- Nelson altered the type of merchandise he auctioned online.

- He used multiple identities, screen names, and e-mail addresses.

- He used several different methods of payment (e.g., via post office boxes, mail drop, or requested payment to one of his numerous PayPal accounts).

Other criminals who committed auction fraud, such as Brian Wildman, Roger Harvey, Gilbert Vartanian, and Timothy Deceuster, requested that buyers pay for items through the U.S. Postal Service or PayPal; like Nelson, they never provided the buyers with the items after they received payment for them.[41]

At other times, defective merchandise might be sent to customers. Customers, not surprisingly, tend to send the product back to obtain a refund for the defective product. However, such refunds are not provided by the seller.

Some online auction frauds mimic real-life auction frauds. At real-life auctions, some bidders in the audience may have been placed there by the seller in an attempt to drive up the bid. These individuals have no intention to buy; they are merely there to influence bidding. The same thing occurs in an online environment. The seller may have others purposely drive up the bid. In addition, the online environment provides a capability that the real-world environment does not: It enables the seller to bid on his or her own product without the other, genuine bidders knowing that they are being manipulated. Specifically, the seller can take on various digital identities by opening up several e-mail accounts and bidding on his or her own items multiple times. By doing so, he or she prompts the genuine bidders to put in much higher bids for the item than they would have done otherwise (i.e., the bid that they would have provided if the seller had not bid on his or her own items to drive up the prices). This tactic is known as "shill bidding," where "shills" are the individuals who bid on their own items to significantly raise bids.

One such example of shill bidding involved Richard Vitrano, who "sold" inauthentic art (paintings, drawings, prints, to name a few) on eBay.[42] To drive up the price of his auctioned items, he bid on his own items through other third-party accounts. His scheme involved contacting the second highest bidder to the item, informing him or her that the deal with the highest bidder fell through, and offering the item to this bidder.

Some bidders are lured off of legitimate auction websites by individuals claiming to offer the auctioned item at a lower cost—a practice known as "bid siphoning." This fraud is similar to those mentioned previously, as the scam artists engaged in the fraud do not send the auctioned item to the buyer once payment is received.

Buyers of items sold on online auctions are well advised to check seller feedback before sending any money.[43] Nevertheless, there is still a risk of falling victim to fraudulent activity even when these precautions are taken. As part of their plan to defraud buyers, unscrupulous sellers may take further steps to encourage people to buy their items on online auction websites—namely, they may pretend to be satisfied customers and provide multiple favorable reviews online for their own items. Individuals reading these positive reviews, which are known as "shill feedback," may be more likely to purchase items from the seller.

At other times, illegal items are posted for sale on online auction websites. For instance, in 1999, a human kidney was being auctioned on eBay.[44] The bidding for this kidney had reached $5.7 million before eBay stopped the auction. The selling of human organs or body parts is strictly prohibited by the **National Organ Transplant Act of 1984**. On occasion, online message boards (e.g., Craigslist) and forums have contained numerous postings of individuals selling their organs and body parts. A search for such items using a search engine such as Google or Bing can bring up several results (although one cannot tell whether these sales are real or a hoax).

A primary difference between real-life auction fraud and online auction fraud is the anonymity that the latter venue affords to the seller. The anonymity of the Internet makes the identification of such criminals much more difficult for law enforcement agencies.

Because the perpetrators of auction fraud use chat rooms, e-mails, message boards, and websites to troll for fraudulent transactions, these forums are the best place to look for electronic evidence of the crime. Other evidence includes the offender's telephone records, financial information, and account data concerning auction websites.[45]

Other Online Sales Fraud

Online sales fraud concerns online sales of counterfeit consumer goods (e.g., clothing, accessories). In this kind of scam, a consumer purchases an item from the seller that appears to be legitimate—let's say a Chanel handbag. When the buyer receives the item through mail, she realizes that it is not an authenticate Chanel handbag as the advertisement stated, but rather a counterfeit product.

A real-life example of this fraud involved Min Zhu and Qiuhui Huang.[46] Zhu and Huang sold what they claimed were authentic Louis Vuitton bags. Upon receiving their bags, buyers soon realized that the bags were fake. Many serious criminals have been known to sell counterfeit merchandise online; among them are organized crime and terrorist groups. For example, terrorist groups, such as al-Qaeda, have been known to encourage their operatives and supporters to sell counterfeit goods online to help finance the operations of the group.

Online sales fraud may also involve nonexistent services. For example, in 2010, an e-mail was sent that provided a website link to users that would supposedly allow them to see the Academy Awards (Oscars). All that was required was one dollar to be paid to a specified account via PayPal. Because the amount was minimal, many did not think twice about paying the fee. Needless to say, when viewers paid the money and clicked on the screen, nothing appeared on the page. They were cheated out of money and, in fact, paid for a link that did not work.

Another instance of online sales fraud involved a woman from New York, Ingrid Dina Levy, who posed as a legitimate retailer of designer clothing through various websites she had set up:[47]

- www.runwayrenegade.com
- www.bubblepopshop.com
- www.shopchic.net
- www.shoplillie.com
- www.shoplily.com
- www.fashionglow.com

After receiving payment from the buyers, Levy either failed to deliver the items or sent items that she had purchased from discount retail stores in New York. Either way, the buyers never received the advertised items that they had paid for. A similar case involved Michael Murgallis, who used websites to sell aquarium equipment.[48] Even though payment for the goods was received, Murgallis never delivered the merchandise.

Some criminals create websites that look identical to known websites. An example of this behavior was detailed by Jonathan Rusch, the Special Counsel for Fraud Prevention of the Department of Justice.[49] In *United States v. Lee*,[50] the defendant purposely designed what appeared to be a legitimate website to defraud individuals. Lee's website (www.hawaiimarathon.com) was specifically created to be similar to an existing, legitimate website that provided information on marathons in Hawaii and allowed users to register for these marathons online (www.hawaiimarathon.org). Individuals who logged on to Lee's website instead of the official Hawaii marathon site would pay Lee the fee to register for the marathons. Perhaps not surprisingly, the fee on Lee's site was higher than the fee charged on the legitimate site.

Investment Fraud

The Internet has afforded criminals with the opportunity to engage in **investment fraud** with much more ease than was previously possible. Instead of calling individuals on the phone or sending out mass letters, a criminal can now enter a chat room or message board and reach more individuals than was possible in real-world investment fraud schemes. Investment fraud involves market manipulation, which in *Ernst and Ernst v. Hochfelder*[51] was defined as "intentional or willful conduct designed to deceive or defraud investors by controlling or artificially affecting the price of securities."

Two acts, the **Securities Act of 1933** (codified at 15 U.S.C. § 77a et seq.) and the **Securities Exchange Act of 1934** (codified at 18 U.S.C. § 78a et seq.), cover securities fraud. According to 18 U.S.C. § 78j(b), it is unlawful

> To use or employ, in connection with the purchase or sale of any security registered on a national securities exchange or any security not so registered, or any securities-based swap agreement (as defined in section 206B of the Gramm-Leach-Bliley Act), any manipulative or deceptive device or contrivance in contravention of such rules and regulations as the Commission may prescribe as necessary or appropriate in the public interest or for the protection of investors.

A person is found to have violated this law if he or she has made "material misstatements or omissions indicating an intent to deceive or defraud in connection with the purchase or sale of a security."[52] Indeed, the **Code of Federal Regulations** (C.F.R.) covers rules and regulations under the Securities Exchange Act. 17 C.F.R. § 240.10b-5 prohibits "any untrue statement of a material fact or [omitting] to state a material fact necessary in order to make the statements made, in the light of the circumstances under which they were made, not misleading."

Two main types of investment fraud exist: the pump and dump strategy and the cybersmear scheme. In a **pump and dump investment scheme**, individuals falsely advertise favorable and misleading representations of a specific stock to convince people to buy that stock. Once the price of the stock has significantly increased, those involved in the scheme sell their stocks. Those not involved in the scheme lose a significant amount of funds. In one notable case, a 15-year-old boy named Jonathan Lebed purchased stocks with low trading activity and posted false and misleading messages touting the stocks.[53] Many investors who bought stocks after reading these messages lost a significant amount of money from the scheme. Lebed made approximately $800,000 in profit. With the help of his lawyer, he was only required to give back $285,000.

Another example of a pump and dump scheme involved Yun Soo Oh Park, who would recommend stock he owned on a website he developed that specialized in stock advice and trading information. Once the value of his stock would increase, Yun Soo would sell his stock. In *Securities Exchange Commission v. Yun Soo Oh Park*,[54] the court held that Yun Soo acted as an investment adviser and his actions violated the **Investment Advisers Act of 1940**, which states that it is illegal for investment advisers to defraud their investors. In *Lowe v. Securities and Exchange Commission*,[55] the court held that investment advisers are those persons who

> engage . . . in the business of advising others, either directly or through publications or writings, as to the value of securities or as to the advisability of investing in, purchasing, or selling, or who for compensation and as part of a regular business, issue or promulgate analyses or reports concerning securities [for compensation].

An example of a **cybersmear investment scheme** is illustrated in *United States v. Jakob*.[56] In this case, the defendant orchestrated a mass distribution of false reports about a company named Emulex. Specifically, the defendant circulated reports stating that Emulex had a loss of earnings and that it was being investigated by the **Securities and Exchange Commission**. When the stock price started to decline, the defendant, among other things, bought it at a lower price and subsequently sold it when it had recovered most of its value.

Two other examples of investment fraud are Ponzi and pyramid schemes. **Ponzi schemes**[57] involve soliciting investors to contribute to investment opportunities that promise high returns for little to no risk. The money provided by new investors is used to pay existing investors, but none of the money that an investor provides is actually invested anywhere. A **pyramid scheme** requires new investors to become involved in marketing a product and recruiting other investors. The emphasis of this scheme is not on the product being marketed, but rather on the recruitment of new investors. The product is purportedly marketed to cover up the pyramid scheme. The fees provided by new entrants are used to pay those who previously entered the pyramid scheme.

Credit Card Fraud

Credit card fraud (or debit card fraud) occurs when an individual obtains, uses, sells, or buys another individual's credit (or debit) card information. An illustrative case of credit card fraud concerns Jeffrey P. Butcher, who obtained and used a Home Depot credit card that was issued in another name (Dureene Kiplinger) to obtain merchandise.[58] Another example concerns a Child Protective Services case worker named Jennifer White who used stolen credit cards numbers she had obtained from a chat room to purchase merchandise online.[59] Serious criminals have also been known to engage in online credit card fraud. For example, al-Qaeda members and supporters have a history of stealing credit card numbers and using them on online stores to buy supplies for jihadists.[60]

The primary statute that deals with credit card fraud is 15 U.S.C. § 1644. An individual violates this section if, among other things, he or she "uses or attempts or conspires to use any counterfeit, fictitious, altered, forged, lost, stolen, or fraudulently obtained credit card to obtain money, goods, services, or anything else of value which within any one-year period has a value aggregating $1,000 or more."[61] For example, in the case known as *United States v. Bice-Bey*,[62] the court held that the defendant, using credit card numbers without the authorization of the card holders to order goods over the phone, violated 15 U.S.C. § 1644(a).

As with other types of fraud, pertinent statutes for credit card fraud include the following:

- 18 U.S.C. § 1343. This section covers fraud committed by wire, radio, or television.

- 18 U.S.C. § 1030. This section covers fraud and related activity in connection with computers. Of particular importance to credit card fraud is 18 U.S.C. § 1030(a)(2)(a), which concerns gaining fraudulent access to a computer and obtaining information from a financial institution, card issuer, or consumer reporting agency.

- 18 U.S.C. § 1029. This section concerns "fraud and related activity in connection with access devices." According to 18 U.S.C. § 1029(e), an access device includes

    any card, plate, code, account number, electronic serial number, mobile identification number, personal identification number, or other telecommunications service, equipment, or instrument identifier, or other means of account access that can be used, alone or in conjunction with another access device, to obtain money, goods, services, or any other thing of value, or that can be used to initiate a transfer of funds (other than a transfer originated solely by paper instrument).

An example of a violation of 18 U.S.C. § 1029 can be seen in the case of Tyson Baird. Baird unlawfully possessed access devices as defined by 18 U.S.C. § 1029(e) and fraudulently obtained and used debit and credit card information to acquire money, goods, and services. He also laundered the money he retrieved from these cards prior to making his purchases in violation of 18 U.S.C. § 1956 (concerning money laundering). Apart from offenses involving credit cards, 18 U.S.C. § 1029 has been successfully applied to cases concerning access devices such as long-distance telephone access codes and mobile phone access numbers.[63]

The scams designed to steal individuals' financial information are fully explored in Chapter 6.

Identity Theft and Fraud

Identity theft occurs when a suspect assumes the identity of the victim by unlawfully using the name, Social Security number, bank account number or other identifying information of that individual with the intent to commit a crime. The **Identity Theft and Assumption Deterrence Act of 1998** made identity theft an offense [see 18 U.S.C. § 1028(a)(7)]. Identity theft occurs when a person

> knowingly transfers or uses, without lawful authority, a means of identification of another person with the intent to commit, or to aid or abet, any unlawful activity that constitutes a violation of Federal law, or that constitutes a felony under any applicable State or local law.

Usually, identity theft is committed to facilitate other crimes:[64]

- **Identification fraud** (18 U.S.C. §1028(a)(1)–(6))
- Credit card fraud (18 U.S.C. §1029)
- Computer fraud (18 U.S.C. § 1030)
- Mail fraud (18 U.S.C. § 1341)
- Wire fraud (18 U.S.C. §1343)
- Financial institution fraud (18 U.S.C. §1344)
- Mail theft (18 U.S.C. § 1708)
- Immigration document fraud (18 U.S.C. § 1546)

A prominent case illustrating this type of fraud involved Chad Hatten. Hatten was charged with—among other criminal activities—promoting and facilitating the distribution and use of credit, debit, and personal information and producing and selling false identification documents.[65] Hatten was a member of the Shadowcrew criminal organization, which maintained one of the largest online sites trafficking in stolen identification and financial information (including bank card, credit, and debit data).[66] In another case, Gregory Kopiloff gained unauthorized access to individuals' computers via file sharing programs to retrieve their personal and financial information.[67] He then used this information to open up credit accounts and make online purchases with these fraudulent accounts. In the same way, Justin A. Perras used multiple methods (including hacking and malware) to obtain usernames and passwords of individuals using a database owned by LexisNexis.[68] The techniques used to steal a person's identity and the scams involving identity theft are explored in more depth in Chapter 6.

When investigating identity theft, the investigator looks for credit card data, checks, money orders, financial records, website transaction records, forged documents, mail sent to the victims, electronic signatures, reproduction of the signatures of the victims, high-quality copiers, scanners, printers and laminators, and photos, birth certificates, and driver's licenses of the victims (or forged copies of these documents).[69]

Telecommunications Fraud

Telecommunications fraud involves the theft of telecommunications services or the use of telecommunications services to commit other forms of fraud. A well-known case of telecommunications fraud involved one of the most notorious organized crime families, the Gambino family. A few members of the Gambino family were involved in a fraud scheme that generated approximately $500 million in gross revenues by placing unauthorized charges on telephone bills of customers. For example, individuals who called 800 (toll-free) numbers for supposedly free samples of psychic readings, horoscopes, phone sex, and dating services were later billed with reoccurring monthly charges disguised as other services, such as voice mail (these charges were at least $40 per month).[70] This scheme also involved Internet fraud, as individuals who took allegedly free tours of porn websites were later subjected to unauthorized debit and credit card charges (these charges reached $90 per month).[71]

Another example of telecommunications fraud is **phone phreaking**. Before the advent of hackers, phone phreakers would trick the telephone system to allow them to make free phone calls. One of the most well-known phone phreaker was John Draper (known as Captain Crunch), who successfully replicated the tones necessary to place free long-distance phone calls. Another famous phone phreaker, known as Captain Zap, changed the clock of AT&T's phone system so that everyone making daytime calls would be charged at a night-time rate (at a lesser charge).

Telemarketing fraud is another example of telecommunications fraud. In this type of scam, a telephone is used to communicate with the prospective victim and try to persuade that person to send money to the fraudulent scheme. A primary example of telemarketing fraud is the operation conducted by David Allen Sussman. In Sussman's fraudulent telemarketing scheme, he had telemarketers present the companies he owned as photography studios that produce photo packages to students, youth clubs, and other organizations.[72] Of course, Sussman never delivered the photo packages for which the customers had paid. In another case, John A. Reece and Patrick J. Soltis used telemarketers to pressure individuals to invest money in fictitious companies. They made approximately $600,000 in profit from this fraudulent scheme.[73]

Embezzlement

Embezzlement is another form of financial fraud. Embezzlement, which is codified in Chapter 31 of Title 18 of the U.S. Code, occurs when an employee misappropriates the funds entrusted to him or her by the company by, for example, making payments to a fictitious employee and then collecting the money. Others simply pocket the company's money outright. In 2009, a U.S. Postal Services worker in Puerto Rico, Arnaldo Cortes-Mestres, was arrested for embezzling passport application and renewal fees.[74] The amount embezzled normally dictates the penalty for engaging in such a criminal activity.[75]

Embezzlement should not be confused with larceny. In fact, as the court stated in *Moore v. United States*,[76] embezzlement "differs from larceny in the fact that the original taking was lawful, or with the consent of the owner, while in larceny the felonious intent must have existed at the time of the taking."

When investigating embezzlement, investigators typically look at the financial records and statements that specific staff have access to and search for discrepancies in accounts, pay, and so on. Overall, investigators look for deviations in normal company spending behavior or statements showing large cash withdrawals.

Intellectual Property Theft and Economic Espionage

Intellectual property is intangible property that the law grants ownership rights to. Federal law protects four distinct areas of intellectual property:[77]

- **Copyright**: A legal document that gives to the copyright holder the right to copy, reproduce, and sell works or authorship such as sound recordings, musical, literary, dramatic, architectural, pantomimes, choreographic, pictorial, graphic, sculptural works, and motion pictures and other audiovisual works.[78]

- **Trademark**: Designed to protect commercial identities or brands. Under 15 U.S.C. § 1127, trademarks include

  > any word, name, symbol, or device, or any combination thereof . . . used by a person, or . . . which a person has a bona fide intention to use in commerce . . . to identify and distinguish his or her goods, including a unique product, from those manufactured or sold by others and to indicate the source of the goods, even if that source is unknown.

- **Patent**: Designed to protect individuals' inventions. Pursuant to 35 U.S.C. § 271(a), patents provide individuals with the rights to make, use, offer to sell, or sell their inventions, products, and processes.

- **Trade secrets**: They include

  > all forms and types of financial, business, scientific, technical, economic, or engineering information, including patterns, plans, compilations, program devices, formulas, designs, prototypes, methods, techniques, processes, procedures, programs, or codes, whether tangible or intangible, and whether or how stored, compiled, or memorialized physically, electronically, graphically, photographically, or in writing (see 18 U.S.C. § 1839(3)).

Trade secrets are fully explored in the next section of this chapter. Two types of intellectual property—copyrights and trademarks—are explored here.

Copyright Infringement

Statutes concerning copyright violations are included in Title 17 and 18 of the United States Code. According to 17 U.S.C. §102(a),

Copyright protection subsists, in accordance with this title, in original works of authorship fixed in any tangible medium of expression, now known or later developed, from which they can be perceived, reproduced, or otherwise communicated, either directly or with the aid of a machine or device.

The main statute criminalizing copyright violation is 17 U.S.C. 506(a). Specifically, this section states that anyone "who willfully infringes a copyright shall be punished . . . , if the infringement was committed . . . for [the] purposes of commercial advantage or private financial gain." For copyright infringement to have occurred, the prosecution must first show that a valid copyright existed.

There are certain exceptions to copyright infringement.[79] The **fair use doctrine** is one such exception. In particular, 17 U.S.C. § 107 states that

> the fair use of a copyrighted work, including such use by reproduction in copies or phonorecords or by any other means specified by that section, for purposes such as criticism, comment, news reporting, teaching (including multiple copies for classroom use), scholarship, or research, is not an infringement of copyright.

Another exception is included in 17 U.S.C. § 108, which allows for the reproduction of a limited number of works by libraries and archives.

No Electronic Theft Act of 1997

Prior to the passing of the **No Electronic Theft Act of 1997**, if an individual did not distribute or reproduce copies of copyrighted works for profit, then no law was violated. For example, in the case of *United States v. La Macchia*,[80] a student at the Massachusetts Institute of Technology (MIT) operated an online bulletin board that distributed copyright software programs. LaMacchia could not be charged under 17 U.S.C. § 506 because he was not engaging in the unauthorized distribution of copyrighted works for profit.

The No Electronic Theft Act sought to close this loophole by making it a criminal offense to distribute or reproduce copies of copyrighted works without authorization, regardless of whether the individual engaged in such activities for profit. The first person convicted under the No Electronic Theft Act was Jeffrey Gerard Levy.[81] Levy illegally posted music, computer software, and movies on his website and allowed the public to download copyrighted work from it.

Digital Millennium Copyright Act of 1998

Due to the rise in circumvention of copyright controls, which seek to control access to copyrighted works, the **World Intellectual Property Organization** (WIPO) responded by adopting the **Copyright Treaty of 1996** (December 20, 1996).[82] According to Article 11 of the Treaty, states need to provide

> adequate legal protection and effective legal remedies against the circumvention of effective technological measures that are used by authors in connection with the exercise of their rights under this Treaty or the **Berne Convention** [for the

protection of literary and artistic works (an international agreement governing copyright)] and that restrict acts, in respect of their works, which are not authorized by the authors concerned or permitted by law.

To implement the **WIPO Copyright Treaty of 1996**, the U.S. Congress passed the **Digital Millennium Copyright Act of 1998** (codified in 17 U.S.C. § 1201–1204). This Act amended the **Copyright Act of 1976** to address digital information. 17 U.S.C. § 1201(a)(1)(A), made it a criminal offense for anyone to "circumvent a technological measure that effectively controls access to" protected works. Moreover, 17 U.S.C. § 1201(a)(2)(A) made it illegal to

> manufacture, import, offer to the public, provide, or otherwise traffic in any technology, product, service, device, component, or part thereof, that . . . is primarily designed or produced for the purpose of circumventing a technological measure that effectively controls access to [protected works].

In *United States v. Elcom*,[83] the court stated that tools designed to avoid or bypass copy restrictions are banned under the Digital Millennium Copyright Act of 1998 because they would most likely be used for unlawful infringement purposes. One such example of someone bypassing restrictions of copyrighted works for profit involved Walter Buchholz, Jr.[84] Buchholz manufactured and distributed technology that was designed to circumvent encrypted technology. Specifically, he reprogrammed DISH Network "smart cards" to circumvent technological measures that were in place to limit access to DISH Network services to legitimate subscribers. Another example involved Mohsin Mynaf, who was found to possess equipment with which to bypass videocassette copyright protections.[85]

Family Entertainment and Copyright Act of 2005

The **Family Entertainment and Copyright Act of 2005** makes it a criminal offense to use a camcorder or other similar recording device to record a movie playing in a movie theater. The majority of pirated copies circulating on the internet are the product of criminals using camcorders to record newly released movies in theaters. Pursuant to 18 U.S.C. § 2319B, if an individual is found with a camcorder or related recording device on his or her person in a movie theater, the device may be used as evidence against the person in a criminal proceeding.

Other Statutes

18 U.S.C. § 2319A

This statute prohibits the creation and trafficking of sound and video recordings of live music performances.

18 U.S.C. § 2318

Under 18 U.S.C. § 2318(a)(1)(A), an individual violates this statute if he or she "knowingly traffics in . . . a counterfeit label or illicit label affixed to, enclosing, or accompanying, or designed to be affixed to, enclose, or accompany" a phonorecord; work of visual art;

documentation, packaging, or copy of a computer program; motion picture (or other audiovisual work); literary work; or pictorial, graphic, or sculptural work. According to 18 U.S.C. § 2318(b)(1), a counterfeit label is defined as "an identifying label or container that appears to be genuine, but is not." The term **counterfeit** should not be confused with terms such as **bootleg**. Specifically, in *United States v. Shultz*,[86] the court

> distinguish[ed] between counterfeit stereo tapes and so-called bootleg . . . stereo tapes. Counterfeit tapes are tapes which are represented to be genuine articles of particular record companies when, in truth, they are not. The process includes reproducing the tape itself and also the recognized label of another record company. A bootleg tape is a reproduction of someone else's recording or recordings marketed under a different label.

Computers have afforded criminals with the opportunity to steal vast amounts of copyrighted material and illicitly distribute them. For instance, Carmine Caridi, a member of the Academy of Motion Picture Arts and Sciences, provided William Sprague with screening copies of movies (many prior to their theatrical or DVD release).[87] Sprague then posted these copies online. Another case of distribution of pirated works involved Robert Thomas and Jared Chase Bowser.[88] In particular, Bowser obtained a copy of an album that had been provided to a music reviewer before the album's release to the public. Bowser sent digital copies of the album to Thomas, who then posted them online. Another case concerned the sale of counterfeit goods online.[89] In this case, Gary Berger advertised and sold what buyers believed to be real Rolex watches. Berger, however, was sending them counterfeit products instead. Likewise, Mark Edward Dipadova and Theresa Gayle Ford operated a website where they sold counterfeit products, such as Rolex watches and Mont Blanc pens.[91]

Some companies have fought back by filing civil suits against violators. For instance, a Dutch file-sharing website, TorrentSpy.com, was ordered to pay more than $110 million in restitution to Viacom Inc.'s Paramount Pictures for distributing pirated films. Other companies have filed civil suits against Internet service providers—albeit unsuccessfully,[91] because the **Online Copyright Infringement Liability Limitation Act**, which is part of the Digital Millennium Copyright Act of 1998, provides immunity to Internet service providers under certain exceptions (e.g., the ISP took action when it became aware of the infringing behavior). Convictions of distributors of pirated software have also increased. Individuals who owned or operated for-profit software piracy websites have received prison sentences and have had their assets forfeited.[92]

Trademark Violations

The **Trademark Counterfeiting Act of 1984** made it a criminal offense, as codified in 18 U.S.C. § 2320, for an individual to traffic in counterfeit goods or services. Specifically, an individual can be found to have violated 18 U.S.C. § 2320 if he or she

> intentionally traffics or attempts to traffic in goods or services and knowingly uses a counterfeit mark on or in connection with such goods or services, or inten-

tionally traffics or attempts to traffic in labels, patches, stickers, wrappers, badges, emblems, medallions, charms, boxes, containers, cans, cases, hangtags, documentation, or packaging of any type or nature, knowing that a counterfeit mark has been applied thereto, the use of which is likely to cause confusion, to cause mistake, or to deceive.

One case concerning the **trafficking of counterfeit goods** involved Kurt Wakefield.[93] Wakefield pleaded guilty to violating 18 U.S.C. § 2320(a) by distributing counterfeit Nike shoes throughout the southeastern United States. Organized crime groups have also been known to engage in trafficking of counterfeit goods. One such group known for its trafficking of counterfeit DVDs and CDs is the Yi Ging organization.

Economic Espionage

Trade secrets are protected under the Economic Espionage Act of 1996 (codified in 18 U.S.C. § 1831–1839) if the trade secret is not generally known or easily accessible to the public, and if the owner of trade secret has taken reasonable measures to keep the information secret.[94] Trade secrets require protection because they are "an integral part of virtually every aspect of U.S. trade, commerce, and business."[95]

According to 18 U.S.C. § 1831, **economic espionage** occurs when an individual steals a trade secret for the benefit of a foreign government, foreign instrumentality, or foreign agent. A case in point is the actions of Fei Ye and Ming Zhong. They violated 18 U.S.C. § 2314 by stealing trade secret information from U.S. companies to help "raise China's ability to develop super-integrated circuit design, and form a powerful capability to compete with worldwide leaders' core development technology and products in the field of integrated circuit design."[96]

The Economic Espionage Act also criminalizes the theft of a trade secret that is not for the benefit of a foreign government, instrumentality, or agent.[97] For instance, William McMenamin violated 18 U.S.C. § 1832 when he illegally accessed a competitor's computer and copied trade secrets.[98] Another case of economic espionage in which the offender sought to benefit an individual other than the owner of the trade secret involved John Berenson Morris. Morris was prosecuted under the Economic Espionage Act for attempting to steal and sell the trade secret information of a New York textile company, Brookwood Companies, to one of its competitors for $100,000.[99] Yet another example of this type of economic espionage involved William Genovese, who stole significant portions of Microsoft Windows source code. This code was subsequently unlawfully released and distributed over the Internet.[100] Genovese was charged with unlawfully distributing a trade secret in violation of the Economic Espionage Act.

Personal Crimes

A wide variety of personal crimes involving a computer angle are possible, including murder, domestic violence, assault, harassment, extortion, drug dealing, and illegal gambling. This section focuses on two crimes that fall under this category: cyberharassment and cyberstalking. Crimes committed against children, such as cyberbullying, child exploitation, and child pornography, which are also considered to be part of this category, are explored in detail in Chapter 6.

Cyberharassment

Cyberharassment is a crime that occurs when an individual uses the Internet, e-mail, mobile phones, fixed telephony, or other forms of communications to intentionally irritate, attack, alarm, or otherwise bother another individual. Consider the actions of Victor Vevea.[101] Vevea gained unauthorized access to the e-mail of the attorney (Michael Kilpatrick) who represented his girlfriend in a lawsuit against him. He then used his access to the attorney's e-mail account to send harassing e-mails to government officials and other lawyers in Kilpatrick's name. Another case involved John Derungs.[102] Derungs sent numerous abusive and insulting (i.e., harassing) e-mails to an employee of the San Francisco Giants who had informed him that someone else was hired for a job for which Derungs had applied.[103]

In one notable cyberharassment case, Shaun Hansen pleaded guilty to committing interstate telephone harassment. Specifically, he was paid to make repeated harassing phone calls to numbers associated with the New Hampshire Democratic Party and the Manchester Professional Firefighters Association on Election Day.[104] Hansen violated 18 U.S.C. § 223, which covers "obscene or harassing telephone calls in the District of Columbia or in interstate or foreign communications." In particular, he was ruled to be in violation of 18 U.S.C. § 223(a)(1)(C), which makes it illegal for someone to "make . . . a telephone call or utilize . . . a telecommunications device, whether or not conversation or communication ensues, without disclosing his identity and with intent to annoy, abuse, threaten, or harass any person at the called number or who receives the communications." It is important to note that 18 U.S.C. § 223 applies only to direct communications between the perpetrator and the victim. Consequently, it is inapplicable to situations in which individuals harass others via public forums such as online bulletin boards.[105]

Cyberharassment and other threats can occur via e-mail, instant messaging (IM), fixed telephony, cell phones, and text messaging. These are the first places to look for evidence of a crime. Subpoenas for information from social networking sites may also be used to find the perpetrators of these crimes.

Cyberstalking

Similarly to cyberharassment, cyberstalking normally occurs via e-mail, social networking sites, newsgroups, bulletin boards, blogs, chat rooms, and websites—so these are great places to look for evidence of this crime. The following federal statutes cover these crimes.

18 U.S.C. § 875

18 U.S.C. § 875 covers interstate threatening communications. An individual violates 18 U.S.C. § 875(c) if he or she "transmits in interstate or foreign commerce any communica-

tion containing any threat to kidnap any person or any threat to injure the person or another"; the punishment for such a violation is a fine, incarceration for up to five years, or both. The courts have held that an Internet transmission can be considered as a transmission in interstate commerce.[106] They have also held that private e-mails are accepted as interstate commerce.[107]

This law does not apply in situations where no real threat has been communicated. The determination of what constitutes a real threat is determined by "whether a reasonable person would foresee that the statement would be interpreted by those to whom the maker communicates the statement as a serious expression of intent to harm or assault."[108] In one case, an anti-abortion group, American Coalition of Life Activists (ACLA), created an online list containing personal information (names and home addresses) of doctors and others who supported or provided abortion services. The website was designed to keep track of those victimized on the list by anti-abortion terrorists by "striking through the names of those who had been murdered and graying out the names of the wounded."[109] According to the court in *Planned Parenthood of the Columbia/Willamette, Inc. v. American Coalition of Life Activists*,[110] the doctors (and other targets of the website) might have interpreted the website and the actions of the ACLA as veiled threats that bodily harm would be inflicted on those listed on the website if they did not stop performing abortions. As such, the contents of the website and actions of the ACLA were interpreted as a form of cyberstalking.

Another case in which the offender's action were found to constitute cyberstalking involved Shawn Memarian.[111] Memarian started harassing a woman after she ended their relationship. He used multiple aliases to send her threatening messages. He further posed as the victim and posted false personal advertisements for sexual encounters on Craigslist and social networking sites (e.g., MySpace and Facebook). As a result of these advertisements, approximately 30 men visited the woman's home (sometimes at night) seeking the sexual encounters she had supposedly advertised.

18 U.S.C. § 2261A

18 U.S.C. § 2261A is also known as the **Interstate Stalking Act of 1996**. An individual violates this statute, if he or she

> travels in interstate or foreign commerce . . . with the intent to kill, injure, harass, or place under surveillance with intent to kill, injure, harass, or intimidate another person, and in the course of, or as a result of, such travel places that person in reasonable fear of the death of, or serious bodily injury to, or causes substantial emotional distress to that person, a member of the immediate family . . . of that person, or the spouse or intimate partner of that person.

The Interstate Stalking Act applies in situations where an individual uses "the mail, any interactive computer service, or any facility of interstate or foreign commerce to engage in a course of conduct that causes substantial emotional distress to that person or places that person in reasonable fear of the death of, or serious bodily injury to" that person, his or her spouse or partner, or immediate family.[112]

Consider a well-known cyberstalking case—that of Amy Boyer.[113] Liam Youens was obsessed with Boyer, with whom he had attended high school. Youens had fantasies about stalking and killing Boyer—fantasies that he had described in detail on his personal website. To find Boyer, Youens requested her date of birth, Social Security number, and employment information from an online investigation and information service, Docusearch. Docusearch could not find Boyer's date of birth, but was able to find her Social Security number and work address. Docusearch retrieved her Social Security number from a credit reporting agency. Then, to obtain her employment address, a subcontractor of Docusearch, Michelle Gambino, placed a "pretext"[114] call to Boyer pretending to be affiliated with Boyer's insurance company. On the phone, Gambino claimed that Boyer needed to verify her work address so that her insurance company could provide her with an overpayment refund that was due to her. Docusearch then forwarded the requested information to Youens. Sometime after receiving the information, Youens went to Boyer's workplace and shot her as she was leaving work.

It is important to note that the Financial Modernization Act of 1999 (otherwise known as the Gramm-Leach-Bliley Act) prohibits pretext calls;[115] however, it applies only to financial institutions and insurance companies. As such, pretext calls made to a victim's family, friends, coworkers, or employers (excluding those from financial institutions and insurance companies) are not covered by this Act. Additionally, the **Telephone Records and Privacy Protection Act of 2006** prohibits the use of **pretexting** to buy, sell, or obtain individuals' telephone records. The first person to be indicted under this statute was Vaden Anderson, who sought to obtain private phone records by serving a fictitious civil subpoena to a phone company (Sprint/Nextel).[116]

Chapter Summary

This chapter focused on the evolution of cybercrime statutes and the crimes they cover. It also illustrated the limitations of the statutes in dealing with specific cybercrimes. The cybercrimes covered in this chapter included computer hacking, website defacement, writing and distribution of malicious code, computer intrusions and attacks, cyberterrorism, different types of fraud, intellectual property theft, electronic espionage, cyberharassment, and cyberstalking. This list of crimes is by no means exhaustive; only the most common cybercrimes and their respective statutes were included in this chapter.[117]

Key Terms

Adware	Copyright
Auction fraud	Copyright Act of 1976
Berne Convention	Counterfeit
Bootleg	Credit card fraud
Code of Federal Regulations	Cybersmear investment scheme

Digital Millennium Copyright Act of 1998
Economic espionage
Executive Order 13133
Fair use doctrine
Family Entertainment and Copyright Act of 2005
Fraud
Identification fraud
Identity Theft and Assumption Deterrence Act of 1998
Intellectual property
Interstate Stalking Act of 1996
Investment Advisers Act of 1940
Investment fraud
National Organ Transplant Act of 1984
No Electronic Theft Act of 1997
Online Copyright Infringement Liability Limitation Act
Online sales fraud
Patent
Phone phreaking
Ponzi scheme
Pretexting

Pump and dump investment scheme
Pyramid scheme
SCADA systems
Securities Act of 1933
Securities and Exchange Commission
Securities Exchange Act of 1934
SYN flood attack
Taxpayer Browsing Protection Act of 1997
TCP handshake
Telecommunications fraud
Telephone Records and Privacy Protection Act of 2006
Trade secrets
Trademark
Trademark Counterfeiting Act of 1984
Trafficking of counterfeit goods
Transmission Control Protocol
Website defacement
WIPO Copyright Treaty of 1996
WIPO Performances and Phonograms Treaty of 1996
World Intellectual Property Organization

Critical Thinking Questions

Choose one cybercrime (it does not have to be included in this chapter). Make up a scenario including this cybercrime or use an existing case (if you choose the latter, make sure to cite your case). As an investigator, which evidence of this crime should you look for when conducting your investigation? Why?

Review Questions

1. What are computer viruses? Worms? Describe the main effects of one virus or worm. Was the perpetrator (or perpetrators) of the virus or worm caught after its release into the wider community?

2. Distinguish between a Trojan horse, a computer virus, and a worm.

3. What is the difference between spyware and adware?

4. Which sections of 18 U.S.C. § 1030 could be used against someone who launched a DoS or DDoS attack? If you believe a section cannot be used to charge someone who engages in these attacks, why do you think this is the case?

5. What is a TCP handshake? How does a SYN flood attack occur?

6. What is cyberterrorism? Is the threat of this crime significant? Why or why not?

7. List the types of fraud that people engage in.

8. What is intellectual property? Why should it be protected?

9. What are trade secrets? Why should the theft of trade secrets be criminalized?

10. What is the main difference between cyberharassment and cyberstalking?

Footnotes

[1] States have also created laws penalizing such unauthorized access. See, for example, New York Penal Law 156.00 to 156.50.

[2] 18 U.S.C. § 1030(a)(1).

[3] *United States v. Czubinski*, 106 F. 3d 1069, 1078 (1st Cir. 1997).

[4] 166 F.3d 1088 (10th Cir. 1999).

[5] See Chapter 1 for more information on these types of crimes.

[6] United States Attorney's Office, Central District of California. (2003, April 18). Ex-employee of airport transportation company guilty of hacking into company's computer. US Department of Justice, Computer Crime & Intellectual Property Section. Retrieved from http://www.justice .gov/criminal/cybercrime/tranPlea.htm

[7] United States Attorney's Office, Central District of California. (2003, June 12). Southern California al Jazeera Website agrees to plead guilty to federal charges. US Department of Justice, Computer Crime & Intellectual Property Section. Retrieved from http://www.justice.gov/criminal/cybercrime/ racinePlea.htm

[8] Web applications allow legitimate website visitors to submit and retrieve information from online databases.

[9] Anderson, K. (2001, October 23). Hacktivists take sides in war. *BBC News* [online]. Retrieved from http://news.bbc.co.uk/2/hi/americas/1614927.stm

[10] Anti-war hackers target Websites. (2003, March 21). *BBC News* [online]. Retrieved from http://news .bbc.co.uk/2/hi/technology/2871985.stm

[11] Kirk, J. (2009, February 2009). Pirate Bay supporters hack Swedish IFPI website. *PCWorld.com*. Retrieved from http://pcworld.about.com/od/securit1/Pirate-Bay-Supporters-Hack-Swe.htm

[12] Hancock, S. (2001, December 22). Hacker boys from Brazil. *BBC News* [online]. Retrieved from http://news.bbc.co.uk/2/hi/science/nature/1723356.stm

[13] In Spain, a long-standing joke exists concerning the similarity of Zapatero with Atkinson. See Mr. Bean replaces Spanish PM on EU presidency site. (2010, January 4). *BBC News* [online]. Retrieved from http://news.bbc.co.uk/2/hi/europe/8440554.stm

[14] *U.S. v. Perez et al.* (S.D. Cal., indicted August 26, 2005). See United States Attorney's Office, Southern District of California. (2005, August 26). Creator and four users of Loverspy spyware program indicted. US Department of Justice, Computer Crime & Intellectual Property Section. Retrieved from http://www.justice.gov/criminal/cybercrime/perezIndict.htm

[15] United States Attorney's Office, Northern District of Florida. (2009, April 2). Individual sentenced after guilty plea to unauthorized access of a computer. Federal Bureau of Investigation, Jacksonville. Retrieved from http://jacksonville.fbi.gov/dojpressrel/pressrel09/ja040209.htm

[16] TCP is responsible for establishing a connection between two host computers and breaking up data into packets.

[17]United States Attorney, Western District of Pennsylvania. (2003, December 2). Former employee of American Eagle Outfitters sentenced to prison for password trafficking and computer damage. US Department of Justice, Computer Crime & Intellectual Property Section. Retrieved from http://www.justice.gov/criminal/cybercrime/pattersonSent.htm

[18]Department of Justice, Western District of Washington. (2005, February 11). Juvenile sentenced for releasing worm that attacked Microsoft website. US Department of Justice, Computer Crime & Intellectual Property Section. Retrieved from http://www.justice.gov/criminal/cybercrime/juvenileSent.htm

[19]United States Attorney, Western District of Pennsylvania. (2003, December 2). Former employee of American Eagle Outfitters sentenced to prison for password trafficking and computer damage. US Department of Justice, Computer Crime & Intellectual Property Section. Retrieved from http://www.justice.gov/criminal/cybercrime/pattersonSent.htm

[20]Dick, R. L. (2001, April 5). Issue of intrusions into government computer networks. Testimony of the Director of the FBI National Infrastructure Protection Center before the House Energy and Commerce Committee, Oversight and Investigation Subcommittee. Retrieved from http://www.fbi.gov/congress/congress01/rondick.htm

[21]Collin, B. C. (n.d.). The future of cyberterrorism: Where the physical and virtual worlds converge. 11th Annual International Symposium on Criminal Justice Issues. Retrieved from http://afgen.com/terrorism1.html

[22]The Working Group on Unlawful Conduct on the Internet was established by **Executive Order 13133** (August 6, 1999) to evaluate the extent of cybercrime and the tools and capabilities of law enforcement to deal with this threat. See Working Group on Unlawful Conduct on the Internet. (2000, March). The electronic frontier: The challenge of unlawful conduct involving the Internet. Retrieved from http://www.justice.gov/criminal/cybercrime/unlawful.htm

[23]Elmusharaf, M. M. (2004, April 8). Cyber terrorism: The new kind of terrorism. Computer Crime Research Center. Retrieved from http://www.crime-research.org/articles/cyber_terrorism_new_kind_terrorism/.

[24]Whitney, L. (2010, March 31). IBM, FAA partner on aviation cybersecurity. *CNET News* [online]. Retrieved from http://news.cnet.com/security/?keyword=FAA

[25]Mills, E. (2009, May 7). Report: Hackers broke into FAA air traffic control systems. *CNET News* [online]. Retrieved from http://news.cnet.com/8301-1009_3-10236028-83.html

[26]Schneier, B. (2003, June 19). The risks of cyberterrorism. Computer Crime Research Center. Retrieved from http://www.crime-research.org/news/2003/06/Mess1901.html

[27]Elmusharaf, M. M. (2004, April 8). Cyber terrorism: The new kind of terrorism. Computer Crime Research Center. Retrieved from http://www.crime-research.org/articles/cyber_terrorism_new_kind_terrorism/

[28]Wilson, C. (2008, January 29). Botnets, cybercrime, and cyberterrorism: Vulnerabilities and policy issues for Congress. *US Congressional Research Report* RL32114, p. 21. Retrieved from http://www.fas.org/sgp/crs/terror/RL32114.pdf

[29]*Ibid.*, pp. 21–22.

[30]Gellman, B. (2002, June 27). Cyber-attacks by al Qaeda feared: Terrorists at threshold of using Internet as tool of bloodshed, experts say. *Washington Post* [online]. Retrieved from http://www.washingtonpost.com/wp-dyn/content/article/2006/06/12/AR2006061200711.html

[31]Elmusharaf, M. M. (2004, April 8). Cyber terrorism: The new kind of terrorism. Computer Crime Research Center. Retrieved from http://www.crime-research.org/articles/cyber_terrorism_new_kind_terrorism/

[32]Public Safety Canada. (2002, January 30). Terrorist interest in water supply and SCADA systems. *Information Note Number: IN02-002.* Retrieved from https://www.publicsafety.gc.ca/abt/index-eng.aspx

[33]Reid, S. T. (2010). *Criminal law* (8th ed.). Oxford, UK: Oxford University Press, p. 401.

[34]Internet Crime Complaint Center. (n.d.). Internet crime schemes. Retrieved from http://www .ic3.gov/crimeschemes.aspx

[35]See Chapter 10 on e-mail investigations, which covers full e-mail header information.

[36]For a more detailed analysis of this example and more personal accounts of auction fraud, visit the following website: http://www.crimes-of-persuasion.com/Crimes/Delivered/internet_ auctions.htm

[37]Hinnen, T. M. (2004). The cyber-front in the war on terrorism: Curbing terrorist use of the Internet. *Columbia Science and Technology Law Review, 5,* 11.

[38]*Ibid.*, p. 11.

[39]United States Attorney's Office, Eastern District of Texas. (2004, June 23). McKinney man sentenced for Internet fraud: Case part of nationwide investigation. US Department of Justice, Computer Crime & Intellectual Property Section. Retrieved from http://www.justice.gov/usao/ txe/news_release/news/kopp_williams.pdf

[40]United States Attorney's Office, Western District of Oklahoma. (2002, July 8). Internet auction scam results in guilty plea. US Department of Justice, Computer Crime & Intellectual Property Section. Retrieved from http://www.justice.gov/criminal/cybercrime/nelsonPlea.htm

[41]United States Attorney's Office, Northern District of Ohio. (2001, December 20). Man pleads guilty to eBay auction fraud. US Department of Justice, Computer Crime & Intellectual Property Section. Retrieved from http://www.justice.gov/criminal/cybercrime/wildmanPlea.htm; United States Attorney's Office, Northern District of Ohio. (2002, June 5). Melvern, Ohio man indicted for eBay auction fraud. US Department of Justice, Computer Crime & Intellectual Property Section. Retrieved from http://www.justice.gov/criminal/cybercrime/harveyIndict.htm; United States Attorney's Office, Eastern District of California. (2005, January 27). Former Sacramento man arrested in eBay auction fraud. US Department of Justice, Computer Crime & Intellectual Property Section. Retrieved from http://www.justice.gov/criminal/cybercrime/vartanian Arrest.htm; United States Attorney's Office, Eastern District of California. (2005, March 28). Fairlawn, Ohio man sentenced in e-Bay auction scam. US Department of Justice, Computer Crime & Intellectual Property Section. Retrieved from http://www.justice.gov/criminal/cybercrime/ deceusterSent.htm

[42]United States Attorney's Office, Southern District of New York. (2006, February 16). Manhattan art dealer sentenced to jail for Internet art fraud. US Department of Justice, Computer Crime & Intellectual Property Section. Retrieved from http://www.justice.gov/usao/nys/pressreleases/ February06/vitranosentencingpr.pdf

[43]Federal Trade Commission. (2000, February). Going, going, gone: Law enforcement efforts to combat Internet auction fraud. Retrieved from http://www.ftc.gov/bcp/reports/int-auction.htm

[44]Associated Press. (1999, September 2). Online kidney auction stopped. Retrieved from http://www9.georgetown.edu/faculty/aml6/econ001/news/kidney.htm

[45]See the following document from the National Institute of Justice for more information on evidence of this crime: Technical Working Group for Electronic Crime Scene Investigation. (2001, July). Electronic crime scene investigations: A guide for first responders. National Institute of Justice, U.S. Department of Justice, p. 37. Retrieved from http://www.ncjrs.gov/pdffiles1/nij/ 187736.pdf

[46]Office of the United States Attorney, Southern District of New York. (2003, February 10). Two men arrested in conspiracy to sell over $200,000 worth of fake Louis Vuitton handbags via the Internet. US Department of Justice, Computer Crime & Intellectual Property Section. Retrieved from http://www.justice.gov/criminal/cybercrime/zhuArrest.htm

[47]United States Attorney's Office, Eastern District of Virginia. (2008, June 12). New York woman sentenced to 46 months in prison for Internet fraud scheme. Federal Bureau of Investigation, Washington. Retrieved from http://washingtondc.fbi.gov/dojpressrel/pressrel08/wf061208a.htm

[48]*Commonwealth v. Murgallis*, 2000 Pa. Super. 167 (2000).

[49]Rusch, J. (2005, May 3). The rising tide of Internet fraud. Department of Justice, Fraud Section of the Criminal Division. Retrieved from http://www.justice.gov/criminal/cybercrime/usamay2001_1.htm

[50]No. 099-00560 SOM (D. Haw. Filed December 9, 1999).

[51]425 U.S. 185, 195 (1976).

[52]*Luce v. Edelstein*, 802 F.2d 49, 55 (2d Cir. 1986); *McMahan and Co. v. Wherehouse Entertainment, Inc.*, 900 F.2d 576, 581 (2d Cir. 1990).

[53]*In re Jonathan G. Lebed*, Securities Act Release No. 33-7891, Securities Exchange Act Release No. 34-43307, 73 S.E.C. Docket 741 (Sept. 20, 2000).

[54]99 F. Supp. 2d 889, 891 (N.D. Ill. 2000).

[55]47 U.S. 181, 203-204 (1985).

[56]No. CR-00-1002-DT (C.D. Cal. indictment filed Sept. 28, 2000; pleaded guilty Dec. 29, 2000).

[57]This type of fraud was developed in 1920 by Charles Ponzi of Boston, Massachusetts. His scheme quickly dissolved when was unable to pay new entrants for their investments.

[58]United States Attorney's Office, Northern District of Ohio. (2004, April 28). Parma, Ohio man indicted for eBay fraud, credit card fraud, and identity theft. US Department of Justice, Computer Crime & Intellectual Property Section. Retrieved from http://www.justice.gov/criminal/cybercrime/butcherIndict.htm

[59]KPHO. (2007, December 20). CPS case worker admits credit card fraud. Retrieved from http://www.kpho.com/news/14901534/detail.html

[60]Hinnen, T. M. (2004). The cyber-front in the war on terrorism: Curbing terrorist use of the Internet. *Columbia Science and Technology Law Review*, 5, 9.

[61]15 U.S.C. § 1644(a).

[62]701 F.2d 1086, 1092 (4th Cir. 1983).

[63]See, for example, *United States v. Brewer*, 835 F.2d 550 (5th Cir. 1987) (long-distance telephone access codes) and *United States v. Bailey*, 41 F.3d 413 (9th Cir. 1994) (mobile phone access codes).

[64]Hoar, S. B. (2001, March). Identity theft: The crime of the new millennium. US Department of Justice. *USA Bulletin, 49*(2) [online]. Retrieved from http://www.cybercrime.gov/usamarch2001_3.htm

[65]US Department of Justice. (2005, November 1). Houston man pleads guilty to federal identity theft charges, says Justice Department. Retrieved from http://www.justice.gov/criminal/cybercrime/hattenPlea.htm

[66]The website was "Shadowcrew.com." US Department of Justice. (2005, November 17). Six defendants plead guilty in Internet identity theft and credit card fraud conspiracy: Shadowcrew organization was called "one-stop online marketplace for identity theft." Retrieved from http://www.justice.gov/criminal/cybercrime/mantovaniPlea.htm

[67]US Department of Justice. (2007, October 6). Seattle man indicted For ID theft using computer file sharing programs: First case in the country involving peer to peer file sharing programs. Retrieved from http://www.justice.gov/criminal/cybercrime/kopiloffIndict.htm

[68]US Department of Justice. (2007, March 5). Defendant sentenced for conspiring to commit computer fraud and identity theft. Retrieved from http://www.justice.gov/criminal/cybercrime/perrasSent.htm

[69]See the following document from the National Institute of Justice for more information on evidence of this crime: Technical Working Group for Electronic Crime Scene Investigation. (2001, July). Electronic crime scene investigations: A guide for first responders. National Institute of Justice, US Department of Justice, p. 39. Retrieved from http://www.ncjrs.gov/pdffiles1/nij/187736.pdf

[70]Brunker, M. (2005, February 15). Alleged mobsters guilty in vast Net, phone fraud: Mafia scheme said to have netted $659 million over 7 years. *MSNBC News* [online]. Retrieved from http://www.msnbc.msn.com/id/6928696/ns/us_news-crime_and_courts//; Freeman, S. (2005, February 15). "Mob nets $650 million from phone and Internet fraud. *Times Online*. Retrieved from http://www.timesonline.co.uk/tol/news/world/article514724.ece

[71]Marzulli, J., & Smith, G. B. (2005, February 15). $650M porn scam Gambino thugs plead guilty to X-rated ring. *Daily News* [online]. Retrieved from http://www.nydailynews.com/archives/news/2005/02/15/2005-02-15__650m_porn_scam__gambino_thu.html

[72]United States Attorney's Office, Western District of Pennsylvania. (2002, May 17). Defendant pleads guilty in three fraud prosecutions for online auction, telemarketing, and computer-generated counterfeit check scams. US Department of Justice, Computer Crime & Intellectual Property Section. Retrieved from http://www.justice.gov/criminal/cybercrime/sussmanPlea.htm

[73]United States Attorney's Office, Northern District of Georgia. (2010, April 7). Defendants sentenced for telemarketing stock fraud scheme: Defendants ran boiler room operation that targeted foreign investors. Federal Bureau of Investigation, Atlanta. Retrieved from http://atlanta.fbi.gov/dojpressrel/pressrel10/at040710.htm

[74]Bureau of Diplomatic Security. (2009, January 30). Diplomatic security arrests a U.S. postal worker for embezzling passport application fees. Retrieved from http://www.state.gov/m/ds/rls/115601.htm

[75]*State v. Savoy*, 205 La. 650, 667, 17 So.2d 908, 914 (La. 1944).

[76]160 U.S. 268, 269 (1895).

[77]Computer Crime and Intellectual Property Section, US Department of Justice. (2006, September). Intellectual Property: An introduction. *Manual on prosecuting intellectual property crimes* [online]. Retrieved from http://www.justice.gov/criminal/cybercrime/ipmanual/01ipma.html

[78]18 U.S.C. § 102(a).

[79]See 17 U.S.C. § 107–§ 112.

[80]871 F. Supp. 535 (D. Mass. 1994).

[81]United States Attorney's Office, District of Oregon. (1999, November 23). Defendant sentenced for first criminal copyright conviction under the "No Electronic Theft" (NET) Act for unlawful distribution of software on the Internet. US Department of Justice, Computer Crime & Intellectual Property Section. Retrieved from http://www.cybercrime.gov/levy2rls.htm

[82]Another treaty was adopted on that day, the **WIPO Performances and Phonograms Treaty of 1996**. Article 18 of this treaty required that signatories "shall provide adequate legal protection and effective legal remedies against the circumvention of effective technological measures that are used by performers or producers of phonograms in connection with the exercise of their rights under this Treaty and that restrict acts, in respect of their performances or phonograms, which are not authorized by the performers or the producers of phonograms concerned or permitted by law."

[83]203 F. Supp. 2d 1111 (N.D. Cal. 2002).

[84]United States Attorney's Office, Northern District of California. (2005, June 10). New York man pleads guilty to reprogramming & selling thousands of satellite TV "smart cards." US Department of Justice, Computer Crime & Intellectual Property Section. Retrieved from http://www.justice.gov/criminal/cybercrime/buchholz.htm

[85]United States Attorney's Office, Eastern District of California. (2003, February 13). First Digital Millennium Copyright Act (DMCA) criminal sentencing in California involving more than 4,500 bootlegged video tapes. US Department of Justice, Computer Crime & Intellectual Property Section. Retrieved from http://www.justice.gov/criminal/cybercrime/mynafSent.htm

[86]482 F.2d 1179, 1180 (6th Cir. 1973).

[87]US Department of Justice. (2004, January 22). Chicago man arrested for criminal copyright infringement. Federal Bureau of Investigation, Los Angeles. US Department of Justice, Computer Crime & Intellectual Property Section. Retrieved from http://www.justice.gov/criminal/cybercrime/spragueArrest.htm

[88]Office of the United States Attorney, Middle District of Tennessee. (2006, August 24). Two men plead guilty to music piracy charges. US Department of Justice, Computer Crime & Intellectual Property Section. Retrieved from http://www.cybercrime.gov/thomasPlea.htm

[89]United States Attorney's Office, District of New Jersey. (2005, July 12). Maple Shade man gets 15 months in prison for selling fake Rolex watches over the Internet. US Department of Justice, Computer Crime & Intellectual Property Section. Retrieved from http://www.cybercrime.gov/berg0712_r.htm

[90]United States Attorney's Office, District of South Carolina. (2001, March 7). Operators of www.fakegifts.com Website plead guilty to selling counterfeit luxury goods over the Internet. US Department of Justice, Computer Crime & Intellectual Property Section. Retrieved from http://www.cybercrime.gov/Dipadova_plea.htm

[91]See, for example, *Hendrickson v. eBay, Inc.*, 165 F. Supp. 2nd 1082 (C.D. Cal 2001); and *Tiffany Inc. v. eBay, Inc.*, No. 04 Civ. 4607 (S.D.N.Y. July 14, 2004), 2004 WL 1413904.

[92]See, for example, US Department of Justice. (2005, September 8). For-profit software piracy Website operator sentenced to 87 months in prison. US Department of Justice, Computer Crime & Intellectual Property Section. Retrieved from http://www.cybercrime.gov/petersonSent.htm; US Department of Justice. (2006, August 25). Operator of massive for-profit software piracy website sentenced to six years in prison. US Department of Justice, Computer Crime & Intellectual Property Section. Retrieved from http://www.cybercrime.gov/ferrerSent.htm

[93]United States Attorney's Office, Southern District of Florida. (2006, August 22). Miramar man pleads guilty to trafficking in counterfeit goods. US Department of Justice, Computer Crime & Intellectual Property Section. Retrieved from http://www.cybercrime.gov/wakefieldPlea.htm

[94]18 U.S.C. § 1839(3).

[95]Kelley, P. (1997, July). The Economic Espionage Act of 1996. *FBI Law Enforcement Bulletin* [online]. Retrieved from http://www2.fbi.gov/publications/leb/1997/july976.htm

[96]United States Attorney's Office, Northern District of California. (2002, December 4). Pair from Cupertino and San Jose, California, indicted for economic espionage and theft of trade secrets from Silicon Valley companies. US Department of Justice, Computer Crime & Intellectual Property Section. Retrieved from http://www.justice.gov/criminal/cybercrime/yeIndict.htm

[97]18 U.S.C. § 1832.

[98]United States Attorney's Office, Northern District of California. (2005, September 29). Software executive admits to conspiring to misappropriate chief competitor's trade secrets. US Department of Justice, Computer Crime & Intellectual Property Section. Retrieved from http://www.cybercrime.gov/mcmenaminPlea.htm

[99]United States Attorney's Office, District of Delaware. (2002, October 17). Guilty plea to economic espionage. US Department of Justice, Computer Crime & Intellectual Property Section. Retrieved from http://www.justice.gov/criminal/cybercrime/morrisPlea.htm

[100]United States Attorney's Office, Southern District of New York. (2005, August 29). Connecticut man pleads guilty in U.S. court to selling stolen Microsoft Windows source code. US Department

of Justice, Computer Crime & Intellectual Property Section. Retrieved from http://www.cybercrime .gov/genovesePlea.htm

[101]United States Attorney's Office, Eastern District of California. (2008, February 13). Bakersfield law student sentenced in email harassment scheme. US Department of Justice, Computer Crime & Intellectual Property Section. Retrieved from http://www.justice.gov/criminal/cybercrime/ veveaSent.pdf

[102]United States Attorney's Office, Northern District of California. (2001, December 18). San Francisco man indicted for selling fake Derek Jeter and Nomar Garciaparra baseball bats on eBay, harassing e-mails. US Department of Justice, Computer Crime & Intellectual Property Section. Retrieved from http://www.cybercrime.gov/derungsIndict.htm

[103]*Ibid.*

[104]US Department of Justice. (2006, November 16). Telemarketing firm official pleads guilty in New Hampshire phone jamming case. Retrieved from http://www.justice.gov/opa/pr/2006/ November/06_crm_769.html

[105]Smith, A. M. (2009, October 19). Protection of children online: Federal and state laws addressing cyberstalking, cyberharassment, and cyberbullying. *US Congressional Research Report* RL34651, p. 7. Retrieved from http://www.ipmall.piercelaw.edu/hosted_resources/crs/RL34651_091019.pdf

[106]*United States v. Alkhabaz,* 104 F.3d 1492 (6th Cir.1997); *United States v. Kammersell,* 196 F. 3d 1137 (10th Cir. 1999).

[107]*United States v. Baker,* 890 F. Supp. 1375 (D.E.D. Mich.1995); *United States v. Kammersell,* 196 F. 3d 1137 (10th Cir. 1999).

[108]*United States v. Orozco-Santillan,* 903 F.2d 1262, 1265 (9th Cir. 1990).

[109]*Planned Parenthood of the Columbia/Willamette, Inc. v. American Coalition of Life Activists,* 290 F.3d 1058 (9th Cir. 2002).

[110]*Ibid.*

[111]United States Attorney's Office, Western District of Missouri. (2009, January 7). KC man pleads guilty to cyberstalking. US Department of Justice, Computer Crime & Intellectual Property Section. Retrieved from http://www.justice.gov/criminal/cybercrime/memarianPlea.pdf

[112]18 U.S.C. § 2261A(2)(B).

[113]Electronic Privacy Information Center. (2006, June 15). The Amy Boyer case: *Resmburg v. Docusearch.* Retrieved from http://epic.org/privacy/boyer/

[114]Pretexting occurs when individuals seek to collect information about an individual using false pretenses.

[115]The primary purpose of this Act is to protect the financial information of individuals that is held by financial institutions.

[116]United States Attorney's Office, Northern District of Ohio. (2008, December 30). News release. US Department of Justice, Computer Crime & Intellectual Property Section. Retrieved from http://www.justice.gov/criminal/cybercrime/andersonIndcit.pdf

[117]Specific cybercrimes are also explored in Chapters 1 and 6 of this book.

Chapter 6

Understanding the Computer-Networking Environment: Beware of the Scam Artists, Bullies, and Lurking Predators!

To be a good computer forensics investigator, you must have sufficient knowledge of existing cybercrime laws and criminal procedure. You must also be familiar with existing technology, computer crimes, and the perpetrators who commit these crimes. Technology is constantly evolving, however, and so are the criminals who commit these crimes. Consequently, as an investigator it is imperative that you stay current in your field, especially with respect to the computer-networking environment in which criminals operate and the types of crimes that criminals are perpetrating within this environment. This chapter explicitly focuses on current online scams, identity theft, cyberbullying, and sexual predators.

Scams and Scam Artists

This section covers a number of different scams and the approaches that a scam artist might use to steal an individual's personal information with the intent to commit fraud. Most scams arrive in the form of unsolicited e-mails, text messages, and phone calls from unknown individuals. The techniques used by scam artists to steal the personal information include, but are not limited to, the following strategies:

- *Dumpster diving.* Criminals may go through the prospective victim's garbage looking for the victim's personal and financial information (e.g., Social Security number, credit card information). Many individuals routinely discard sensitive documents and financial information without shredding them first. One can also dumpster dive at financial institutions. For example, Jonah

Hanneke Nelson stole more than 500 identities by dumpster diving behind banks and other businesses and retrieving sensitive material and blank checks. Sometimes information that is retrieved from the dumpster of a company can be used to obtain sensitive information from employees of that company. For example, the Phonemasters, a criminal group, tricked employees of companies (e.g., AT&T, MCI, Sprint, Equifax) into revealing their usernames and passwords by using the materials they had extracted from the dumpster, which included old phone and technical manuals for the computer systems of these companies.[1]

- *Shoulder surfing.* Criminals may watch prospective victims at automatic teller machines (ATMs) to "steal" the personal identification number (PIN) that each victim enters into the machine. Usually, after watching someone enter the PIN, the offender either distracts the victim and steals the victim's debit or credit card or pickpockets the victim after he or she leaves the ATM.

- *Skimming device.* A skimming device reads the magnetic strip on credit and debit cards. These devices have been surreptitiously placed on ATMs to collect this information unbeknownst to users. Some criminals have handheld devices they use to swipe unsuspecting users' cards so as to steal their data. In one case, employees of the Cheesecake Factory used card-skimming devices to steal the credit card numbers of restaurant customers.[2]

- *Stealing a prospective victim's wallet or breaking into the home or car of the prospective victim to steal documents that contain the victim's personal information.* A recent example of this kind of scam occurred on June 15, 2010, when a famous English playwright, Alan Bennett, had his wallet stolen by thieves. The thieves—two women and one man—threw ice cream at his coat and pretended to clean it up. By distracting him in such a manner, they managed to steal Bennett's wallet, which he claimed contained 1500 pounds.[3]

Various scams that have been used to steal money, financial data, and/or personal information are described next.

Online Auction Scams

In one type of online scam with auction fraud (covered in Chapter 5), the scam artist provides overpayment for an item that is being auctioned. The buyer sends too much money for the item (probably a few hundred dollars over the price of the item) via international money order. The seller, in an effort to maintain a good business relationship with buyer, sends the excess money back to the buyer after the bank has accepted the money order. Several weeks later, the seller is informed by the bank that the money order was a forgery. Consequently, the seller loses both the few hundred dollars refunded to the buyer and the goods that were sold.

Online Rental and Real Estate Scams

According to the website of the Federal Bureau of Investigation (FBI), recent online scams have included rental properties and real estate (**rental and real estate scam**).[4] The technique used in such cases is similar to the scam mentioned previously involving auction sites. The scammer, who disguises himself or herself as an interested buyer or renter, agrees with the seller or landlord on a specific price. The offender then sends the seller or landlord a check for the amount agreed upon. At this point, the scammer backs out of the deal and asks for a full refund. After sending the scammer the amount refunded, the seller or landlord finds out that the check was counterfeit. Not only has the individual lost the money he or she forwarded to the scammers, but now the person has to pay the bank that same amount because the check was counterfeit.

Online Dating Scams

Online **dating scams** are conducted in more or less the same manner by most offenders. The offender poses as an attractive person (and shows the victim pictures to illustrate his or her supposed appearance) on an online dating site and develops an online relationship with the victim. After courting the victim for a few months, the offender informs the victim that he or she must travel abroad for business. While abroad, the offender notifies the victim that an unexpected tragedy has occurred; either the person is the victim of a street crime (e.g., robbery), is detained by authorities, or cannot pay a hotel or hospital bill. The offender then asks the victim for financial assistance. Sometimes the offender will come up with more scenarios that require further financial assistance from the victim. Either way, after one or more payments are made by the victim, the offender vanishes and the victim never hears from him or her again.[5]

Online Lottery Scams

Other ways to fraudulently solicit money from a victim include claiming that the victim won a prize—typically a foreign lottery. The **lottery scam** artist informs the victim that he or she must supply the lottery agency with a bank account number and pay certain fees and taxes in advance to obtain the winnings. The victim is further informed that he or she must take these steps immediately because the deadline to claim the prize will expire soon. After the victim provides the bank account number and money to the "foreign lottery agency," the person never hears from the "agency" again. The bank account for which the victim provided access information will probably be emptied as well.

Participation in a foreign lottery is a violation of U.S. law. Accordingly, the victim is unlikely to report the fraud. Even if a victim does report the crime to the authorities, the money the victim sent to the scam artist cannot be retrieved because he or she has engaged in unlawful activity.[6] An example of a lottery scam, retrieved from the author's own inbox is included in the nearby box. Notice the errors in the original text—such spelling and grammatical errors are characteristic of scams.

Lottery Scam

From: 2010 WORLD CUP AWARD <claimingza@sify.com>
Subject: Call Urgently on +27 839470181
To: xxxxx@xxxxxx.com
Date: Wednesday, June 9, 2010, 7:53 AM

WINNING NOTIFICATION
(2010 WORLD CUP LOTTERY AWARD)

We happily announce to you the draw of South African 2010 World Cup Bid Lottery Award International programs held in U.K your "email address" was attached to ticket number; B9665 75604546 199 serial number 97560 This batch draws the lucky numbers as follows 60/84/27/17/36, bonus number 2, which consequently won the lottery in the second category. Congratulations your email is among the three lucky winning that won **$2,000,000.00{Two million United State Dollars}**in the just concluded south Africa world cup bid 2010 promotion sponsored by Coca-Cola British American tobacco companies south Africa.

COMPLETE THIS INFORMATIONS BELOW:
NAME:................................
ADDRESS:...........................
NATIONALITY:.......................
SEX:................................
AGE:................................
PRIVATE PHONE/MOBILE NO

PRIVATE FAX NO:....................
OCCUPATION:........................
BATCH/WINNING 60/84/27/17/36

The lottery program took place to promote South African 2010 world cup award.

His contact details are as follows . . .
CONTCAT YOUR CLAIMING AGENT:
Contact: MR.STEPHEN VALE
TEL: + 27-839470181
EMAIL: claimingza@sify.com

Yours Faithfully,
Management.
JOHN CLARK

Charity Scams

Many e-mails are purportedly sent on behalf of charities in the aftermath of a disaster. This rush to capitalize on people's desire to help others in distress happened after the terrorist attacks in the United States on September 11, 2001; the tsunami in Indonesia in December 2004; Hurricane Katrina in September 2005; and the earthquake in Haiti in 2010; to name but a few. For instance, following the 2004 tsunami and Hurricane Katrina, numerous fake websites were set up asking for donations from unsuspecting individuals via PayPal, credit card, or bank account, thereby misdirecting funds from legitimate disaster relief efforts. Additionally, in the immediate aftermath of the earthquake in Haiti, numerous fake e-mails and messages on social networking sites (such as Facebook, MySpace, and Twitter) were sent soliciting funds.

Some scams even claimed to be from legitimate organizations, such as the United Nations International Children's Emergency Fund (UNICEF). Here is how the scam worked: An individual would receive an e-mail that appeared to be from UNICEF. When the individual would click on the link provided in the e-mail, he or she would be diverted to an official-looking UNICEF Web page. This page would ask for personal information (e.g., name, home address, and Social Security number), credit card information, and bank account information. The information provided by the victim would subsequently be used by the offender to commit other crimes (e.g., identify theft).

Another form of scam was developed to divert funds that were obtained from text message short codes. For example, a legitimate short code set up for donations in the aftermath of the Haiti earthquake was "90999". If an individual typed the word "Haiti" and sent the message to this code, $10 was donated to the Red Cross in the United States. Scammers created a text message number to which donors could send money that mimicked the legitimate one. Specifically, variations of the legitimate "90999" code were provided by scam artists (e.g., "99099") to retrieve some of the money that was being donated to Red Cross for Haiti. Indeed, many such false codes were provided following the earthquake in Haiti. Both the FBI and the Federal Trade Commission (FTC) issued warnings to the public soon after the earthquake concerning which charities were legitimate. The FBI and FTC also warned the public about e-mail and text message charity scams.

Government E-mail Scams

There are several ways in which scammers try to steal personal information from an individual's computer. One of the most common ways is the distribution of malicious software through what appear to be legitimate **government e-mails**—at least on the surface. For instance, e-mails have been distributed claiming to be from the FBI, some of which come with an attachment. The e-mail in the nearby box was found among many similar spam emails in the author's inbox; it included an attachment titled "FBI Report.txt". One should be extremely wary when opening attachments to suspicious e-mails as they (more often than not) carry a malicious payload, which is intended to download spyware or key-logging software to retrieve information from a user's computer.

Fake FBI E-mail

From: FBI OFFICE <drdwilson@btconnect.com>
Subject: REPLY NOW
To: xxxxxxxx@xxxxxxx.com
Date: Wednesday, June 23, 2010, 2:37 AM

See attachment below for a current report on our investigation. You are advised to Contact me asap for further clarification. This has to be cleared! You are warned!

In reality, the FBI always uses its official e-mail addresses for communications. Besides, an FBI official would not send such a communication if a matter was truly urgent: It is much more likely that an FBI agent would visit the individual in person. In these types of scams, the e-mail addresses of the supposed FBI officials are typically from free website service accounts, such as Hotmail, Yahoo, or Gmail. In the case of the "Example of Nigerian Scam" e-mail, the message was sent from a communications carrier in the United Kingdom known as BT (British Telecommunications).

One example of this kind of scam can be seen in the e-mails that were distributed on behalf of the FBI purportedly claiming to include an Intelligence Bulletin concerning "New Patterns in Al-Qaeda Financing." This e-mail was accompanied by an attachment of the supposed bulletin, which was actually an executable file ("bulletin.exe") that contained malicious software. Upon being opened, a malicious payload was downloaded that was designed to steal information from the user's computer.[7] Another e-mail masquerading as an official urgent message from the Department of Homeland Security (DHS) and the FBI was distributed that claimed to contain a recording of a speech Osama bin Laden gave directed at Europe.[8] This email also had an executable file ("audio.exe") attached to it; as with the "bulletin" attachment, the attachment contained malicious software designed to steal the user's personal information.

As a general rule, e-mails with attachments from unknown senders should not be opened.

Nigerian Scams

The **Nigerian scam** is also known as the "419 scam," named after the section of the Nigerian criminal code that this scam violates.[9] Many variations of the Nigerian scam exist. Some claim that the victim has received an inheritance from a long-lost relative from Nigeria. Others, which constitute the majority of this type of scam, involve e-mails sent by individuals fraudulently claiming to be government, business, or banking officials.

In the typical Nigerian scam, a victim is contacted by someone impersonating one of the previously mentioned officials. This official informs the victim that he or she has been

Example of Nigerian Scam

From: Mrs. Debbie Anderson <nelson_welter@secreatrias.com>
Subject: Stop Contacting Scam Alert. . . . !
To: xxxxx@xxxxxxx.com
Date: Tuesday, June 8, 2010, 5:55 AM
Attn: My Dear Good Friend

I am Mrs. Debbie Anderson, I am a US citizen, 48 years Old. I reside here in New Braunfels Texas 78132. My residential address is as follows. 108 Crockett Court. Apt 303, New Braunfels Texas, United States, am thinking of relocating since I am now rich. I am one of those that took part in the Compensation in Nigeria many years ago and they refused to pay me, I had paid over $20,000 while in the US, trying to get my payment all to no avail.

So I decided to travel down to Nigeria with all my compensation documents, And I was directed to meet Mr Michael Bolts, who is the member of COMPENSATION AWARD COMMITTEE, and I contacted him and he explained everything to me. He said whoever is contacting us through emails are fake.

He took me to the paying bank for the claim of my Compensation payment. Right now I am the most happiest woman on earth because I have received my compensation funds of $1,600,000.00 Moreover, Mr Nelson Oboh, showed me the full information of those that are yet to receive their payments and I saw your name as one of the beneficiaries, and your email address, that is why I decided to email you to stop dealing with those people, they are not with your fund, they are only making money out of you. I will advise you to contact Mr Nelson Oboh.

You have to contact him directly on this information below.

COMPENSATION AWARD HOUSE
Name : Nelson Oboh
Email: nelson_oboh@secretarias.com
Phone: +234-808-286-2330

You really have to stop dealing with those people that are contacting you and telling you that your fund is with them, it is not in anyway with them, they are only taking advantage of you and they will dry you up until you have nothing.

The only money I paid after I met Mr Nelson Oboh was just $95 for the paper works, take note of that.

Once again stop contacting those people, I will advise you to contact Mr Nelson Oboh so that he can help you to Deliver your fund instead of dealing with those liars that will be turning you around asking for different kind of money to complete your transaction.

Thank You and Be Blessed.
Mrs. Debbie Anderson.

selected to partake in the sharing of a percentage of millions of dollars. To obtain the money, the victim is required to allow the "official" to deposit these millions into the individual's bank account. To do so, the victim is required to provide the "official" with all of the relevant personal and banking information to complete the transaction. The victim is also required to pay certain legal fees, taxes, and bribes to government officials. Of course, the victim is reassured by the "official" that the money provided for the fees, taxes and bribes will be paid back in full. The victim, of course, never sees any money from this scheme, nor will he or she ever get back the money given to the scammers for the fictitious fees, taxes, and bribes. Additionally, if the victim's bank account had any money in it, the scammer will have withdrawn all of it.

Other victims, thinking that they are dealing with legitimate government, business, or bank officials, have been lured to travel to Nigeria to supposedly receive their funds.[10] Before they set out for Nigeria, victims are purposely told that they do not need a visa to enter the country. Entering the country without a visa, however, is illegal in Nigeria. As such, the scammers use the fact that the victim entered the country illegally as leverage to compel the victim into paying the scammers the money that they are demanding. This fraud does not only result in financial loss to the victim; some individuals who have traveled to Nigeria in pursuit of the money purportedly offered to them have died or gone missing. As certain websites have reported, an American was murdered while pursuing such a scam in Lagos, Nigeria, in 1995.[11]

E-mails claiming to need the target's assistance to receive money from Nigeria should be sent to the following e-mail address of the FTC: spam@uce.gov.

Work-at-Home Scams

Many fraudulent advertisements exist concerning work-at-home opportunities. The individuals who are usually victims of **work-at-home scams** are stay-at-home mothers, disabled individuals, and persons who are unemployed and are desperately seeking some form of income. Most offers contain exaggerated claims of potential earnings such as "Make $7000 a week working part-time; no experience required" or "Make $1000 in as little as 4 hours; no experience needed." To convince victims of the scheme's legitimacy, scammers will create false websites containing news articles claiming that such opportunities are real and that many have benefited from them. Either this scam will advertise opportunities for individuals to create their own home business or it will advertise positions in which employers are seeking individuals who can process e-mails, transfer funds, assemble products, reship products, or process payments (to name a few bogus jobs).

When the scam focuses on establishing a home business, the victim is required to pay an advance fee—usually a few hundred dollars—for the materials needed to set up this business. The materials the victim actually receives are useless, meaning the victim has been scammed out of the money paid for the advance fee.

When the scam focuses on working at home for an invisible employer, either the victim never receives money for the job performed or receives very little money in return. Either way, the person never receives the large sum of money originally promised as compensation. At times, those offering work-at-home jobs solicit individuals to engage in illegal activities by having the victims receive and cash fraudulent checks, transfer the offender's illegally obtained funds, or receive and ship stolen merchandise to the offender.[12]

Virus Protection Scams

Users browsing the Internet may receive pop-up security warnings falsely informing them that their computer has been infected with numerous computer viruses.[13] This type of scam is known as **scareware**, and the goal is to persuade the user to buy fake software and/or to download malicious software on the user's computer—a type of fraud known as a **virus protection scam**.

Many different types of scareware or rogue antivirus software are available online. One very well-known scareware program is XP Protection Center, which may be downloaded on a user's computer through fake, malicious websites and Trojan horses. XP Protection Center was designed to hijack certain websites and redirect users to a website where the program could be purchased.

This type of scareware is also designed to bombard the user's computer with repeated security messages warning of an infection of the system by multiple viruses and malware. If a user is redirected to a website that provides security warnings claiming that the user's computer has multiple infections, the user should not click on the "X" on the upper-right hand corner of the Web page, nor should he or she click on a box on the page that says "No Thanks" to exit the program. Taking either of these actions may cause execution of malicious code. Instead, the user should close his or her browser by clicking on the "File" tab, and then choosing "Exit." The alternative is to right-click on the program icon on the task bar and then click "Close."

Spammers

Terrorists have been known to secure their communications by sending fake streams of e-mail **spam** to disguise a single targeted message.[14] Spam can be sent via both e-mail and cell phones. Spam can be annoying and time-consuming because it fills up individuals' e-mail accounts with junk e-mail that these individuals have to sort through to distinguish it from their legitimate mail. Spammers tend to distribute unsolicited e-mails with the intent to defraud users by, for example, offering deals on products or services. Most often, the advertised products or services the buyer sends money for are never received. At other times, the products received are of very low quality and value—worth much less than what the victim paid for them.

In recent years, spam has become more dangerous because it includes—more often than not—links in the e-mails that may contain malicious software that can install spyware onto an unsuspecting victim's computer system. Sometimes just opening the e-mail can deliver a malicious payload onto a person's computer. For this reason, if an e-mail is suspected to be of the spam variety, the user should just delete it. Additionally, if an individual responds to spam, his or her personal information and e-mail address are usually sold to other companies. In turn, the victim is subjected to even more spam.

For the recipient of spam, it can be extremely difficult to distinguish between legitimate advertisements and those masquerading as advertisements. The Controlling the Assault of Non-Solicited Pornography and Marketing Act of 2003 (CAN-SPAM Act) (codified in 15 U.S.C. 7701, et seq.) was developed to remedy this situation and deal with the growing problem of spam. If a glimpse of the author's inbox is any indication as to its effectiveness, however, this measure has definitely fallen short of its stated purpose.

The CAN-SPAM Act was designed to, among other things, afford the user with the option to opt out of services. Not all spammers, however, have abided by these rules. Most spam includes a message at the bottom of the e-mail that states that an individual opted in to receive certain updates and special offers through a partner website (which is not named). It further states that if an individual believes that he or she has received the e-mail in error or does not wish to receive additional updates in the future, the user should click on a link that is provided and unsubscribe. Sometimes those links contain malicious software. At other times, an individual who attempts to unsubscribe to such e-mails may end up receiving more spam because he or she was tricked into entering the e-mail address in a website that sends the user's information to other spammers.

Phishers

Phishers are known for posing as legitimate companies and government agencies and using misleading or disguised hyperlinks and fake e-mail return addresses to trick Internet users into revealing their personal information. Normally, such e-mails claim that a customer's account information needs to be verified to protect the user against identity theft. To ensure that many individuals reply to the phishing email, the message also warns users that if they do not reply in a timely manner, their accounts will be terminated. Some scam artists have even pretended to be part of the Financial Crimes Enforcement Network (FinCEN); in so doing, they have tricked their victims into providing them with personal information, to be later used by these scam artists to commit other crimes.[15]

In another case, phishers targeted senior victims by creating a phishing scam that involved Social Security benefits.[16] They sent e-mails to senior citizens claiming that if they did not update their information by responding to the e-mail, their accounts would be suspended indefinitely and they would lose their Social Security benefits. Many people responded to the e-mail by sending their personal and financial information to the website the phishers provided, which looked like the authentic Social Security Administration website.

Yet another example of phishing involved fraudulent e-mails sent on behalf of the Internal Revenue Service (IRS) concerning the economic stimulus tax rebate implemented during the administration of President George W. Bush.[17] These messages instructed recipients of the e-mail to provide direct deposit information so that they would receive the rebate faster. Any personal information sent to the phishers was subsequently stolen and used for other fraudulent purposes.

A more recent phishing scheme also involved the IRS. Specifically, Mikalai Mardakhayeu and his co-conspirators participated in a scheme to defraud taxpayers across the United States out of their income tax refunds by luring them to websites that purportedly offered lower-income taxpayers free tax return preparation and electronic filing services.[18] After the unsuspecting victims uploaded their tax return information, Mardakhayeu and his cohorts altered the information they provided and had the tax refunds sent to bank accounts controlled by them.

As this brief survey suggests, phishing schemes abound. In all cases, however, phishing is a form of identity theft. It seeks to dupe individuals into giving away their personal information and passwords. If an individual's password has been compromised, the offender may use the victim's e-mail account to send requests for money to individuals listed in the victim's address book. Specifically, the offender—posing as the victim—may claim that the individual needs money, immediately. The e-mail might state that the victim was in a terrible accident or stuck in a foreign country without any money and needs financial assistance.

Phishing attacks have also occurred by means of social networking sites. One such example involves Twitter. On January 3, 2009, a message with a link in it was sent out via Twitter, disguised as being from a friend or someone who was allowed to follow the victim's tweets. If the victim clicked on the link, he or she was redirected to another website that looked identical to the official Twitter page. When prompted to enter his or her username and password, the victim typed it in. This information was subsequently stolen by phishers, giving them access to the victim's Twitter account.

Vishing

Vishing (voice phishing) is another well-known type of scam. Vishing can occur via voice or text messages sent to cell phones. In one variant of this scheme, an offender poses as a representative of the victim's bank and notifies the victim that his or her account has been compromised. The offender then provides the victim with a number to call that specifically deals with these issues. When the victim calls the number, an official-sounding recording is heard that instructs the victim to provide the account number and password. Once the account number and password are provided, the call is terminated—and the victim's bank account is quickly emptied.

Pharming ✳ test

Pharming is a type of scam in which the offender creates a website that looks identical to an authentic website. However, the mirror website carries a malicious payload. According to McLean and Young:

Pharming uses Trojan horse programs to redirect people to counterfeit banking or e-commerce sites (sometimes called "page hijacking"). The compromised computer or server redirects consumers to fraudulent websites . . . The fraudulent sites are formulated to look like authentic, legitimate sites (and may even include a bogus "secure site" logo indicating that the site is genuine). The site may install spyware or prompt the consumer to enter personal information, including user name and password.[19]

Pharming has occurred primarily with pornographic websites. In addition, pharmers have targeted municipal and government websites.

Identity Theft

Identity theft usually occurs when someone steals an individual's identity by obtaining his or her personal information or by accessing an individual's account with the intention of using it to commit illegal activities. For instance, an employee of the New York State Tax Department stole thousands of taxpayers' identities and made more than $200,000 in fraudulent charges with credit card accounts and lines of credit opened with those identities.[20] In another case, a computer technician stole the identities of more than 150 employees of the Bank of New York Mellon Corporation.[21] He subsequently used these identities to conduct criminal activities, stealing more than $1.1 million from charities, nonprofit organizations, and other related groups. A third case involved a former U.S. military contractor who pleaded guilty to exceeding his authorized access to a computer of the Marine Corps Reserve Center and obtaining and selling the names and Social Security numbers of 17,000 military employees.[22]

Terrorists have also engaged in identity theft for a variety of reasons. In fact, investigations of the terrorist attacks on September 11, 2001, revealed that the terrorists responsible for the attack "repeatedly committed acts of identity theft to advance their destructive goals."[23] Indeed, intelligence has shown that terrorists have obtained genuine

---

Interesting Fact About Pharmed Websites

Suppose you are searching for the link to Best Buy so that you can buy electronics. You type the phrase "Best Buy" into a search engine, such as Google or Bing. Numerous results are found. To determine if a link has been pharmed, you can move your mouse so that it hovers over the title of the first search result (actually, any search result could be used). If the Web address that shows up is www.bestbuy.com, then this is the legitimate site. In contrast, if the Web address revealed is www.bestby.com, then the website is pharmed. Of course, you can avoid pharmed websites altogether if you know the website you are seeking to access and type its locator correctly into the address browser bar.

passports by using falsified personal identity information and fraudulent supporting documents.[24] Because such documents are issued by the appropriate agency, they are indistinguishable from legitimate passports.[25] For example, an al-Qaeda terrorist cell in Spain used "false passports and travel documents . . . to open bank accounts where money for the mujahidin movement was sent to and from countries such as Pakistan, Afghanistan, etc."[26] A cloak of anonymity is provided to those who steal identities. Thus stolen identities afford terrorists with the opportunity to enter and exit the United States (and other countries) undetected.

Terrorists have also engaged in identity theft by bribing employees of the Department of Motor Vehicles (DMV) into providing them with driver's licenses. For example, the terrorists responsible for the September 11 attacks bribed DMV employees so that they could receive driver's licenses. Indeed, examples abound of corrupt DMV employees illegally providing licenses for a fee. One such case involved an identity theft ring that was caught by counterterrorism investigators.[27] A member of this ring, a Pakistani woman named Shamsha Laiwalla, had provided individuals who came to the United States illegally from Pakistan with driver's licenses, Social Security numbers, and birth certificates. To obtain the driver's licenses, Laiwalla bribed DMV workers. Laiwalla is also suspected of providing some of the proceeds of her illicit gains to finance a Lebanon-based terrorist group known as Hezbollah.

Types of Identity Theft

According to the Berkshire, Massachusetts District Attorney's Office, one's personal information is stolen to commit four major types of crimes:[28]

1. *Financial identity theft.* This occurs when an offender uses a victim's identity to obtain money, goods, or services. For example, an offender may take the following actions:
   - Open up a bank account in the victim's name.
   - Obtain debit and credit cards in the victim's name.
   - Take out mortgage loans in the victim's name.
   - Buy an automobile by taking out a loan in the victim's name.

2. *Criminal identity theft.* With this type of identity theft, an offender poses as the victim to commit a crime or claims to be the victim when apprehended for a crime.

3. *Identity cloning.* This occurs when an offender assumes the identity of the victim in his or her daily life. To do so, the offender usually retrieves duplicates of the victim's driver's license, birth certificate, passport, and other personally identifying records. The offender subsequently takes over all of the victim's existing accounts (e.g., bank, phone).

4. *Business/commercial identity theft*. Using this form of identity theft, offenders use another business' or organization's name to obtain credit, funds, goods, or services.

How Can Someone Steal an Identity?

According to the Congressional testimony of Grant Ashley, even one stolen document can assist someone in taking over a person's identity and using it for fraudulent purposes.[29] Consider the following example: If an offender has the name and date of birth of an individual, he or she can go to bureaus that have open records policies and obtain the birth certificate of the individual. This information can be then used to contact the Social Security Administration and obtain the victim's Social Security number. With a Social Security number, credit reports can be obtained. These credit reports allow the offender to determine two things:

- If the victim has good credit with which the offender can open up accounts and apply for credit cards
- Which bank accounts and credit cards the victim has

Using the information the offender has retrieved from the credit report, he or she can place pretext calls to the victim's bank and obtain the victim's bank account number. This information can subsequently be used for a variety of fraudulent purposes.

Criminals can steal a person's identity in many different ways:

- By stealing the victim's personal information (e.g., Social Security number, home address, bank account numbers where your salary is deposited) from an employer
- By bribing an employee from the human resources department who has access to the aforementioned information
- By stealing your mail (especially pre-approved credit card forms) and filling out a change of address form to divert the victim's mail to the criminal
- By hacking into computer systems and stealing personal information

The last tactic is particularly important to note because it is easily forgotten that individuals may become victims of identity as a result of the theft of their personal information from company or government databases. Most governments and private companies provide users with one solution to identity theft: Protect yourself (by, for example, shredding documents containing personal information before putting them in the garbage). This advice distracts the public from one very important fact: Most of their personal data is stored in remote databases beyond the individuals' control and reach.

Thousands of incidents in which these kinds of databases were breached by criminals have been recorded by the Privacy Rights Clearinghouse. Specifically, this organization's website details cases of data breaches that have resulted from hacking incidents, theft of data from databases, the theft and loss of computers and related devices that store personal data, accidental disclosure of personal data, and careless discarding of documents containing personal data into trash and dumpsters without shredding them first. Each of these possibilities is explored in further detail here.

Numerous examples exist of theft of data from the databases of government agencies and businesses:[30]

- On April 28, 2009, a hacker illegally accessed a database of the West Virginia State Bar that contained lawyers' identification numbers, home addresses, e-mail addresses, and some of the lawyers' Social Security numbers.

- On March 12, 2009, individuals gained unauthorized access to a U.S. Army database that contained the personal data of approximately 1600 soldiers.

- On January 23, 2009, someone gained unauthorized access to a database of Monster.com (a website where people post résumés and search for employment) that contained the usernames, passwords, names, dates of birth, e-mail addresses, gender, and ethnicity of the users.

- On May 12, 2007, a hacker gained unauthorized access to a computer containing approximately 7300 Social Security numbers, dates of birth, names, home addresses, and phone numbers of the students of Goshen College in Indiana.

- On May 8, 2007, a hacker accessed databases at the University of Missouri that contained the names and Social Security numbers of employees who were former or current students of the campus in Columbia, Missouri.

Laptops and related electronic devices of businesses, military, state, and government agencies have also been stolen:[31]

- On June 10, 2008, the billing records of 2.2 million patients at the University of Utah Hospitals and Clinics were stolen from a vehicle.

- On January 29, 2008, a laptop that an employee of Horizon Blue Cross Blue Shield was taking home with him was stolen. This laptop contained the personal information of more than 300,000 members of the health plan.

- On March 30, 2007, three laptops containing sailors' personal data were stolen from the Navy College Office at the San Diego Naval Station.

- On October 20, 2006, a laptop containing the personal information (names, Social Security numbers, and medical data) of 1600 veterans who received care at the Manhattan Veterans Affairs Medical Center was stolen.

- That same day, a laptop containing 130,500 Social Security numbers was stolen from the Los Angeles County Child Support Services.

Many examples also exist of employees who have lost laptops, flash drives, and other electronic storage devices containing sensitive personal information:[32]

- On May 14, 2010, a laptop was stolen from the Department of Veterans Affairs in Washington, D.C., that contained Social Security numbers and other personal data of more than 600 veterans.

- On March 24, 2009, an employee of Massachusetts General Hospital took records home to do work over the weekend and ended up losing confidential medical records by leaving them on a train.

- On June 5, 2006, an IRS employee lost a laptop that contained fingerprints, Social Security numbers, and other personal data of current employees and job applicants during transit on an airline flight.

- On June 1, 2006, an employee of Miami University lost a computer that contained the personal information of students enrolled in the university between June 2001 and May 2006.

- On January 12, 2006, People's Bank lost a computer tape containing customers' personal information and checking account numbers.

Incidents of exposure of personal data as a result of an employee error or accidental disclosure have also been noted:[33]

- On September 23, 2009, someone observed that the names and Social Security numbers of more than 5000 employees and students of Eastern Kentucky University were posted online. The data remained online for about a year before someone reported that fact and had them taken down.

- On January 30, 2009, the Social Security numbers of current and former employees of the Indiana Department of Administration were accidentally posted on a website for approximately two hours before someone noticed them and had them removed.

- On January 21, 2009, at Missouri State University, e-mails sent soliciting help with language tutoring included, by mistake, an attachment of a spreadsheet with the names and Social Security numbers of international students.

- On August 29, 2008, the personal data (including Social Security numbers) of 16,587 individuals who applied for benefits appeared on two privately owned websites of the Federal Emergency Management Administration (FEMA). FEMA was unsure how long this information was posted online.

- On April 7, 2008, a spreadsheet with a hidden column that contained the Social Security numbers of officers and civilian employees at the Army Acquisition Center was posted on a website for approximately 5 months before it was discovered.

- On March 20, 2008, the voter registration website of the Pennsylvania Department of State was shut down after it was discovered that it allowed anyone visiting the website to view voters' names, dates of birth, political parties, and driver's license numbers.

- On June 1, 2007, the Jax Federal Credit Union accidentally posted client information—Social Security numbers and bank account numbers—online.

Businesses have also carelessly discarded their clients' personal information:[34]

- On April 9, 2010, an individual in Sparks, Nevada, observed that thousands of documents including customers' personal data, credit card information, and signatures had been discarded in dumpsters (without shredding them) by the local Hollywood Video store when it closed.

- On May 6, 2008, news reporters from Channel 8 found Social Security numbers, bank account numbers, and canceled checks in the dumpster of Northeast Security in New Haven, Massachusetts. When Northeast Security closed and moved out of the store, the company threw its clients' data out without shredding it.

- On May 1, 2007, Healing Hands Chiropractic threw away hundreds of patients' medical records containing names, Social Security numbers, home addresses, and (in certain records) credit card information, without first shredding them.

- That same day, garbage bags containing financial data of customers of J. P. Morgan were found outside five branch offices in New York City.

- Similarly, on that day, documents that included personal data, medical records, and the results of police background checks from the Maine State Lottery Commission were found in a dumpster.

The theft, loss, or accidental disclosure of personal data from such databases is the responsibility of the companies or agencies that hold this information—not the responsibility of the individuals whose data are held within the database. To effectively protect individuals' identities and prevent identity theft, both systemic solutions (security of personal data held in databases) and individual responses (actions that individuals take to protect themselves from becoming victims of identity theft) are required.

Where Can Someone Find Information to Steal?

A plethora of information can be found about individuals online. websites that provide anything from background checks to individuals' photographs, home addresses, phone numbers, e-mail addresses, age, date of birth, work history, education, alumni information, family members, average income, criminal records, and so on include, but are not limited to, the following:

- Yahoo! People Search: http://people.yahoo.com/
- U.S. Search: www.ussearch.com

Things to Remember

To protect yourself from becoming a victim of identity theft, you should take the following precautions:

- Avoid storing your personal information on computers and related electronic devices.

- Shred documents containing personal information before you discard them.

- Check your bank accounts frequently for unusual activity.

- Get a credit report and check your credit score.

- Do not use the same password for all of your accounts. If an offender hacks into one of the accounts of the victim, a similar password will provide the hacker with access to his or her other accounts.

The U.S. Department of Justice provides information on identity theft and identity fraud on its website: http://www.usdoj.gov/criminal/fraud/websites/idtheft.html. Specifically, this site includes information on what identity theft and identity fraud are, what the most common ways to commit these crimes are, which actions the Department of Justice is taking to combat these crimes, how individuals can protect themselves from these crimes, and what actions individuals should take if they become a victim of these crimes. This site also provides links to other sites that have information on identity theft and identity fraud.

- U.S.A. People Search: www.usa-people-search.com
- Pipl: www.pipl.com
- Intelius: http://www.intelius.com/people-search.html
- People Finder: http://www.peoplefinders.com/
- Wink: http://wink.com/
- PeekYou: http://www.peekyou.com/
- PeopleLookup: http://www.peoplelookup.com/

Social networking websites (e.g., Facebook, Twitter, MySpace) also store a vast amount of individuals' personal information. Many individuals holding accounts on social networking sites do not password-protect their accounts and routinely accept strangers' "friend requests," which provides strangers with access to all of the photographs, videos, and information (e.g., home address, e-mail address, and phone number) in the individual's account. Of course, these sites do have privacy protection controls that allow users to limit access to their accounts and even allow users to restrict access to specific areas in the account to a limited number of friends (or users whom the individual has allowed access to his or her account). Many users, however, do not choose these options, even though these measures would enhance the security of their personal information. To verify this fact, just visit these sites and search through the numerous open accounts.

Most personal and financial information can be obtained online if the price is right. Chat rooms, bulletin boards, and advertisements exist that openly sell such data. Indeed, identity thief Robbin Shea Brown bought approximately 4500 credit card numbers, PINs, and personal data relating to the customers' accounts—such as the expiration dates of credit cards, passwords, and Social Security numbers—in online chat rooms from sellers who claimed to have retrieved these data from "phishing, pharming, and spamming unsuspecting victims."[35]

Cyberbullying

With cyberbullying, bullies no longer need to confront their victims face-to-face. Instead, young cyberbullies use communications technologies to annoy, embarrass, humiliate, abuse, threaten, stalk, or harass other children or teenagers. Generally, two types of cyberbullying are distinguished:[36]

- *Direct cyberbullying*. As the name clearly indicates, this type of cyberbullying occurs when an individual attacks a victim directly.
- *Cyberbullying by proxy*. With this type of cyberbullying, an individual enlists the help of others to assist him or her with bullying the victim. Sometimes this occurs without the knowledge of the "helpers." For instance, to alienate a student from her classmates, a girl might pose as the victim and post insulting comments to her fellow students. The students then retaliate against the victim, thinking that she initiated the verbal attack. Depending on where the information is posted and who views it, a victim may be cyberstalked subsequently to fake postings.

There are many different ways in which a cyberbully can either directly attack the victim or orchestrate an attack by proxy. Shielded by anonymity, children (or teenagers) can use websites, tweets (messages sent via Twitter), instant messages, e-mails, blogs, polls, and posts on social networking sites to belittle, verbally attack, stalk, or otherwise threaten other children or teenagers. Cyberbullying can also occur when perpetrators upload embarrassing photos or videos of the victim to websites (e.g., YouTube) or e-mails, and then send them or make them available to numerous people so that they can collectively bully the victim. The harassers can further bully their victims by using text messages or multimedia messages (to send pictures and videos) via cell phones. In particular, cyberbullying may take any of the following forms included in the next subsections.

Stalking the Victim

First, the harasser may stalk the victim, by sending him or her repeated rude, threatening or harassing e-mails, instant messages, or text messages. To see how this works, consider what happened to Amanda Marcuson of Birmingham, Michigan.[37] Marcuson had reported some girls for stealing a pencil case she owned with makeup in it. In retaliation, the girls bombarded her with instant messages calling her a tattletale and a liar. Marcuson had set up her phone to have e-mails and instant messages sent to her phone. As a result, she received so many messages that her mobile phone inbox was filled to capacity—a phenomenon known as a text war. Text wars occur when a few people get together to send the victim hundreds of messages (instant, e-mail, or text), causing e-mail stress and significantly increasing the costs to the victim (by increasing the cell phone charges from the sheer number of message received on the victim's phone).[38]

Another case involved Lauren Newby of Dallas, Texas. In her situation, a cyberbully started posting offensive and vicious comments about Newby on a website. Other individuals joined in, making insulting comments about her weight and a health issue she had (multiple sclerosis).[39] However, the bullying did not stop there: It moved offline and into the real world. Specifically, "Lauren's car was egged, 'MOO BITCH' was scrawled in shaving cream on the sidewalk in front of her house, and a bottle filled with acid was thrown at her front door."[40]

The communication of violence or threats is prohibited online. In the majority of cases, this behavior violates the terms of use of websites. Indeed, social networking sites (e.g., Facebook and MySpace) typically have terms of service agreements that prohibit any type of verbal abuse, homophobic, racist, or otherwise offensive remarks.[41] For example, according to its website, when you use or access Facebook you agree to its "Statement of Rights and Responsibilities."[42] Under Section 3(6) of Facebook's "Statement of Rights and Responsibilities," an individual cannot bully, intimidate, or harass another user. MySpace has developed similar terms of use that prohibit the users of its site from harassing other users.[43] Other sites, such as Twitter, prohibit the communication of violence. In particular, Twitter's terms of use explicitly forbid the publishing or posting of "direct, specific threats of violence against others."[44]

Sharing the Victim's Personal Information

A cyberbully may share personal information about the victim or post the passwords of the victim online. This type of information may place the victim in harm's way. For example, if someone posts a victim's address online, the individual may be stalked, harassed, or harmed in real life. If the victim's password is posted online, individuals can steal the victim's personal information or can pose as the victim. For instance, upon gaining access to their victim's account, they may steal personal photographs and videos of the victim and send them to others en masse to humiliate the target. If they steal their victim's password, they may access the target's online account and post insulting messages to others to provoke retaliation. This behavior is prohibited by most online sites. In fact, the terms of use of most websites are violated if an individual discloses another person's username and password to a third party or uses "the account, username, or password of another [m]ember at any time."[45]

Posing as the Victim and Soliciting Sex

Third, the cyberbully may pose as the victim and post solicitations for sex in the victim's name. With this approach, the cyberbully typically includes the personal contact information of the victim in the solicitations. Sometimes cyberbullies can intentionally seek to harm a victim by posting solicitations for sex online on behalf (and unbeknownst) to the victim along with the victim's telephone number (fixed telephony and mobile) and home address. Sometimes these advertisements are posted on child predator websites. The transmission of material or content that promotes the physical harm or injury of any individual is explicitly forbidden by most online sites.[46]

Posing as the Victim to Send Messages

A cyberbully may send e-mail messages, instant messages, or text messages to others while disguising himself or herself as the victim. Cyberbullies may pose as their victims by using similar screen names (with minor changes that will go unnoticed by most online users) and post inappropriate things to other online users. Those users, thinking it is the victim who is saying such things about them, will most likely retaliate and make similarly offensive remarks about the victim in return. To be cruel, sometimes bullies pose as other victims to embarrass them and alienate them from other classmates. A real life example of this involved Kylie Kenney. Her cyberbullies had chosen screen names that were similar to Kenney's screen name to make sexual remarks about and advances on her female hockey teammates, alleging that the remarks were made by Kenney herself.[47]

Moreover, a cyberbully usually poses as the victim as a form of retaliation for something the victim has done (whether real or perceived) to the cyberbully. Consider the following example: Becky is angry at Maggie for going out with her ex-boyfriend, Mark. She poses as Maggie online and posts messages to the accounts of Mark's friends—Ryan, Tristan, and John—that include insults about Mark and ask his friends to go out with her. Ryan, Tristan, and John respond by insulting Maggie for her inappropriate behavior. They also inform Mark about the messages from Maggie. Mark calls the relationship off with Maggie as a result of the messages he thought she sent to his friends.

Some online sites specifically forbid impersonation of others. For instance, MySpace prohibits "impersonating or attempting to impersonate MySpace or a MySpace employee, administrator or moderator, another [m]ember, or person or entity (including, without limitation, the use of email addresses associated with or of any of the foregoing)."[48] Twitter also prohibits the impersonation of others "through the Twitter service in a manner that does or is intended to mislead, confuse, or deceive others."[49] Therefore, using a false identity or otherwise attempting to mislead others as to a user's identity or the origin of messages is expressly prohibited.[50] Furthermore, Formspring.me, an online forum that allows users to ask questions and give answers to just about anything, prohibits the transmission of "any material or content that attempts to falsely state or otherwise misrepresent" the user's identity or affiliation with a person or entity.[51]

Creating Polls to Humiliate the Victim

Cyberbullies may create embarrassing, insulting, and oftentimes offensive polls to bully their victims online, such as "Who is the ugliest person in the class?", "Who is the biggest slut in the class?", "Who is Hot? Who is Not?", and so on. Additionally, cyberbullies may encourage others to participate in Internet polls and share offensive or insulting comments about classmates (e.g., their top 10 ugliest girls or boys in the class). This behavior also violates the terms of use of most websites, because the transmission or encouragement of "harassing, libelous, abusive, threatening, harmful, vulgar, obscene or otherwise objectionable material of any kind or nature" is strictly prohibited.[52]

Creating Websites to Humiliate the Victim

Cyberbullies may create websites that ridicule, humiliate, or intimidate others. This involves uploading or disseminating embarrassing or inappropriate videos or pictures of the victim. Cyberbullies sometimes write about their victims on online blogs as well. Other children can also take part in the quest to humiliate or harass the target of the cyberbullying by creating websites dedicated to verbally attacking and embarrassing the victim.

For example, a website was created about David Knight, titled "Welcome to the website that makes fun of David Knight."[53] This page provided individuals with the opportunity to post hateful comments about Knight, which they did in abundance. Knight was not alone. Jodi Plumb, a 15-year-old girl from Mansfield, England, also discovered a website containing abusive comments concerning her weight and even had the date of the day she was supposedly going to die posted on it.[54] Yet another example of this type of cyberbullying involved Kylie Kenney. Her classmates created websites with names such as "Kill Kylie Incorporated" that contained offensive and homophobic remarks about Kenney.[55]

Sexting Distribution

Another humiliating and degrading form of cyberbullying is sexting. **Sexting** "is the act of sending, receiving, or forwarding sexually explicit messages, photos, or images via cell phone, computer, or other digital device."[56] The original messages and photos are then forwarded beyond the intended recipient, resulting in widespread humiliation and ridicule of the subject of the photo.

An example of this kind of cyberbullying involved Hope Witsell.[57] Witsell sent a photo of her breasts to a boy she liked in her school. This photo was subsequently distributed to others in the school. The distributor of these photos was not the boy to whom Witsell had sent the photos, but rather a girl who had borrowed his phone. When the girl noticed the picture of Witsell, she forwarded the photo to her own phone and to the phones of other students. The picture was viewed by the majority of students in Witsell's school. Witsell was taunted about the picture relentlessly. When the taunting proved too much for her to bear, she hung herself. She was only 13 years old when she died.

In a similar case, an 18-year-old teenager, Jesse Logan, killed herself after an ex-boyfriend forwarded naked pictures of her after they ended their relationship.[58] She, too, could not handle the humiliation she felt as a result of his actions.

It is important to note that pornographic material cannot be posted or distributed on certain websites. Doing so may result in the deletion of the account by the site administrators, as this behavior violates the terms of use of most online sites.[59]

Notification Wars

"Notification wars" (or "warning wars") may occur online. Sometimes cyberbullies will falsely report their victims for terms of use violations on websites on more than one occasion. When a site receives notice of a violation of the terms of service, the victim's account may be temporarily suspended while an investigation ensues or the account may be permanently deleted.[60]

The High Costs of Cyberbullying

Cases involving cyberbullying have led to the death of many of the victims. A recent case involved Phoebe Prince, who was subjected to intolerable ridicule and torment at school. People would insult her and call her names on Facebook, Formspring, and Twitter.[61] Prince was not alone; numerous cases like hers have occurred in the United States. Prince's cyberbullying case is just the most recent. Before her death by suicide, there was the case of Carl Joseph Walker-Hoover, who hung himself after enduring constant bullying at school and being repeatedly subjected to anti-gay slurs.[62]

Sadly, the suicides from cyberbullying do not stop there. Another widely recognized case involved Megan Meier.[63] Meier was a former friend of Lori Drew's daughter, Sarah. Lori Drew and her daughter decided to set up a false identity, "Josh Evans." The elder Drew stated that this was done to see if Meier was spreading rumors about Sarah. Meier believed that "Josh" was a real person. "Josh" complimented Meier and told her she was beautiful; Meier believed that they had formed an online relationship. One day, "Josh" broke off the relationship, claiming that he had heard that she was a cruel person. Meier committed suicide after "Josh" told her that the "world would be a better place without you." She was 13 years old when she took her life.

As a result of the lack of adequate laws on cyberbullying at that time, Lori Drew was charged only with violating the terms of service agreement of MySpace. In 2009, she was acquitted of all charges on appeal. A cyberharassment law was enacted in Missouri in response to the Megan Meier case.

One individual charged with violating this law was Elizabeth Thrasher. Thrasher posed as a 17-year-old girl (the daughter of her ex-boyfriend's girlfriend) and posted sexual advertisements on the "Casual Encounters" section on Craigslist.[64] She also posted a picture, the cell phone number, and e-mail address of the teenager. As a result, the 17-year-old girl received numerous offensive e-mails, calls, text messages, and pornographic photographs from individuals responding to her supposed advertisement on Craigslist.

Another cyberbullying victim was Ryan Patrick Halligan.[65] Rumors about his sexual orientation were spread through the Internet by people he believed to be his friends. He started talking with one of the most popular girls in school over the summer of 2003. From their conversations, he had believed that they had grown quite close. When the school year started, Halligan approached her to speak with the girl in person. In front of her friends, she ridiculed him and informed him that their online relationship was a hoax. She also informed him that she shared everything that he had confided with her to her friends. Halligan was unable to cope with the rejection and humiliation he felt and ended his life. His family later lobbied successfully to have an anti-cyberbullying statute passed.[66]

Some harassers have even continued bullying individuals even after the death of the victims. Specifically, after Phoebe Prince and Alexis Pilkington died following bullying incidents, their tormentors posted vicious comments on their Facebook memorial pages.[67]

Some children's parents have sought to fight their children's bullies in court. A successful example of such an occurrence involved the parents of Ghyslain Raza, who became known as the "*Star Wars* kid." Ghyslain's classmates uploaded a humiliating video of him pretending to be a *Star Wars* character online. The boy was subjected to extensive ridicule worldwide because of this video. His parents successfully sued the parents of the bullies who posted Raza's video online.

It is clear that technology has made cyberbullying far worse than bullying by conventional (in person) means, as rumors may now be spread much faster to a larger number of individuals. Once the embarrassing or offensive material has been posted, it is extremely difficult (if not impossible) to completely remove it. For instance, in the now famous case of Ghyslain Raza, his video was viewed more than 1 million times approximately one month after it was posted. Anyone can post this kind of information online. Only if these cyberbullies are caught will their accounts be deleted—but by then the damage to the victim is already done. Yet, cyberbullies may continue on, even after their accounts have been deleted, because it is very easy for them to open up accounts under a false name, as long as a different e-mail address is provided. This is possible because the majority of websites do not authenticate the personal information that individuals provide.

Child Exploitation Online

Computer forensics investigators often use Internet resources to gather evidence on **child exploitation**. Specifically, the Internet can assist investigators and members of the public in identifying predators and child molesters.

The **National Sex Offender Registry** is one such website that can provide investigators with information on **sexual predators**. The Family Watchdog website (http://www .familywatchdog.us) provides access to such a registry. This site offers many essential tools for parents, concerned citizens, and law enforcement agencies with which to protect children. Parents and concerned citizens can search this site for registered sex offenders in their neighborhood. Investigators can type in the victim's address and locate any sex offenders who live close by using these types of websites. If an incident occurs at a school, an investigator can locate the school on the map of the residences as well as areas of local employment of sex offenders, thereby determining whether a sex offender is in close proximity to that school.

Another very important tool for investigations is the reverse e-mail lookup. This tool scans various social and picture sites to see where e-mail addresses have been registered. Investigators can use this information to track down user profiles created by the sex offenders on these social sites and determine whether these individuals have contacted children using the sites by looking at their "friends" lists. Many sites are available online that provide reverse e-mail searches, including the following sites:

- AnyWho (http://www.anywho.com/rl) provides reverse phone lookup.

- People Search Pro (http://www.peoplesearchpro.com/resources/email-search/reverse-email-lookup/) provides reverse e-mail lookup.

- Reverse Records (http://www.reverserecords.org/?hop=haskinsmic) provides reverse e-mail, phone, cell phone, and home address lookups.

The Dru Sjodin National Sex Offender public website provides real-time access to state and territory sex offender registries, including registries from all 50 states and the U.S. territories of Guam, Puerto Rico, the District of Columbia, and participating tribes.[68] This registry was named after Dru Sjodin, who was kidnapped, brutally raped, and murdered by a three-time convicted sex offender, Alfonso Rodriguez, Jr. The public outcry in the wake of this crime led to "Dru's law," which called for establishment of a national sex offender registry.

Dru's law became part of the **Adam Walsh Protection and Safety Act of 2006**, which was passed in response to the abduction of Adam Walsh, the son of former *America's Most Wanted* host John Walsh, from a mall in Florida and his subsequent murder. Adam Walsh's killer was never caught. Title I of the Adam Walsh Child Protection and Safety Act is known as the **Sex Offender Registration and Notification Act of 2006** (SORNA). SORNA created a national sex offender registry, where the offenders' photographs, full names and any aliases, home and work addresses, dates of birth, and offenses committed are posted. The Adam Walsh Children Protection and Safety Act also strengthened the national sex offender registry by requiring more data to be included about sex offenders, such as their physical description, fingerprints, palm prints, DNA samples, criminal history, and a detailed summary of their crimes. This law was intended to standardize state sex offender laws nationwide. It further provided for a three-tier system to be used when categorizing sex offenders for the purposes of the registry:

- Tier 1 offenders must be registered for 15 years and update their information annually.

- Tier 2 offenders must be registered for 25 years and update their information biannually.

- Tier 3 offenders—the most dangerous offenders—must register for life and are required to update their information in the database every three months (to ensure that it remains up-to-date).

Moreover, the **Campus Sex Crimes Prevention Act of 2000** requires that sex offenders who enroll or work at a community college, college, university, or trade school must notify local law enforcement agencies.

Prior to the passage of the Adam Walsh Children Protection and Safety Act, legislators at both the federal and state levels had passed several other laws that paved the way for national registration of sex offenders. These laws were triggered by the abduction and murder of children by sexual offenders. In particular, in 1990, Megan Kanka was raped and murdered by a neighbor who was a sex offender. Her parents were unaware that a sex offender lived in their neighborhood. The young girl's death triggered the creation of state and federal laws that focused on sex offender registration and notification of residents of neighborhoods when sex offenders move in. The **Jacob Wetterling Crimes Against Children and Sexually Violent Offender Act of 1994** (codified in 42 U.S.C. § 14071), named after another victim, was enacted as part of the **Violent Crime Control and Law Enforcement Act of 1994** and required states to implement a sex offender registry.

Some state sex offender registries provide users with the option to track sexual predators. For instance, Florida has an offender alert system comprising a free service that provides e-mail alerts when an offender or predator moves close to any address in Florida that an individual chooses to monitor.[69]

Furthermore, America's Missing: Broadcasting Emergency Response (AMBER) alerts were developed to give notice to the public of recent child abductions. The **AMBER alert** system was named after Amber Hagerman, who was abducted outside of a Winn-Dixie store in her hometown of Arlington, Texas, as she was riding her bike. Four days after her abduction, the child's lifeless body was found in a drainage ditch. According to the U.S. Department of Justice, as of 2009, AMBER alerts have helped rescue 495 children.[70] For an AMBER alter to be issued, the following conditions must be met:

- Law enforcement agents must confirm that the child has been abducted.
- The abducted child must be at risk of serious harm or death.
- A sufficient description of the child, the child's abductor, or the abductor's vehicle must exist.
- The abducted child must be 17 years old or younger.

Apart from appearing on televisions, radios, websites, and highway traffic boards, AMBER alerts can be issued through messages sent to wireless devices, should users opt to receive such alerts relative to their geographical location.

Online Sexual Predators

Sexual predators have been known to frequent chat rooms and monitor ongoing conversations in search of their next victims. Once they find a potentially suitable target, they engage in conversation with him or her. Sometimes they attempt to engage the target in a private conversation. Slowly they develop a relationship of trust with their victim, encour-

---

The Child Pornography Protection Act of 1996

The sections of the **Child Pornography Protection Act** that explicitly criminalize the exploitation of children online include, but are not limited to, the following:

- 18 U.S.C. § 2251: prohibits the sexual exploitation of children.
- 18 U.S.C. § 2251A: provides severe penalties for persons who buy and sell children for sexual exploitation.
- 18 U.S.C. § 2252: covers activities relating to material involving the sexual exploitation of minors.
- 18 U.S.C. § 2252A: prohibits activities relating to material constituting or containing **child pornography**.

---

aging the victim to confide in them by convincing the victim to share his or her personal information (e.g., age, home address, phone number, relationship status). They may also try to convince the minor to engage in sexually explicit conduct. For example, Ivory Dickerson persuaded and enticed "female minors into engaging in sexually explicit conduct for the purpose of manufacturing child pornography."[71] His computer also contained more than 600 child pornography images. Mark Wayne Miller posed as a young male to persuade his victims—young girls to engage in sexually explicit conduct.[72] He recorded these sessions and even distributed some of these sessions to third parties via active Webcams.

Eventually, sexual predators try to set up a meeting with their victims. Some even travel across states to meet these youths. In one case, an individual violated 18 U.S.C. § 2423(b), which prohibits a "person from traveling in interstate commerce, or conspiring to do so, for the purpose of engaging in criminal sexual activity with a minor," when he traveled across state lines (from Minnesota to Wisconsin) to meet with a minor (a 14–year-old girl) he met in an online chat room and engage in sexual acts.[73]

Adults may identify themselves as children on websites to lure children into dangerous situations. One cannot definitively know that the person with whom an individual is communicating in chat rooms and social networking sites or through e-mails and instant messages is not a sexual predator. In an attempt to remedy this problem, the **Keeping the Internet Devoid of Sexual Predators Act of 2008** (KIDS Act) was enacted. The KIDS Act requires sex offenders to submit their e-mail addresses and online screen names to the national sex offender registry.

The Internet also provides these predators with anonymity. They can pretend to be any age—some have even entered chats among 10-year-olds pretending to be the same age. When these adults pretend to be children or teenagers, their victims are more likely to develop friendships and online relationships with these individuals. At other times, offenders have pretended to be a different gender. William Ciccotto, a 51–year-old male, posed as a young girl (between the age of 13 and 14 years old).[74] Specifically, he opened a Hotmail account and a MySpace account under the name "Cindy Westin" through which he proceeded to send friend requests to young girls and chat with them within this site and via AOL instant messages. Under this persona, Ciccotto convinced the girls to send nude photos of themselves, sometimes convincing his correspondents to take photos of themselves engaging in sexually explicit conduct and send them to "Cindy." Ciccotto then distributed the photos through a private peer-to-peer (P2P) network.

Child Pornography Images and Videos

The Internet provides sexual predators with easy, fast, convenient, and anonymous access to vast quantities of pornographic material featuring children worldwide for a low cost. Sometimes child pornography images can be stored in servers beyond U.S. law enforcement agencies' reach. Thus child pornography poses unique challenges to law enforcement agencies. Coordinating responses with multiple jurisdictions may be required to track even one child pornography distributor.

In the United States, case law and statutes have outlawed the manufacture, possession, and distribution of child pornography. For instance, the **Sexual Exploitation of Children Act of 1978** made it illegal for someone to manufacture and commercially distribute obscene materials that involve minors younger than 16 years old. Ten years later, the **Child Protection and Obscenity Enforcement Act of 1988** was enacted. This Act made it illegal for an individual to use a computer to depict or advertise child pornography. Later, in 1990, the *Osborne v. Ohio*[75] ruling stated that private possession of child pornography was illegal.

Child pornography can be found in the following areas on the Internet:[76]

- *Newsgroups.* Members may share child pornography images and information on child pornography websites through this medium to avoid unwanted attention by Internet service providers (ISPs) and law enforcement agencies. They may use code to discuss child pornography websites or hide the images of child pornography among legal adult pornographic images.

- *Bulletin boards.* Discussions on child pornography often occur in this forum, and websites containing child pornography are frequently rated and shared among child pornographers. These forums may be password protected to avoid infiltration by undercover law enforcement agents and individuals who oppose child pornography collection and distribution.

- *Chat rooms.* These areas are used to exchange child pornography images and find minors to sexually exploit and victimize.

- *Peer-to-peer networks.* These networks enable the sharing of child pornography images (files) to closed groups to avoid detection.

- *E-mails.* This method is not often used by seasoned sexual predators to share images because of their fear that they might unwittingly transfer such material to undercover law enforcement agents.

In a landmark, but controversial case, *Ashcroft v. Free Speech Coalition*,[77] the U.S. Supreme Court held that virtual (i.e., computer-generated) images of child pornography were legal—as long as a real child was not used to produce the image. This ruling raised significant issues for prosecutors of child pornography cases because the burden of proof is now on prosecutors to show that the images depict actual children, rather than adult models made to look like children. The **Child Victim Identification Program**, which houses the largest database of child pornography images for the purpose of identifying victims of child exploitation and abuse, was developed in the aftermath of the Supreme Court's ruling in the *Ashcroft* case.

In 1998, the Child Protection and Sexual Predator Punishment Act was enacted, which required ISPs) to report incidents of child pornography to authorities when they come across them. ISPs are not actively required to monitor their websites or customers, however. Additionally, Section 508 of the **Prosecutorial Remedies and Other Tools to End the Exploitation of Children Today Act of 2003** (PROTECT Act) amended the **Victims of Child Abuse Act of 1990** to

> authorize a provider of electronic communication or remote computing services that reasonably believes it has obtained knowledge of facts and circumstances indicating a State criminal law child pornography violation to disclose such information to an appropriate State or local law enforcement official.

In addition, the PROTECT Act prohibited the use of misleading domain names with the intent to deceive a minor into viewing online material that is harmful to minors. For example, John Zuccarini used the Internet domain names of a famous amusement park, celebrities, and cartoons to deceive minors into logging into pornography websites.[78] Specifically, he intentionally misspelled versions of a domain name (www.dinseyland .com) owned by the Walt Disney Company (www.disneyland.com). Zuccarini also used 16 misspellings and variations of the name of the legitimate website of the popular female singer Britney Spears (www.britneyspears.com). In addition, Zuccarini used misspellings and variations of the domain names of legitimate websites depicting two popular cartoon characters, Bob the Builder and Teletubbies—for example, www.bobthebiulder.com and www.teltubbies.com.[79]

The **Securing Adolescents from Exploitation Online Act of 2007** (SAFE Act) expanded the definition of an ISP to include wireless hot spots such as libraries, hotels, and municipalities. It also required such service providers to "provide information relating to the Internet identity of any individual who appears to have violated a child exploitation or pornography law, including the geographic location of such [an] individual and images of any apparent child pornography." Moreover, the SAFE Act required these ISPs to preserve images of child pornography that were observed for evidentiary purposes. Failure to do so or to report instances of child pornography will result in significant penalties to an ISP. If civilians find websites that exhibit child pornography, the Child Exploitation and Obscenity Section (CEOS) website states that these individuals should contact the **National Center for Missing and Exploited Children** (NCMEC).

Internet service providers can help fight against the proliferation of child pornography by taking the following steps:[80]

- Removing illegal sites wherever and whenever they encounter them.

- Establishing websites and hotlines where individuals can complain about any child pornography images or websites encountered.

- Taking responsibility for the content of sites they host and notifying authorities when child pornography is encountered.

- Preserving their records to make them available for law enforcement agencies during investigations of child pornography creation, collection, and distribution.

- Requiring the verification of personally identifying information given by individuals to open up accounts. Individuals may use fake names, home addresses, and phone numbers to open up accounts, for example, because this information is not authenticated by ISPs. This factor makes it extremely difficult for law enforcement authorities to trace and find those persons who are responsible for these illegal activities.

Law enforcement responses to child pornography should include the following measures:[81]

- Locate and take down child pornography websites.

- Find out who is signed up to download and post child pornography from these websites.

- Conduct undercover sting operations. For instance, an undercover FBI agent downloaded child pornography images from P2P networks from a user known as "Boys20096."[82] The undercover agent traced the IP address of "Boys20096" to a residence in Wheaton, Illinois. A laptop seized at the residence was believed to belong to a live-in nanny, Lubos Albrecht. The laptop contained approximately 6000 images or videos of child pornography. The FBI eventually linked Albrecht to the laptop and the P2P sharing network from which the child pornography images were obtained.

- Create "honeypots" to lure child pornographers in an attempt to identity them. Honeypots also discourage other child pornographers from visiting these websites for fear that they are not real child pornography websites, but rather fake websites designed to bait and catch them.

- Publicizing captures of individuals engaging in sexual conduct with minors and the creation, collection, and distribution of child pornography. This publicity is meant to instill fear and uncertainty in the offender or potential offenders that their online illegal actions may be brought to the attention of authorities—for which they will certainly be punished.

Chapter Summary

This chapter explored the computer-networking environment in which scam artists, identity thieves, cyberbullies, and sexual predators operate in and the crimes they commit. Through auction, rental, retail, dating, lottery, employment, and virus protection scams, individuals' personal, financial, and banking details can be stolen. Individuals posing as international or national government officials may also seek to take advantage of unsuspecting victims and steal their money and/or identities. Identity thieves can steal victims' identities for profit, so as to avoid detection or deceive law enforcement authorities and to commit other crimes. Indeed, individuals' personal information is readily available online from a variety of sources.

Cyberbullies may verbally attack their victims through text messaging on their cell phones and in cyberspace through websites, instant messages, e-mails, posts on social networking sites, or blogs. Their victims are exposed to demeaning text messages, embarrassing photos, humiliating videos, and crude opinion polls that make this type of bullying particularly disturbing. By providing bullies with the ability to post their hurtful messages anonymously, the Internet has created a breeding ground for cyberbullying.

Children may attract unwanted attention from online sexual predators as well. Sexual predators use the Internet to stalk their victims under the cloak of anonymity afforded to them by the online environment. The predator may use multiple personas to lure the victim into a false sense of security. Many laws have been passed to prohibit the sexual exploitation of children and the creation, collection, and distribution of child pornography online. Because of problems related to the Internet's lack of boundaries and single jurisdictions, the regulation of child pornography, prevention of child exploitation, and investigation of child sexual predators pose unique challenges to law enforcement agencies worldwide.

Key Terms

Adam Walsh Protection and Safety Act of 2006

AMBER alert

Campus Sex Crimes Prevention Act of 2000

Child exploitation

Child pornography

Child Pornography Protection Act of 1996

Child Protection and Obscenity Enforcement Act of 1988

Child Victim Identification Program

Dating scam

Dru's law

Government e-mail scam

Identity theft

Jacob Wetterling Crimes Against Children and Sexually Violent Offender Act of 1994

Keeping the Internet Devoid of Sexual Predators Act of 2008

Lottery scam

National Center for Missing and Exploited Children

National Sex Offender Registry

Nigerian scam

Pharming

Prosecutorial Remedies and Other Tools to End the Exploitation of Children Today Act of 2003

Rental and real estate scam

Scareware

Securing Adolescents from Exploitation Online Act of 2007

Sex Offender Registration and Notification Act of 2006

Sexting

Sexual Exploitation of Children Act of 1978

Sexual predator

Spam

Victims of Child Abuse Act of 1990

Violent Crime Control and Law Enforcement Act of 1994

Virus protection scam

Vishing

Work-at-home scam

Practical Exercise

Consider the cyberbullying case that resulted in the death of 13-year-old Megan Meier. After Meier's suicide in 2006, Missouri revised its harassment statutes to include bullying, stalking, and harassment via telecommunications and electronic communications. Why do you think this happened?

In considering this question, refer *United States v. Drew*[83] to provide a brief description of the case. In your answer, indicate which law the suspect was charged with violating. Given that these charges were later overturned, also explain why this happened in your answer.

Review Questions

1. Which types of scams do scam artists use to steal the personal, financial, and banking information of victims?

2. What is the "419 scam," and what does it involve?

3. Why is spam dangerous to the victim?

4. What are the similarities and differences between vishing, phishing, and pharming?

5. What are four types of identity theft?

6. How have existing practices in storing individuals' personal information facilitated identity theft?

7. Where can offenders find the personal information of a victim?

8. What is cyberbullying, and where does it occur?

9. What is sexting?

10. Where can child pornography be found online?

11. How have ISPs contributed to the fight against child pornography?

12. How have law enforcement agencies sought to combat child pornography?

Footnotes

[1]Pethia, R. (2006). Information assurance and computer security incident response: Past, present, future. CERT, Software Engineering Institute, Carnegie Mellon University. Retrieved from http://search.first.org/conference/2006/papers/pethia-richard-slides.pdf

[2]White, J. (2010, May 24). District food servers charged in theft of patron's credit card numbers. *Washington Post* [online]. Retrieved from http://www.washingtonpost.com/wp-dyn/content/article/2010/05/23/AR2010052302921.html?hpid=newswell

[3]For further information, see Winterman, D. (2010, June 15). How not to get scammed like Alan Bennett. *BBC News* [online]. Retrieved from http://news.bbc.co.uk/2/hi/uk_news/magazine/8740984.stm

[4]Federal Bureau of Investigation. (2010, March 12). Rental and real estate scams. *New E-Scams & Warnings*. U.S. Department of Justice, Investigative Programs: Cyber Investigations. Retrieved from http://www.fbi.gov/cyberinvest/escams.htm

[5]For more information on this scam, see US Department of State. (2007, February). International financial scams: Internet dating, inheritance, work permits, overpayment, and money-laundering. Retrieved from http://travel.state.gov/pdf/international_financial_scams_brochure.pdf

[6]For further information on this scam, see US Department of State. (2009, August 5). Lottery scams. Retrieved from http://travel.state.gov/travel/cis_pa_tw/cis/cis_2475.html

[7]Federal Bureau of Investigation. (2009, October 5). Fraudulent email claiming to contain FBI "Intelligence Bulletin No. 267." *New E-Scams & Warnings*. US Department of Justice, Investigative Programs: Cyber Investigations. Retrieved from http://www.fbi.gov/cyberinvest/escams.htm

[8]Federal Bureau of Investigation. (2009, October 5). Fraudulent email claiming to be from DHS and the FBI Counterterrorism Division. *New E-Scams & Warnings*. US Department of Justice, Investigative Programs: Cyber Investigations. Retrieved from http://www.fbi.gov/cyberinvest/escams.htm

[9]Internet Crime Complaint Center. (n.d.). Internet crime schemes. Retrieved from http://www.ic3.gov/crimeschemes.aspx

[10]See Fraud Watch International. (n.d.). Nigerian 419 scams. Retrieved from http://www.fraudwatchinternational.com/nigerian-419/

[11]See, for example, http://www.scam-info-links.info/; http://www.internetscamswatch.com/; http://www.fraudaid.com/ScamSpeak/Nigerian/nigerian_scam_letters.htm

[12]Federal Bureau of Investigation. (2009, February 9). Work-at-home scams. *New E-Scams & Warnings*. US Department of Justice, Investigative Programs: Cyber Investigations. Retrieved from http://www.fbi.gov/cyberinvest/escams.htm

[13]Federal Bureau of Investigation. (2009, December 11). Pop-up advertisements offering anti-virus software pose threat to Internet users. *New E-Scams & Warnings*. US Department of Justice, Investigative Programs: Cyber Investigations. Retrieved at: http://www.fbi.gov/cyberinvest/escams.htm

[14]Thomas, T. L. (2003). Al Qaeda and the Internet: The danger of "cyberplanning." *Parameters*, p. 115.

[15]Financial Crimes Enforcement Network. (2008, July 18). FinCEN reminds public to be aware of financial scams. Retrieved from http://www.fincen.gov/news_room/nr/pdf/20080718.pdf

[16]Social Security Administration. (2006, November 7). Public warned about email scam. News release. Retrieved from http://www.internetcases.com/library/statutes/can_spam_act.pdf

[17]Federal Bureau of Investigation. (2008, May 8). Phishing related to issuance of economic stimulus checks. *New E-Scams & Warnings*. US Department of Justice, Investigative Programs: Cyber Investigations. Retrieved from http://www.fbi.gov/cyberinvest/escams.htm

[18]US Department of Justice, (2010, June 24). Nantucket man arrested and charged with operating international online "phishing" scheme to steal income tax refunds. Retrieved from http://www.justice.gov/criminal/cybercrime/mardakhayeuIndict.htm

[19]McLean, P. S., & Young, M. M. (2006, March/April). Phishing and pharming and Trojans: Oh my!. *Utah Bar Journal, 19*, 32.

[20]New York State Attorney General. (2009, April 22). Attorney General Cuomo announces arrest of former State Tax Department employee for using position to steal taxpayer's identities. Retrieved from http://www.ag.ny.gov/media_center/2009/apr/apr22a_09.html

[21]Sandholm, D. (2009, October 28). NY computer repairman accused of identity theft in $1.1M scheme. *ABC News* [online]. Retrieved from http://abcnews.go.com/Blotter/man-allegedly-orchestrated-11m-fraud-scheme/story?id=8940921

[22]Gross, G. (2008, May 2). Military contractor convicted on ID theft charges. *IDG News Service* [online]. Retrieved from http://www.networkworld.com/news/2008/050208-military-computer-contractor-convicted-on.html

[23]Brown seeks identity theft law. (2002, June 2). *New York Law Journal*, p. 9.

[24]Rudner, M. (2008). Misuse of passports: Identity fraud, the propensity to travel, and international terrorism. *Studies in Conflict & Terrorism, 31*(2), 103.

[25]*Ibid.*

[26]Lormel, D. M. (2002, July 9). Hearing on S.2541, "The Identity Theft Penalty Enhancement Act." Testimony of Chief of Terrorist Financial Review Group, FBI Before the Senate Judiciary Committee Subcommittee on Technology, Terrorism and Government Information. Retrieved from http://www.fbi.gov/congress/congress02/idtheft.htm

[27]ID theft ring allegedly bribed DMV. (2010, January 7). *NBC News, Los Angeles* [online]. Retrieved from http://www.nbclosangeles.com/news/local-beat/ID-Theft-Ring-Allegedly-Bribed-DMV-Employees.html

[28]Berkshire District Attorney. (2010). Identity theft. Commonwealth of Massachusetts. Retrieved from http://www.mass.gov/?pageID=berterminal&L=2&L0=Home&L1=Crime+Awareness+%26+Prevention&sid=Dber&b=terminalcontent&f=awareness_prevention_identity_theft&csid=Dber

[29]Ashley, G. D. (2002, September 19). Preserving the integrity of Social Security numbers and preventing their misuse by terrorists and identity thieves. Testimony of Assistant Director, Criminal Investigation Division, FBI, Before the House Ways and Means Committee, Subcommittee on Social Security. Retrieved from http://www.fbi.gov/congress/congress02/ashley091902.htm

[30]Privacy Rights Clearinghouse. (2010, June 25). Chronology of data breaches. Retrieved from http://www.privacyrights.org/ar/ChronDataBreaches.htm#CP

[31]*Ibid.*

[32]*Ibid.*

[33]*Ibid.*

[34]*Ibid.*

[35]United States Attorney's Office, District of Arizona. (2008, June 3). Tucson man sentenced to over 5 years in prison for aggravated identity theft. US Department of Justice, Computer Crime & Intellectual Property Section. Retrieved from http://www.cybercrime.gov/brownSent1.pdf

[36]WiredKids, Inc. (n.d.). How cyberbullying works. Retrieved from http://www.stopcyberbullying.org/how_it_works/direct_attacks.html

[37]Jackson, D. (2005). Examples of cyberbullying. Retrieved from http://www.slais.ubc.ca/courses/libr500/04-05-wt2/www/D_Jackson/examples.htm

[38]Division of Criminal Justice Services. (2007). Cyberbullying. New York State. Retrieved from http://www.criminaljustice.state.ny.us/missing/i_safety/cyberbullying.htm

[39]Jackson, D. (2005). Examples of cyberbullying. Retrieved from http://www.slais.ubc.ca/courses/libr500/04-05-wt2/www/D_Jackson/examples.htm

[40]*Ibid.*

[41]See Section 8.1, MySpace.com. (2009, June 25). Terms of use agreement. Retrieved from http://www.myspace.com/index.cfm?fuseaction=misc.terms

[42]Statement of Rights and Responsibilities. (n.d.). *Facebook.* Retrieved from http://www.facebook.com/terms.php?ref=pf

[43]Section 8.2, MySpace. (n.d.). Terms of use agreement. Retrieved from http://www.myspace.com/index.cfm?fuseaction=misc.terms

[44]See The Twitter rules. (2009, January 14). Retrieved from http://twitter.zendesk.com/forums/26257/entries/18311

[45]Section 8.28, MySpace. (n.d.). Terms of use agreement. Retrieved from http://www.myspace.com/index.cfm?fuseaction=misc.terms

[46]See, for example, Formspring.me terms of service. (n.d.). Retrieved from http://about.formspring.me/terms

[47]Chaker, A. M. (n.d.). Schools move to stop spread of "cyberbullying." *Pittsburg-Post Gazette* [online]. Retrieved from http://www.post-gazette.com/pg/07024/756408-96.stm

[48]Section 8.27, MySpace. (n.d.). Terms of use agreement. Retrieved from http://www.myspace.com/index.cfm?fuseaction=misc.terms

[49]The Twitter rules. (2009, January 14). Retrieved from http://twitter.zendesk.com/forums/26257/entries/18311

[50]Formspring.me terms of service. (n.d.). Retrieved from http://about.formspring.me/terms

[51]*Ibid.*

[52]*Ibid.*

[53]Division of Criminal Justice Services. (2007). Cyberbullying. New York State. Retrieved from http://www.criminaljustice.state.ny.us/missing/i_safety/cyberbullying.htm

[54]*Ibid.*

[55]Chaker, A. M. (n.d.). Schools move to stop spread of "cyberbullying." *Pittsburg-Post Gazette* [online]. Retrieved from http://www.post-gazette.com/pg/07024/756408-96.stm

[56]Berkshire District Attorney. (2010). Sexting. Commonwealth of Massachusetts. Retrieved from http://www.mass.gov/?pageID=berterminal&L=3&L0=Home&L1=Crime+Awareness+%26+Prevention&L2=Parents+%26+Youth&sid=Dber&b=terminalcontent&f=parents_youth_sexting&csid=Dber

[57]Murrhee, K. C. (2010, Winter). Cyber bullying: Hot air or harmful speech?. *University of Florida, Levin College of Law.* Retrieved from http://www.law.ufl.edu/uflaw/10winter/features/hot-air-or-harmful-speech

[58]Inbar, M. (2009, December 2). "Sexting" bullying cited in teen's suicide. *MSNBC News* [online]. Retrieved from http://today.msnbc.msn.com/id/34236377/ns/today-today_people/

[59]See terms of use of social networking websites such as Facebook, MySpace, and Twitter.

[60]See, for example, the terms of use of Twitter and MySpace.

[61]Ellis, R. (2010, March 30). 9 teens indicted as a result of bullying Phoebe Prince to suicide. *NY Parenting Issues Examiner*. Retrieved from http://www.examiner.com/examiner/x-29163-NY-Parenting-Issues-Examiner~y2010m3d30-9-teens-indicted-as-a-result-of-bullying-Phoebe-Prince-to-suicide

[62]James, S. D. (2009, April 14). Carl Joseph Walker-Hoover commits suicide after anti-gay slurs. *Huffington Post* [online]. Retrieved from http://www.huffingtonpost.com/2009/04/14/carl-joseph-walker-hoover_n_186911.html

[63]*United States v. Drew*, No. 08-00582 (C.D. Cal. May 15, 2008).

[64]Harvey, M. (2009, August 19). American woman Elizabeth Thrasher faces jail over "cyber-bullying." *The Times* [online]. Retrieved from http://technology.timesonline.co.uk/tol/news/tech_and_web/the_web/article6802494.ece

[65]For more information on Ryan Patrick Halligan's story, see the website that is dedicated to him: http://ryanpatrickhalligan.org/. It also contains important information and links on cyberbullying.

[66]VT. STAT. ANN. Title 16, § 1161a(a)(6) (2007).

[67]Ellis, R. (2010, March 25). Long Island teen commits suicide: Is cyberbullying to blame? *Cyber Safety Examiner*. Retrieved from http://www.examiner.com/x-39476-NY-Cyber-Safety-Examiner~y2010m3d25-Long-Island-teen-commits-suicide-Is-cyberbullying-to-blame

[68]US Department of Justice. (n.d.). Dru Sjodin National Sex Offender Public Website. Retrieved from http://www.nsopw.gov/Core/Conditions.aspx?AspxAutoDetectCookieSupport=1

[69]Florida Department of Law Enforcement. (n.d.) Florida Offender Alert System. Retrieved from http://www.nsopw.gov/Core/ResultDetails.aspx?index=3&x=0AD7C0B4-EA7E-41C9-AD1D-C0BDF0167096

[70]US Department of Justice. (2010, January). Amber alert timeline. Retrieved from http://www.ojp.usdoj.gov/newsroom/pdfs/amberchronology.pdf

[71]United States Attorney, Middle District of Florida. (2007, November 30). North Carolina man sentenced to 110 years for computer hacking and child pornography. US Department of Justice, Computer Crime & Intellectual Property Section. Retrieved from http://www.justice.gov/criminal/cybercrime/dickersonSent.pdf

[72]United States Attorney, Southern District of Ohio. (2006, January 19). Dayton man pleads guilty to sexual exploitation crimes involving minors. US Department of Justice, Computer Crime & Intellectual Property Section. Retrieved from http://www.justice.gov/criminal/cybercrime/millerPlea.htm

[73]Working Group on Unlawful Conduct on the Internet. (1999, August 5). Appendix C: Online child pornography, child luring, and related offenses. Retrieved from http://www.cybercrime.gov/append.htm

[74]United States Attorney's Office, Middle District of Florida. (2010, April 30). Brevard man pleads guilty to producing child pornography. Federal Bureau of Investigation Tampa. Retrieved from http://tampa.fbi.gov/dojpressrel/pressrel10/ta043010.htm

[75]495 U.S. 103 (1990).

[76]Wortley, R., & Smallbone, S. (2006, May). Child pornography on the Internet. US Department of Justice, Office of Community Oriented Policing Services, *Problem-Oriented Guides*, Series No. 41. Retrieved from http://www.cops.usdoj.gov/files/RIC/Publications/e04062000.pdf

[77]535 U.S. 234 (2002).

[78]United States Attorney, Southern District of New York. (2004, February 26). "Cyberscammer" sentenced to 30 months for using deceptive Internet names to mislead minors to X-rated sites. US Department of Justice, Computer Crime & Intellectual Property. Retrieved from http://www.justice.gov/criminal/cybercrime/zuccariniSent.htm

[79]The legitimate websites were www.bobthebuilder.com and www.teletubbies.com.

[80]Wortley, R., & Smallbone, S. (2006, May). Child pornography on the Internet. US Department of Justice, Office of Community Oriented Policing Services, *Problem-Oriented Guides,* Series No. 41. Retrieved from http://www.cops.usdoj.gov/files/RIC/Publications/e04062000.pdf

[81]*Ibid.*

[82]US Department of Justice. (2010, March 16). Wheaton nanny arrested for distribution of child pornography. Federal Bureau of Investigation Chicago. Retrieved from http://chicago.fbi.gov/pressrel/pressrel10/cg032610.htm

[83]No. 08-00582 (C.D. Cal. May 15, 2008).

Chapter 7

Where Is the Electronic Evidence and Which Tools Can We Use to Find It?

Before we look at how to manage crime scenes and incidents (a topic that we consider in Chapter 8), we first need to explore the various hardware and software computer components, electronic devices, and media that may contain electronic evidence. The first part of this chapter explores certain hardware and software computer components and electronic devices that might be the source of evidence for computer forensics investigations. The chapter then describes the tools required to search and collect electronic evidence.

The Location of Electronic Evidence

Evidence is most commonly found on hard drives. Data within the hard drives of computers consist of volatile and nonvolatile data. **Volatile data** disappear when the computer is powered off, whereas **nonvolatile data** are stored and preserved in the hard drive when the computer is powered off. Evidence in the hard drives of computers may be found in files created by the computer user (e.g., e-mails, spreadsheets, and calendars), files protected by the computer user (e.g., encrypted and password-protected files), files created by the computer (e.g., log files, hidden files, and backup files), and other data areas (e.g., metadata).

Files Created by Computer Users

Files created by the user include document (e.g., Word; file extensions of either ".doc" or ".docx"), text, spreadsheet (e.g., Excel), image, graphics, audio and video files. These files contain **metadata** (i.e., data about data). Metadata can provide the following kinds of information:

- The name of the author of the document and the company the document belongs to
- The owner of the computer
- The date and time the document was created
- The last time the document was saved and by whom it was saved
- Any revisions made to the document
- The date and time the document was last modified and accessed
- The last time and date the document was printed

Consider the embarrassment of the British government when the embedded metadata of a Word document on Iraqi intelligence that it published revealed that the document contained plagiarized text.[1] The identity of the BTK killer (Dennis Rader) was also revealed by embedded metadata in a deleted Word document included on a floppy disk that the suspect had sent to a TV station. Specifically, this information enabled law enforcement agents to trace the document back to a church to which Dennis Rader belonged. As these cases suggest, metadata can yield substantial evidence related to an incident or crime.

The files of the Windows operating system contain metadata that can be accessed in Windows by right-clicking on the icon of the desired file and selecting "Properties."[2] Investigators should keep in mind that whenever they access a file, they modify system metadata.

Timestamp data (i.e., the time of events recorded by computers) also may provide valuable information to an investigation. This was demonstrated in *Jackson v. Microsoft Corporation*,[3] where the timestamp data on confidential files in the defendant's possession provided evidence of intellectual property theft. Programs are available online, such as Timestomp, that try to delete or modify timestamp information. When changing timestamp information, a suspect may have one of the following aims:

- Validate a statement or testimony the suspect made
- Provide the suspect with an alibi
- Eliminate the person from consideration as a possible suspect

The use of Timestomp can frustrate computer forensics investigations if the offender has taken specific measures to conceal his or her use of this software. For instance, an investigator will be alerted to the use of this type of software if the suspect clears the timestamp of the entire system. Additionally, the dates the offender modifies must be believable and must not draw unwanted attention from authorities. For example, if an offender changes the timestamp of a Microsoft Office Publisher (".pub") document to 1980, the examiner would consider this document to be suspicious—".pub" documents were not available in 1980. The investigator would suspect that the offender had modified timestamp information, and would proceed to look at other documents to see if their timestamps had been modified as well.

A website called "Anti-forensics," which claims to render computer investigations irrelevant, warns users that if they use Timestomp, they should take a few steps to hide their use of this software. One such step is to rename the "timestomp.exe" file (which loads the program to a user's computer) as "RUNDLL32.exe," which is a legitimate file responsible for loading dynamic link library (DLL) files to memory and running them.[4] An investigator may still be able to find this file with a simple word search because it contains strings featuring the term "timestomp." Another recommendation made by the "Anti-forensics" website is to delete the Timestomp prefetcher (".pf") file, which is stored in a directory (c:\windows\prefetch directory) when an executable file is run in Windows.[5] Regardless of the efforts the suspect takes to conceal his or her use of this software, one thing remains certain: The use of Timestomp can significantly disrupt the forensic timeline that the investigator is seeking to establish.

An investigator should also check the computer to see if the suspect has created a calendar. Calendars may hold appointment information and other data that can reveal important clues about the suspect's whereabouts on a given date and time. It can also reveal the contacts of a suspect. Questioning these contacts may provide investigators with valuable information about the case.

Additionally, investigators should examine Web browsers for any files created by the user. In particular, they may look at particular websites that a user may have bookmarked or added to his or her favorites folder in the Web browser (different versions of Internet Explorer, Mozilla Firefox, Netscape Navigator, and Chrome, to name a few). **Figure 7–1** shows the Bookmarks menu in Mozilla Firefox; **Figure 7–2** shows the Favorites menu in Internet Explorer.

Evidence can also be retrieved from e-mail accounts. For instance, address books in e-mail accounts can include the contacts of the suspect. Other pertinent information to a criminal or civil case under investigation can be retrieved from e-mails in the inbox, sent, delete, draft, and spam folders of an account, which reveal the content of communications and the persons with whom the suspect was communicating. Of particular note are the draft and spam folders.

Members of al-Qaeda have reportedly used a technique known as an electronic or virtual "dead drop" to communicate.[6] This technique involves opening up an account, creating a message, saving this message as a draft, and then transmitting the account user name and password to the intended recipient (or recipients). The receiver of this information then logs on to the account and reads the message that was saved as a draft. This technique was used by the perpetrators of the Madrid train bombings in 2004, for example. These terrorists communicated with other members of their cell by saving messages for one another in the draft folders of preselected email accounts.[7] By using this

Figure 7–1 Depicts the location of bookmarks in the Mozilla Firefox web browser.

technique, terrorists avoid the possibility that their e-mails might be intercepted by eaves-droppers. Spam folders may also need to be checked: Al-Qaeda terrorists have been known to hide their messages in spam and then send these seemingly junk e-mails to their intended recipients.[8]

Evidence retrieved from e-mail investigations is explored in more depth in Chapter 10.

Files Protected by Computer Users

There are many different ways in which a user can protect his or her files:

- An individual can modify files or folders within the computer to look like something else.

- He or she can add a password to the file or folder and/or encrypt it to ensure that no one will be able to see what is in the file or folder.

- An individual can make the file or folder invisible.

These same tactics are also used by criminals to hide evidence of their crimes.

Figure 7–2 Depicts the location of the "Favorites" button in the Internet Explorer web browser.

Renamed Files and Files with Changed Extensions
Individuals can hide files in plain sight by renaming or changing the file extensions. As an example of the former technique, a file that contains sexually explicit material involving minors might be given an innocuous label such as "Thanksgiving" or "Mom's Birthday." Using the latter technique, a file extension might be changed by a criminal to hide files that contain incriminating evidence. For instance, a drug dealer who has a spreadsheet that lists clients, their drugs of choice, telephone numbers, payment for drugs, and any money owed by clients to the dealer might change the extension of the spreadsheet from ".xls" to ".jpeg". The drug dealer would probably change the extension of the spreadsheet to that of an image file because investigators looking for information concerning the dealer's sales would most probably look for it in spreadsheet and document files.

The change of file extension corrupts the file. Such a file may be viewed only in its original format. Investigators, however, can use forensics tools (which are explored later in this chapter) to reveal the changed file's original file type. As such, this is not a very effective way to hide data from a computer forensics investigator.

Deleted Files

Evidence can be found in files deleted by a computer user, although deleted files can typically be recovered by investigators. The first thing an investigator should do when searching for a deleted file is to check the Recycle Bin. When a file is deleted, it is moved to the Recycle Bin. More often than not, the files in the Recycle Bin will have been emptied (i.e., deleted) by the offender. When this occurs, any file that has been deleted from the Recycle Bin is removed from the file allocation table.

Depending on which Windows operating systems the investigator is dealing with, the file allocation table can be in the FAT, FAT32, or NFTS format. The first two file allocation table types—**FAT** and **FAT32**—are used in Windows 3.1, 95, 98, and ME editions. **NFTS**—more formally, New Technology File System—is used in the more recent versions of Windows (NT, 2000, XP, 2003, 2008, Vista, and 7). These file allocation tables keep track of and store the names and physical locations of every file on a computer hard drive.

Once a file is removed from the file allocation table, the space where the deleted file resided is marked as free space. In particular, when a file is deleted by the user, the operating system indicates that the space occupied by that file is now available for use by another file. However, the contents of the original file remain in that space until the space is overwritten with new data. Even if a file has been deleted and partially overwritten, it is still possible to recover the file fragment that has not been overwritten. Software is commercially available that allows the user to not only delete the selected files, but also write over all of the free space that is available. This action ensures that the space where the deleted information was will be used and, therefore, that space will not contain any old information.

When a disk is wiped with a single, nonrandom bit pattern, it is possible that data may still be found by a computer forensics investigator using specialized tools. U.S. Department of Defense standards hold that files and hard drives need to be overwritten at least seven times with random binary bit patterns to ensure that the original data are completely deleted. Many software programs are available online that claim to permanently delete files and other data on hard drives and related storage devices (e.g., CDs, DVDs, flash drives). Among them are data shredders, which overwrite files on a hard drive or other storage device with a random series of binary data. This tactic ensures that the data stored on these drives or devices are obliterated beyond recognition.

Encrypted Files

To protect his or her files, an individual may use **encryption** to physically block third-party access to them, either by using a password or by rendering the file or aspects of the file unusable. How does encryption render a file unusable? Encryption basically scrambles the data and makes it unreadable. It does so by transforming plaintext into ciphertext, which is essentially gibberish. A decryption key is required to transform the ciphertext back into plaintext.

Individuals have used encryption to protect their privacy and secure the data in their computer and related electronic devices. Encryption also gives criminals a powerful tool with which to conceal their illegal activities. In 2000, U.S. Director of Central Intelligence George Tenet testified that terrorist groups such as Hezbollah, Hamas, and al-Qaeda routinely use encryption to conceal files and communicate undetected.[9] In respect to al-Qaeda, Tenet noted that convicted terrorist Ramzi Yousef, the mastermind of the bombing of the World Trade Center in 1993, "had stored detailed plans to destroy United States airliners on encrypted files on his laptop computer."[10]

The United Kingdom has a law that grants law enforcement agencies the power to compel suspects to provide their decryption keys. Specifically, consider Part III of the **U.K. Regulation of Investigatory Powers Act of 2000** (RIPA). Section 49 (notices allowing disclosure) provides agencies with the power to monitor individuals' e-mails by allowing law enforcement authorities to serve written notice to demand that either an encryption key is handed over or a communication is decrypted when such action is deemed to be in the interests of national security, for the purpose of preventing or detecting crime, and in the interests of the economic well-being of the United Kingdom. As of October 1, 2007, Part III of RIPA was activated; now individuals using encryption technology can no longer refuse to reveal keys to U.K. law enforcement agencies if served with a notice. Failure to hand over encryption keys upon request is an offense, even if the material is completely legal. Section 53 of RIPA, which deals with failure to comply with notices to hand over encryption keys, is particularly problematic. It assumes the guilt of the accused by potentially criminalizing individuals with poor memories or reversing the burden of proof in the case of those who claimed to have forgotten (which can easily occur because these keys are normally long passwords) or lost keys to their data.[11]

The United Kingdom is not alone in demanding access to encrypted files. France has a similar law in effect, the Daily Safety Law (*Loi sur la sécurité quotidienne*), which provides the government with access to private encryption keys, restricts import and export of encryption software, and imposes strict sanctions for using cryptographic techniques when committing a crime.[12]

By contrast, the United States does not have a law that grants law enforcement agencies the power to compel suspects to provide their decryption keys. In fact, in *In re Grand Jury Subpoena to Sebastien Boucher* (hereafter *In re Boucher*), the court ruled that a hard drive could not be accessed because it was protected by **Pretty Good Privacy** (PGP).[13] PGP is software for encrypting files and messages, which provides cryptographic privacy and authentication for users, but also hampers law enforcement agencies' ability to access the content of files or communications. The *In re Boucher* ruling stated that compelling an individual to provide a password to an encrypted hard drive violated that individual's privilege against self-incrimination under the Fifth Amendment to the U.S. Constitution. An appellate court, however, reversed this decision on different grounds.

In the *In re Boucher* case, border agents had reviewed some of the contents of the defendant's encrypted Z drive that the defendant enabled them to see, which contained child pornography.[14] In *Fisher v. United States*,[15] the court had ruled that if the information the government seeks to compel the defendant to provide adds "little or nothing to the sum total of the Government's information" on the defendant, then no constitutional rights of the defendant are touched. In line with this argument, the appellate court in the *In re Boucher* case reasoned that providing law enforcement agents with access to the Z drive "adds little or nothing to the sum total of the Government's information" about the existence and location of potentially incriminating files in the drive. Boucher was thus ordered to provide an unencrypted version of the Z drive to authorities.[16]

As matters now stand, the courts have not taken an affirmative stand on whether to grant law enforcement agents unrestricted access to encrypted files. So far, they have allowed this type of access only when the defendant has provided agents with information about the contents of the encrypted files or hard drive during questioning or border searches. If this information has not been disclosed by the defendant, then the agents cannot have access to the encrypted files. The suspect may decide to provide law enforcement agents with the password to these files. However, investigators should take care lest the offender use this opportunity to destroy evidence rather than provide authorities with it.

Some publishers of encryption programs leave backdoors in their software through which to enter their systems. If so, investigators might be able to contact the publishers to gain access to the files protected by these programs. Indeed, encryption programs contain key escrows, which enable the recovery of data in the event that a user forgets his or her passphrase.[17] When issues of national security arise, these keys must be provided to government agencies.

Password cracking software should be used cautiously, as an individual may have set up the computer to recognize unauthorized access to the system (e.g., repeated failed attempts to access the system) and created a "booby trap" to delete the data that the investigator is trying to retrieve. The suspect may have also created a kill switch—in the form of a command or the pressing of a particular key on the mouse or keyboard—to erase the data on the hard drive.

Hidden Data: Steganography

Something that needs to be protected from prying eyes can be "camouflaged in sound, pictures or other routine content in ways analogous to hiding a pebble on a shingle beach."[18] This technique is known as **steganography** (information hiding). Steganography seeks to make data and messages invisible by hiding them in various files. For example, individuals who collect and distribute child pornography may hide child pornography images and videos in files to avoid being caught by law enforcement agents. Using steganography, these individuals may store child pornography for their own personal use on their computers and send images or other files that have child pornography images or videos hidden in them to others via emails, chat rooms, discussion boards, social networking sites, peer-to-peer networks, blogs, and websites.

To determine if an image contains steganography, an investigator usually makes a visual, side-by-side comparison of the original image and the processed image to identify any differences between them. Unfortunately, computer forensics investigators may not have this luxury; they often have only the processed image available, making visual detection of steganography extremely difficult. To find it, investigators would basically need to know what they are looking for and in which type of file it is possibly hidden. With steganography, only those individuals with the appropriate software can see the hidden information. Nevertheless, certain tools exist that can assist an investigator in determining whether steganography has been used. Specifically, investigators can use steganalysis to detect if steganography was used on a particular file—that is, to determine whether hidden data exist in the file.

Steganography software is sold commercially and can be downloaded for free over the Internet. Even the most inexperienced user can learn how to hide data using steganography. Free online tutorials that show individuals how to hide data using this technique are readily available from YouTube and websites that offer steganography software.

Files Created by the Computer

Files that are created by the computer may also have evidentiary value. Files that may assist a computer forensics specialist in his or her investigation include event logs, history files, cookies, temporary files, and spooler files.

Event Logs

Event logs automatically record events that occur within a computer to provide an audit trail that can be used to monitor, understand, and diagnose activities and problems within the system.

To find the event logs in a Windows operating system, for Windows 7,[19] click on the Start menu on the lower-left corner of the desktop and select "Control Panel." Once in the "Control Panel" page, click on "System and Security." On this page, click on "Administrative Tools" (**Figure 7–3**).

Next, select "View event logs" from the right side of the screen (see Figure 7–3). In the next screen (**Figure 7–4**), click on "Windows logs" on the upper-left corner of the screen.

Several event logs are now displayed on the screen (**Figure 7–5**), including the following:

- **Application logs.** These logs contain the events that are logged by programs and applications. Errors of these applications and programs are also recorded in this log.

- **Security logs.** These logs record all log-in attempts (both valid and invalid) and the creation, opening, or deletion of files, programs, or other objects by a computer user.

Figure 7–3

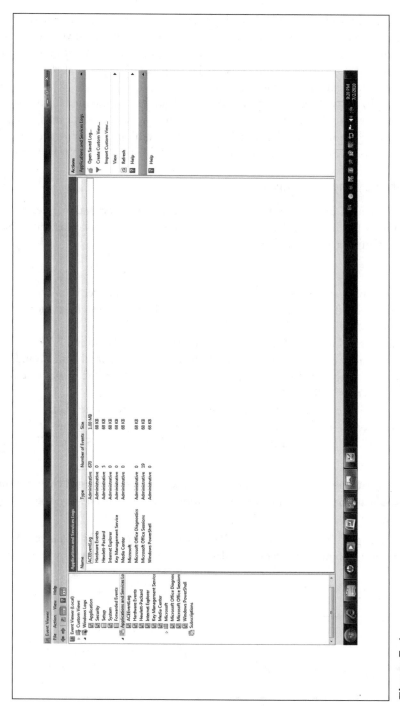

Figure 7–4

Figure 7–5

- **Setup logs.** These logs provide data on applications that are installed on a computer.
- **System logs.** These logs provide information on Windows system components. For example, they record any failure of a component to load during the startup process.
- **Applications and services logs.** These are new event logs in Windows 7. Instead of recording events that may affect the system as a whole, each log stores events from a single application or component.

Critical information and evidence of a crime may be lost if the logs are full. Depending on your settings, either new events will be added by overwriting older events or newer events will not be added at all. To fix this problem, the size of the logs should be adjusted (i.e., increased) to allow more events to be logged, the logs should be manually emptied, or the logs should be set to overwrite older events automatically.

The most important event log of those mentioned previously is the security log, which records all log-in attempts and activities of the computer user. As such, this log can indicate that malicious activity or other forms of cybercrime have been or are being committed. For instance, numerous failed log-in attempts in the security log may indicate that someone is trying to access the computer without authorization. Moreover, this log can reveal a suspect's attempt to delete data from the computer.

History Files

The operating system also collects data about the websites visited by a user. In *United States v. Tucker*,[20] computer forensics investigators found important electronic evidence of the crime—namely, deleted Internet cache files showing that Tucker had visited child pornography websites—on the suspect's hard drive.

Internet cache files have also served as important evidence in murder cases. For example, the Web searches of Melanie McGuire provided critical evidence about the brutal murder of her husband, William.[21] McGuire shot her husband, drained his body of blood, dismembered him, placed his body parts in various garbage bags and suitcases, and later disposed of them in the Chesapeake Bay.[22] When investigators searched McGuire's home computer, the Internet search history showed that she had typed into the Google search bar, among other things, "undetected poisons," "how to commit murder," and "how to buy a gun in Pennsylvania." The results of her Internet search for weapons led investigators to the Pennsylvania store where McGuire had purchased the weapon with which she killed her husband. Investigators also found an e-mail McGuire had sent to a friend requesting information on how to buy a gun quickly. Based on the information retrieved from McGuire's home computer and the leads it produced, this defendant was found guilty of first-degree murder and other charges. Clearly, Internet history can provide invaluable information to an investigation.

The toolbars of most Web browsers save the browsing history of the computer user (**Figure 7–6** and **Figure 7–7**). While the majority of cybercriminals erase their browser history, it is important to check this location in case its contents have been overlooked by the offender. The address bar of a Web browser should also be checked, as it is often overlooked by offenders. This area does not provide information on all websites viewed, but only those whose addresses were explicitly typed or copied and pasted into the address bar by the user.

Most online chat room software temporarily stores chat session logs. It also affords users with the opportunity to permanently save logs of chat sessions. The default settings of certain chat room software (e.g., Yahoo! Messenger) are set to temporarily save messages until the user signs out of the application (**Figure 7–8**).

Users may actually set this software to save all of the messages sent or received. With these settings, the logs can provide details of the discussions the suspect had. Using these forums, individuals can forward calls; forward instant messages to their **mobile phones**; make phone calls to cell phones, landlines, or other computers; send text messages; engage in live conference calls; archive messages; and receive and save voice mails, videos, photos, and other files. Accordingly, chat room software provides a wealth of information of potential value to investigations.

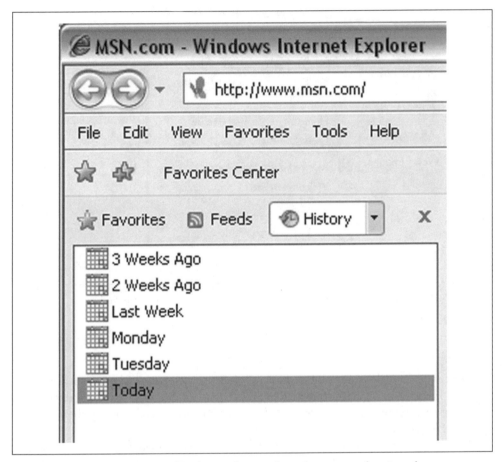

Figure 7-6 Area in Internet Explorer where website history can be viewed.

Cookies

Cookies are files created by websites that are stored on a user's computer hard drive when he or she visits that particular website. As such, by viewing cookies, the investigator can determine which websites the user has visited. Certain cookies are used by websites to gather information about an individual's activities, interests, and preferences. Others are used to store credit card information, user names, and passwords. Some cookies do both. The type of information an investigator finds depends on the cookies stored on the suspect's computer.

Temporary Files

Some files are created by the computer unbeknownst to the user. Specifically, the operating system collects and hides certain information from the user. One example of this

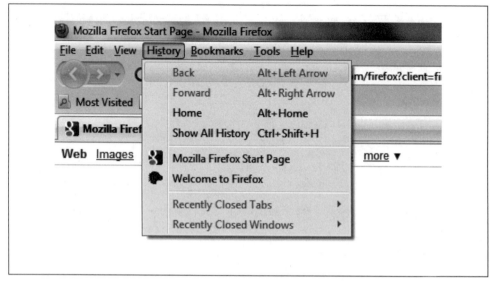

Figure 7–7   Area in Mozilla Firefox where website history can be viewed.

kind of temporary file is unsaved documents. In the case of the "Gap-Toothed Bandit," who was involved in 12 bank robberies in San Diego in 1999, unsaved Word documents including threatening demand notes used for the robberies were saved by the operating system of his computer in a temporary location.[23]

The computer also stores information about websites browsed, items searched online, user names, and passwords. This material is stored in temporary Internet files or cache. To delete this information, click on the "Start" button, and then select "Control Panel." At the upper-right corner of the screen, click on "Category" and choose "Large Icons." Then, in the "Control Panel" screen, choose "Internet Options." When the page shown in **Figure 7–9** appears on your screen, under "Browsing history," click the "Delete" button. You can then choose to delete cookies, temporary Internet files, Web browser history, items searched online, and passwords (**Figure 7–10**).

An investigator should check the temporary files because criminals may forget to delete the information that the computer stores. Some are not even aware that this information is stored. Other criminals take additional steps to delete these data (i.e., beyond those described previously and depicted in Figures 7-9 and 7-10). In particular, they may use software to delete browser cache, cookies, and other files. One such software package, Evidence Shredder Pro, claims to permanently delete this information and even provides the user with a panic button, which will close all browser windows and wipe the computer

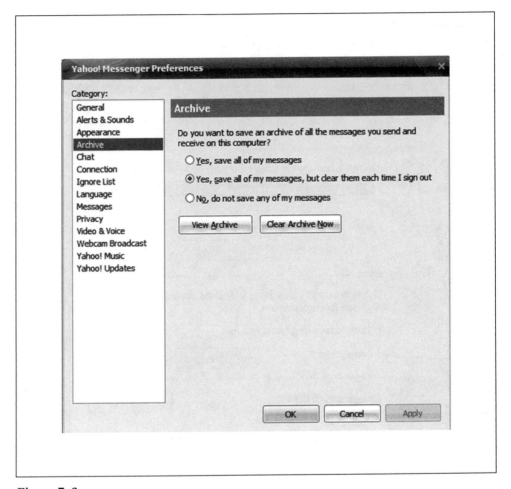

Figure 7–8

Source: Reproduced with permission of Yahoo! Inc. ©2010 Yahoo! Inc. YAHOO! and the YAHOO! logo are registered trademarks of Yahoo! Inc.

if the user clicks on it.[24] Others, such as Fixcleaner and CCleaner, offer to delete browser history, temporary files, and activity logs at the user's command.[25]

Spooler Files
As a default setting, most Microsoft Windows operating systems have print jobs "spool" to the hard drive before they are sent to the **printer**. Accordingly, a copy of the printed item is stored on the hard drive of the computer. This copy can be recovered and could provide vital evidence in the case under investigation.

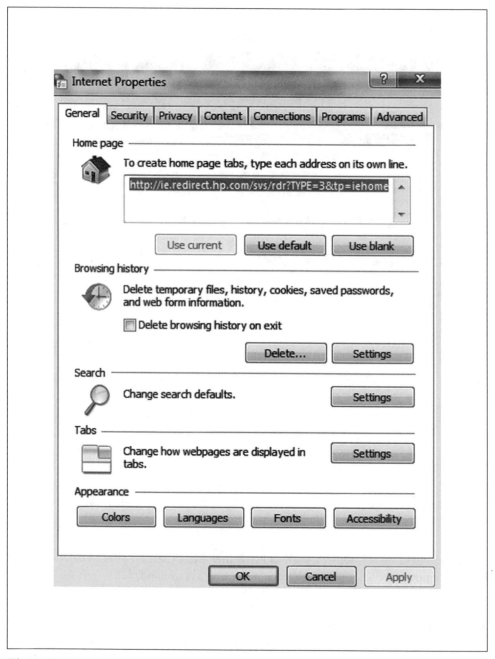

Figure 7–9

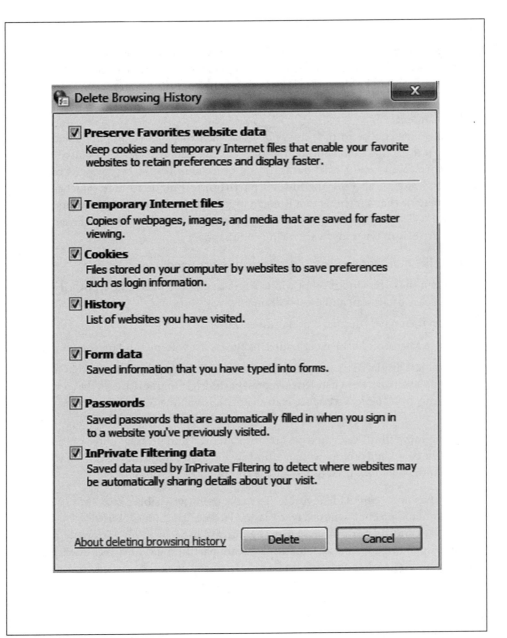

Figure 7–10

Other Data Areas on Computers

Unallocated space—space available because it was never used or because the information in it was deleted—may also contain important evidence of a crime or incident. Evidence may also be found in hidden partitions, bad clusters, and slack space.

Hidden Partitions

Individuals may choose to hide drives or files when they share computers with others, especially if these files hold confidential and sensitive data (e.g., Social Security numbers, bank information, credit card data). It is a quite simple and easy way to hide data. Criminals, however, also use the **hidden partition** technique to hide evidence of their crimes. Imagine that a member of a Russian organized crime group, Ivan, wants to hide a hard drive "G", which contains evidence of the group's money laundering transactions. To hide the drive, Ivan types in the following commands:

1. In the "Run" box on the "Start" menu, he types "cmd".

2. When the command prompt window appears on the screen, he types "cd \" (i.e., cd space backslash) and presses "Enter."

3. Ivan then types "diskpart" and presses "Enter."

4. Once Microsoft DiskPart is loaded, he types "list volume" and presses "Enter."

5. A table is displayed that contains several items, among them columns that include volumes and letters, where letters represent the letters of the drives in the computer. In this hypothetical scenario, Ivan's G drive is volume 3. Ivan then types "select volume 3" and presses "Enter."

6. To remove the drive, Ivan types "remove letter G" and presses "Enter." According to the command prompt window, the drive has been successfully removed (at times, an individual may need to reboot the computer to see this effect).

The drive has now been made invisible. To make the drive visible again, Ivan needs to go to the command prompt window, type "assign letter g", and press "Enter."

This data hiding technique can make an investigator's job extremely difficult. Windows operating systems automatically create partition gaps between each partition. Therefore, incriminating evidence can be hidden in any one of these gaps. To find the evidence, an investigator needs to search each of these gaps.

Slack Space and Bad Clusters

Certain programs (e.g., Slacker) exist that can help users hide files from computer forensics investigators in slack space. Specifically, the program breaks up the file that a user wants hidden and places parts of that file into the slack space of other files.[26]

Other areas of the computers that investigators should look at are clusters—that is, areas of the operating system where data are stored. In particular, **bad clusters**, which are not accessed and thus overlooked by the operating systems, should be examined.[27] Criminals have been known to mark clusters as bad and hide data within them.

Peripheral Devices

Peripheral devices are devices that are not essential parts of a computer system, such as **scanners**, copiers, printers, and **fax machines**. Such devices can contain valuable information about the case being investigated. Suppose the crime being investigated is child pornography, and images of child pornography have been found on the suspect's computer. These digital images could have been generated from a variety of sources. It is possible for investigators to determine the particular device that generated the image and the make and model of the device.

For instance, the scanner may leave potential markings on scanned items that may link pictures or documents to the particular scanner. Specks or marks on the scanned item may result from dirt or scratches on the glass window where the original document was placed in and scanned. Copied documents may also be linked to a particular copy machine. Irregularly shaped characters on a copied document may be the result of imperfections on the drum in the copier that is forming the image. Similarly to scanners, if the glass window where the document is placed in is dirty or has certain imperfections, the copied document will have distinguishing markings on it that can link it to the copier.

Printers may contain logs of printer use, time and date information for printed items, and a memory card. Investigators can also determine whether a particular printer generated the image in question. For instance, banding imperfections, where some areas of a printed document look lighter or darker than other areas, constitute one way in which investigators can match a particular document to a specific printer.[28] These imperfections may or may not be visible to the naked eye. Additionally, a dot encoding mechanism, which is invisible to the naked eye, exists in many laser printers.[29] This mechanism is used by some printer companies to encode the serial number and the manufacturing code into every document that is processed by their color laser printers. A person can determine if such a mechanism exists in his or her own color laser printer by flashing a blue LED light on a printed page from the printer and viewing it under a magnifying glass.[30]

For convenience and to save space in an office or home, many individuals have bought multifunctional ("all-in-one") machines. These machines typically contain printer, scanner, and copier capabilities. In addition, these multifunction machines usually contain a hard disk that captures all of the records and images processed by the machine. This disk is of great forensics value—as long as the user or owner of the machine has not prevented the recording of such data, either by setting it not to store the data or by deleting the contents of the disk. Some devices offer users the opportunity to encrypt the data processed by the device with the installation of specific software programs.

Sometimes, multifunction machines may have fax capabilities. Fax machines may also hold important evidence. Types of data that they retain include incoming and outgoing fax numbers, documents sent and received, and transmission logs.

Evidence may also be found on peripheral storage devices such as CDs, DVDs, backup tapes, external hard drives, thumb or flash drives, and Zip drives.

Telecommunications Devices

Evidence can be retrieved from devices other than computers. **Telecommunications devices** where evidence can be found include fixed telephony, mobile phones, and answering machines.

The following types of evidence are available from **fixed telephony**:

- Calls made, received, and missed
- Voice mails
- Messages
- Favorite numbers

In mobile phones, the following types of evidence may be found:

- Names and numbers of contacts
- Calls made, received, and missed
- Date, time, and duration of calls
- Text messages—that is, **Short Message Service (SMS)** data
- Messages with a combination of text, images, videos, and sound—that is, **Multimedia Messaging Service (MMS)** data

Nowadays, mobile phones have vast storage capacities and can hold even more information than that listed above that can be used by investigators. In particular, mobile phones may be able to send e-mails, take photographs, download music, send instant messages, record and play videos, open application files (e.g., documents, spreadsheets, and presentations), and browse the Internet. Mobile phones may even store **global positioning system (GPS)** coordinates when photographs are taken, along with the time and date when the photo was created. Additionally, mobile phones may contain GPS navigation systems. Thus an investigator can pull up the GPS history and any addresses programmed into the GPS and determine which places an offender visited. Moreover, some mobile phones can link to work and home computers, thereby providing investigators with access to even more potential evidence. Further information about mobile phones and the evidence retrieved from them is provided in Chapter 12.

Finally, **answering machines** can contain, among other things, recorded voice messages (current or deleted), missed calls, caller identification information, and the last number called or dialed on the device.

Handheld Computing and Wireless Devices

Examples of handheld computing and wireless devices include **pagers** and **personal digital assistants** (PDAs). Extra care must be taken when seeking evidence from these devices because these devices lose their evidentiary value if power is lost. The information on these devices is also easily destroyed. For instance, incoming messages can delete a pager's stored information.

Pagers may contain messages that are of interest to the investigator seeking forensic evidence. Many different kinds of pagers are available on the market, and the type of pager will determine the data that may be retrieved from it:

- A tone-only pager alerts the user that an individual has tried to contact him or her. To hear a message that an individual may have left for the user, the recipient must contact the paging service.

- Numeric pagers, as the name implies, provide data in the form of a numeric code or telephone number.

- Alphanumeric pagers can handle both text and numeric messages.

- Voice pagers actually transmit the voice messages directly to the user.

Pagers have largely fallen out of favor today, and many companies have discontinued making them. However, they are still used by some people (e.g., emergency services and medical personnel).

The procedure for handling a PDA is similar to that for handling a pager. PDAs may contain evidence of a crime or incident in its documents. Previously, these devices were limited to a single function—acting as a personal information organizer. These days PDAs can be used not only as organizers but also, among other things, to browse the Internet and send and receive text messages and e-mails; all of these actions may produce information that is pertinent to an investigation. PDA investigations are explored in further detail in Chapter 12.

Miscellaneous Electronic Devices

Another type of electronic device that may be of interest to computer forensics investigators is the **digital camera**. Evidence of images, sounds, and date and timestamps may be retrieved from the memory cards of digital cameras. Digital cameras contain a wealth of metadata in Exchangeable Image File Format (EXIF). EXIF can provide the following kinds of information:

- Date and time when a picture was taken (assuming that this capability has been set properly by the user)

- Make and model of the camera used

- Latitude, longitude, altitude, and Universal Time Coordinates (UTC) of the location where the picture was taken

The location information is obtained from the GPS capability that many cameras support. If the camera was registered after its purchase, then the manufacturer (e.g., Panasonic, Nikon, Olympus, Canon) also could have information about the consumer who purchased the camera in question.

Other Devices?

Conduct independent research and answer the following questions:

1. Which other devices might contain evidence of the crime?
2. Which types of evidence might they contain?

Tools Used to Search and Collect Electronic Evidence

Specialized tool kits are required for computer forensics investigations. A computer forensics investigator must be equipped with the appropriate kits to collect, store, preserve, and transport forensic evidence. The tools the investigator uses will depend on the operating system (e.g., Windows or BlackBerry) and type of electronic device (e.g., computer or mobile phone) to be examined. Choosing the right tools with which to examine computer system components and electronic devices for evidence is extremely important.

Before examining the evidence with the chosen forensics tool, an investigator must ask himself or herself if the chosen software is appropriate for the computer system or electronic device in question. The software used for this purpose (e.g., Encase) must be sound. To be forensically sound, a computer forensics tool must not modify the data on a computer when it is used. A **write blocker** device is used to prevent anything from being written to the hard drive or other data source. Specifically, this device blocks the wires that would communicate the data to be written to the drive.

The National Institute of Standards and Technology (NIST) has established certain requirements for computer forensics imaging tools. **Imaging** is the process by which a duplicate copy of the entire hard drive is created. Once an exact copy of a suspect's hard drive has been made, the investigator must verify that it is an exact copy. An investigator verifies this fact by computing an **MD5 hash algorithm** for both the original hard drive and the copy. If the MD5 values are exactly the same, then the copy is an exact copy of the original drive. Hashing using the MD5 or **SHA** hash algorithm has a standard of certainty even higher than that of DNA evidence. Indeed, it has been validated by many courts. For example, in *State v. Morris*,[31] prosecutors not only validated the MD5 hash process but also validated the forensic imaging process.

Once a write blocker has been set up to prevent any data from being written to a suspect's hard drive, the computer forensics investigator clones or copies the drive by writing each and every part of the drive to a blank hard drive. That is, the investigator will make a bitstream copy of the hard drive. The process by which a duplicate copy of the entire hard drive is created is known as imaging. With imaging, all data on the hard drive are copied, including metadata, system-created files, and deleted files. Sometimes, an investigator might only create a copy of an individual file; at other times, he or she might also print a hard copy of particular files. While these methods may be quick and simple, they may result in a substantial loss of information (e.g., metadata).

The integrity of an investigation is ensured by imaging an exact copy of the hard drive or other media using the proper software. In fact, in *Gates Rubber Co. v. Bando Chemical Industries, Ltd.*,[32] the court held that creating a mirror image copy of the hard drive is considered as the most complete and accurate method for processing evidence. When an investigator creates a duplicate electronic copy of the entire storage device, it prevents changes or damages to the original hard drive. By contrast, if a computer forensics investigator copies individual files on the hard drive, data on the hard drive may be unintentionally altered or destroyed. For instance, opening a file in the computer changes the time and date stamp indicating when the file was last accessed.

According to NIST, the computer forensics tools that an investigator uses must meet the following requirements:[33]

- Make a bitstream duplicate or an image of an original disk or partition
- Not alter the original disk in any way
- Log input and output errors and offer a resolution to fix such errors
- Keep correct documentation

These standards apply to software tools that copy or image hard drives. NIST has not used these standards to test other media such as mobile phones and PDAs.

When considering the use of computer forensics tools, two important questions need to be considered:

- Should examiners possess significant experience to use computer forensics tools?
- Alternatively, are computer forensics tools "idiot-proof," such that they will provide accurate information regardless of who uses them?

While some forensics tools do not require specialized training for their use, the investigator needs at least some basic level of computer expertise or at least some practice with the tool to understand how it works. As such, computer forensics investigators must have knowledge of forensics hardware and software tools and investigative methods using these tools.

The computer forensics tool that is chosen must have been successfully used in court cases. Forensics tools that meet this criterion include FTK, Encase, ILook, and e-Fense Helix and Live Response. Of these tools, the most commonly employed are FTK and Encase. In fact, the court verified this contention in *United States v. Gaynor*,[34] when it explicitly stated that FTK and Encase are the most widely used tools by computer forensics investigators.

Forensic Toolkit

Forensic Toolkit (**FTK**) is a product sold by AccessData. This software has many capabilities, including the ability to create images of hard drives, analyze the registry, scan slack space for file fragments, inspect e-mails, and identify steganography. Unlike other computer forensics tools on the market, FTK can crack passwords. This tool can also be used to decrypt files. Indeed, FTK was used to decrypt files seized from a safe haven of a Bolivian terrorist organization that had assassinated four U.S. Marines.[35] Furthermore, this tool is quite beneficial because even if a computer crashes while using FTK software, the information will not be lost. FTK has been widely recognized by U.S. courts as a valid computer forensics tool.[36]

Encase

Encase is a computer forensics tool that is widely used by law enforcement agencies. It allows the user to create an image of a drive without altering its contents and calculates the hash value for further authentication. It can locate hidden drives or partitions within a drive, as well as other hidden files or media that some other programs would not be able to discover. Encase can search multiple file locations and devices simultaneously. In doing so, it creates an index of what is found on the computer, such as e-mails and deleted files. Overall, it is an incredibly useful tool for investigations and for law enforcement personnel.

Encase has been used by law enforcement agencies worldwide. A case in point is that of the terrorists who bombed the Indian Parliament on December 13, 2001. Investigators in India used Encase software to find evidence that the laptop of a terrorist suspect, Mohammed Afzal, was used to make the fake identification cards that were found on the dead bodies of the terrorists responsible for the bombings.[37]

The use of Encase by computer forensics investigators has been validated by U.S. courts as well.[38] For example, consider *State v. Cook*.[39] In this case, the defendant was convicted of possessing and viewing child pornography materials. Cook attempted to appeal the decision on the grounds that the forensics tool used to obtain the evidence against him (Encase) was neither valid nor reliable. The court disagreed, and upheld the validity and reliability of Encase.

ILook

ILook is another tool that is used to forensically examine computer media. Its capabilities include imaging, advanced e-mail analysis, and data salvaging (to recover files that have been deleted by the user).[40] This tool is used by the Criminal Investigation Division (CID) of the Internal Revenue Service (IRS), U.S. Department of Treasury. It is not available to the general public, but rather is provided to law enforcement agencies, government intelligence agencies, military agencies, and government, state, and other regulatory agencies with law enforcement missions.

E-fense Helix and Live Response

E-fense offers cybersecurity and computer forensics software such as Helix3Pro and Live Response. **Helix3Pro** software can be used on multiple operating systems (Windows,

---

Other Tools?

Numerous computer forensics tools are available on the market. Search online and find a computer forensics tool that is not mentioned in this chapter. After a forensics tool has been chosen, answer the following questions:

1. Is this tool forensically sound?

2. Why or why not?

Macintosh, and Linux). This tool is carefully designed to ensure that data are not altered during the imaging process. Local, state, and federal law enforcement agencies, along with private practitioners, have used this computer forensics tool. **Live Response** is a **Universal Serial Bus (USB)** key that is designed to be used by first responders, investigators, information technology professionals, and security professionals to collect non-volatile and volatile data (which will be lost if the computer is shut down) from live running systems.[41]

Chapter Summary

This chapter explored evidence that may be contained in electronic devices and the tools used to collect this evidence. Electronic evidence can be found in computer systems, peripheral devices (e.g., printers, scanners, copiers, and fax machines), telecommunications devices (e.g., fixed telephony and mobile phones), handheld computing and wireless devices (e.g., PDAs and pagers), and other miscellaneous devices (e.g., digital cameras). Many hurdles arise when investigators try to extract electronic evidence, especially when criminals have employed measures to conceal evidence of their unlawful activities.

This chapter also covered the process of extracting electronic evidence from computers and the problems that investigators run into when doing so. When investigators need to examine a computer hard drive for evidence, the original disk is not used. Instead, a write blocker is used to prevent any data from being written to the drive, and then a forensics tool is used to capture a bit-by-bit image of the drive. To show the validity of the image, the original drive is hashed and then the image is hashed. If the hash values match, the copy is considered authentic.

Several computer forensics tools are commercially available that can acquire both volatile and nonvolatile data. The validity of some of these tools, such as Encase and FTK, has been upheld by U.S. courts. Some experience is required to use these tools, and the type of tool that is ultimately used depends on capabilities of the software and the operating system of the computer or electronic device that will be examined for evidence.

Key Terms

Answering machine
Application log
Applications and services log
Bad clusters
Cookie
Digital camera
Encase
Encryption
Event log
FAT
FAT32
Fax machine
Fixed telephony
FTK
Global positioning system (GPS)
Helix3Pro
Hidden partition
ILook
Imaging
Live Response
MD5 hash algorithm
Metadata
Mobile phone

Multimedia Messaging Service (MMS)
NFTS
Nonvolatile data
Pager
Peripheral devices
Personal digital assistant (PDA)
Pretty Good Privacy
Printer
Scanner
Security log
Setup log
SHA hash algorithm
Short Message Service (SMS)
Steganography
Systems log
Telecommunications devices
U.K. Regulation of Investigatory Powers
    Act of 2000
Unallocated space
Universal Serial Bus (USB)
Volatile data
Write blocker

Practical Exercises

1. Locate a court case that either challenges a computer forensics tool or establishes its validity. Summarize the facts in this case, and briefly explain why the defendant challenged the tool used in an investigation or how the prosecution established its validity.

2. Search online for a tool or technique that criminals can use to hide evidence of their activities. Provide one example here. Include in your answer how the technique or tool works and how it makes an investigator's job more challenging.

Critical Thinking Question

Are deleted files ever really gone?

Review Questions

1. In which devices can electronic evidence be found?

2. List five types of files that may be created by the computer user.

3. What are metadata? Why is this type of data important?

4. What are event logs? Why are they important?

5. What are volatile data?

6. What is encryption?

7. What is steganography?

8. Name five types of files that are created by the computer.

9. What is imaging?

10. What is a write blocker?

11. How can an investigator validate that he or she made an exact copy of a hard drive?

12. Which computer forensics tools are widely used in the United States?

Footnotes

[1] For further information, see http://www.casi.org.uk/ discuss/2003/msg00457.html

[2] This is the easiest way to access the file metadata of the majority of the versions of Windows. There are other ways to access "Properties." For instance, with ".doc" files, you can click on "File," then "Properties" to view metadata.

[3] 211 F.R.D. 423 (W.D. Wash. 2002).

[4] Modify NTFS timestamps and cover your tracks with Timestomp.exe. (2009, March 5). Anti-Forensics. Retrieved from http://www.anti-forensics.com/modify-ntfs-timestamps-and-cover-your-tracks-with-timestomp

[5] *Ibid.* Note that the prefetcher directory can also alert users to malicious software that masks itself as legitimate ".pf" files. An example of this kind of malicious software is the "i.explore.exe.pf", which masks itself as the legitimate "IEXPLORE.EXE" (which is always in capital letters).

[6] Coll, S., & Glasser, S. B. (2005, August 7). Terrorists turn to the Web as base of operations. *Washington Post* [online], p. A01. Retrieved from http://www.washingtonpost.com/wp-dyn/content/article/2005/08/05/AR2005080501138_pf.html

[7] McLean, R. (2006, April 28). Madrid suspects tied to e-mail ruse. *International Herald Tribune* [online]. Retrieved from http://www.iht.com/articles/2006/04/27/news/spain.php

[8] See, for example, spammimic.com, a website that offers such tools. Thomas, T. L. (2003). Al Qaeda and the Internet: The danger of "cyberplanning." *Parameters*, p. 115.

[9] Hezbollah is a Lebanon-based terrorist group that was founded in 1982. Hamas is a Palestinian terrorist group that was founded in 1987. Freeh, L. J. (2000, March 28). Statement for the record of . . . the Director of the Federal Bureau of Investigation on cybercrime before the Senate Committee on Judiciary Subcommittee for the Technology, Terrorism, and Government Information in Washington, D.C. U.S. Congress. Retrieved from http://www.cybercrime.gov/freeh328.htm

[10]A truck laden with explosives was driven by Ramzi Yousef and Eyad Ismoil into the garage under the North Tower of the World Trade Center in New York City on February 26, 1993. After igniting the fuse, they fled. Six individuals died and more than 1000 people were injured. Freeh, L. J. (2000, March 28). Statement for the record of . . . the Director of the Federal Bureau of Investigation on cybercrime before the Senate Committee on Judiciary Subcommittee for the Technology, Terrorism, and Government Information in Washington, D.C. U.S. Congress. Retrieved from http://www.cybercrime.gov/freeh328.htm

[11]House of Lords, U.K. Parliament. (2007, July 17). Hansard vol. 694 col. GC5; Donohue, L. K. (2006). Criminal law: Anglo-American privacy and surveillance. *Journal of Criminal Law and Criminology*, *96*(3), 1180.

[12]See Loi n° 2001–1062 du 15 novembre 2001 relative à la sécurité quotidienne; Privacy International. (2007, December 18). PHR 2006: French Republic. Retrieved from http://www .privacyinternational.org/article.shtml?cmd[347]=x-347-559537

[13]2007 WL 4246473 (D. Vt. Nov. 29, 2007).

[14]*In re Grand Jury Subpoena to Sebastien Boucher*, 2007 WL 4246473 (D. Vt. Nov. 29, 2007), rev'd, 2009 WL 424718 (D. Vt. Feb. 19, 2009).

[15]425 U.S. 391, 408 (1976).

[16]*In re Grand Jury Subpoena to Sebastien Boucher*, 2009 WL 424718 (D. Vt. Feb. 19, 2009).

[17]Electronic Privacy Information Center. (1998, April 14). Key escrow. Retrieved from http://epic .org/crypto/key_escrow/

[18]Bowden, C. (2002). Closed circuit television for inside your head: Blanket traffic data retention and the emergency anti-terrorism legislation. *Computer and Telecommunications Law Review*, *8*(2), 23.

[19]The events logs of Microsoft Vista and XP can be found in a similar manner. Event logs of older versions of Windows are found more or less in the same way.

[20]150 F. Supp. 2d 1263 (D. Utah 2001).

[21]James, S. (2007, May 16). Did Melanie McGuire dismember her husband. *MSNBC* [online]. Retrieved from http://www.msnbc.msn.com/id/18688528/

[22]Culora, J. (2007, April 29). Inside cheating wife's gruesome "suitcase murder." *New York Post* [online]. Retrieved from http://www.nypost.com/p/news/regional/item_iRDscNdoDdlHo XRYMDT8tJ/0

[23]Hassell, J., & Steen. S. (December 2002/January 2003). Demystifying computer forensics. *Louisiana State Bar*, *50*, 279.

[24]For more information, see the Evidence Shredder Pro website: http://evidenceshredder.com/ product_info.html

[25]For FixCleaner, see http://fixcleaner.com/; for CCleaner, see http://www.piriform.com/ccleaner

[26]Berghel, H. (2007). Hiding data, forensics, and anti-forensics. *Communications of the ACM*, *50*(4), 18.

[27]*Ibid.*

[28]Viegas, J. (2004). Computer printers can catch terrorists. *Discovery Channel News* [online]. Retrieved from http://dsc.discovery.com/news/briefs/20041011/printer.html

[29]Esguerra, R. (2008, October 24). EFF's yellow dots of mystery. Electronic Frontier Foundation. Retrieved from http://www.eff.org/deeplinks/2008/10/effs-yellow-dots-mystery-instructables

[30]Tuohey, J. (2004, November 22). Government uses color laser printer technology to track documents. *PC World* [online]. Retrieved from http://www.pcworld.com/article/118664/government_ uses_color_laser_printer_technology_to_track_documents.html

[31]2005 WL 356801 (Ohio App. 9 Dist. Feb. 16, 2005).

[32]9 F.3d 823 (10th Cir. 1993).

[33]National Institute of Justice. (n.d.). Test results for disk imaging tools: dd GNU fileutils 4.0.36, provided with Red Hat Linux 7.1. US Department of Justice (NCJ 196352), p. 1. Retrieved from http://www.ncjrs.gov/pdffiles1/nij/196352.pdf

[34]2008 WL 113653.

[35]Denning, D. E., & Baugh, W. E. (1999). Hiding crimes in cyberspace. *Information Communication and Society*, *2*(3). Retrieved from http://all.net/books/iw/iwarstuff/www.infosoc.co.uk/00107/feature.htm

[36]See, for example, *Commonwealth v. Koehler*, 914 A. 2d 427 (Pa. Super. 2006); *United States v. Luken*, 515 Supp. 2d 1020 (D.S.D, August 21, 2007); *United States v. Graziano*, 558 F. Supp. 2d 304, 75 Fed. R. Evid. Serv. 1220 (E.D.N.Y., March 20, 2008); *United States v. Richardson*, 583 F. Supp. 2d 694 (W.D. Pa. October 31, 2008).

[37]Negi, S. S. (2005, August 4). Afzal to die; Shaukat gets 10-year jail term: SC acquits Geelani in Parliament attack case. *The Tribune* [online]. Retrieved from http://www.tribuneindia.com/2005/20050805/main1.htm

[38]See, for example, *Williford v. State*, 127 S.W.3d 309, 311 (Tex. App 2004); *Fridell v. State*, 2004 WL 2955227 (Tex. App. Dec. 22, 2004); *State v. Morris*, 2005 WL 356801 (Ohio App. 9 Dist. Feb. 16, 2005); *United States v. Bass*, 411 F.3d 1198 (10th Cir. 2005); *State v. Howell*, 609 S.E.2d 417,419 (N.C.App. 2005).

[39]777 N.E.2d 882 (Ohio Ct. App. 2002).

[40]See U.S. Internal Revenue Service, Criminal Investigative Division, Electronic Crimes. (n.d.). ILook investigator. Retrieved from http://www.perlustro.com/wp-content/uploads/oldwebpage.pdf

[41]See the following website for further information: http://www.e-fense.com/live-response.php

Chapter 8

Crime and Incident Scene: What Should an Investigator Do?

The previous chapters provided information about different cybercrimes, cybercrime laws, the computer-networking environment, searches and seizures, laws regulating access to electronic evidence, the types of electronic evidence, places to find this evidence in computers and related electronic devices, and tools used to discover electronic evidence. This chapter explores how to conduct a computer forensics investigation. Specifically, it examines the types of physical evidence retrieved from the scene of a crime (or incident), the steps in processing a crime (or incident) scene, and the steps in searching the scene. Furthermore, it considers the procedures for handling evidence and documenting the investigation.

Conducting an Investigation

Beginning Steps

An investigator must first ensure that the search of the crime or incident scene that he or she is about to conduct is lawful. The investigation of a suspect must be based on probable cause and the search that is about to be conducted must occur pursuant to a warrant. If no warrant has been acquired, then the search must be executed pursuant to one of the exceptions that allow investigators to conduct warrantless searches (e.g., exigent circumstances).[1] Investigators should try—to the best of their abilities—to avoid giving advance notice to the suspect of an upcoming search (if, of course, they have obtained permission to do so with a no-knock search warrant).[2] Advance notice may lead to the destruction of evidence by a suspect. If a no-knock warrant cannot be obtained, the investigators are required to announce their presence and intentions to the suspect and provide him or her with the search warrant (if present) before entering the suspect's premises.

Processing a Crime (or Incident) Scene

The first step in processing a crime (or incident) scene is to secure the scene. As an investigator, you are responsible for controlling the scene and the individuals on or near the scene. To secure the crime scene, all nonessential personnel must be removed and the boundaries of the scene need to be established. The area to be secured should encompass the entire crime scene. If it does not, any potential evidence outside the boundaries can be contaminated by, for example, pedestrian traffic. After the boundaries are marked off, a single path of entry and exit needs to be established to protect the crime scene and preserve the evidence. The personal information (e.g., name, phone number, work and home addresses) of all individuals who enter and exit the crime scene must be recorded. This can be done in a separate log or else documented in an investigator's notebook.

After the scene is secured, a clear search pattern for the evidence must be established to ensure that all areas of the crime scene are thoroughly searched and vital evidence within these areas is not overlooked. Once this step is complete, the crime scene must be documented in its entirety; that is, the investigator must document the conditions of the scene and identify all relevant physical evidence within it. The evidence an investigator looks for will depend on the crime (see Chapter 5). At this stage, the investigator must decide which evidence can be lawfully collected from the crime (or incident) scene. The evidence that can be lawfully seized must then be documented.

Documentation

Documentation is important for the following reasons:[3]

- It helps police officers corroborate or refute the testimony of a witness, suspect, or victim.
- It helps investigators corroborate, refute, or modify an existing hypothesis as to what happened at the crime scene.
- It helps forensic scientists understand how the evidence found relates to the crime.
- It helps prove that data were not altered during the investigation.
- It helps illustrate to the court what the conditions of the crime scene were, how the crime occurred, and who was involved.

There are five ways to document the crime scene:[4]

- Notes
- Sketches
- Photographs

- Videos
- Reports

Notes

Documentation must include a description of the crime scene and evidence. The crime scene must be documented in its entirety as part of the investigator's **notes**. Individuals at the crime scene and their locations at the time of an investigator's arrival should be noted. The location of each piece of evidence must be documented, along with the time and date it was found. The details of the individual who discovered the evidence must also be documented. If anything (e.g., a piece of evidence) or anyone (e.g., a dead body) was moved prior to the investigator's arrival, it must be documented as well.

The physical conditions of the evidence should also be noted. For instance, any distinguishing marks on the evidence should be included in the notes. Additionally, if evidence, such as computers or phones, has been damaged, that fact should be documented. For example, the suspect may have tried to damage the evidence contained in these items to prevent the extraction of incriminating evidence. Regardless of why the evidence was damaged, its condition should be recorded to protect the investigator and his or her department from possible allegations that the suspect's property was mishandled or damaged by the investigator or other individuals working on the case.

In summary, anything relevant to the investigation at hand must be included in the investigator's notes. When taking notes, an investigator should keep in mind the following points:[5]

- Notes must be made in ink in a bound notebook with its pages numbered sequentially.
- No spaces should be left to go back and make any additions to previous entries.
- If a mistake is made, it needs to be crossed out.
- Under no circumstance should an entry be erased, nor should another entry be written over the erased item.

Notes are important because if an investigator is called to testify, he or she must be able to recall what was observed and which actions were taken at the crime scene. Notes should be supplemented by sketches, photographs, and, wherever possible, video.

Sketches

Sketches include the measurements taken by an investigator to provide accurate dimensions of the crime scene and the location of the evidence in relation to that scene. A compass direction should also be included in each sketch. Moreover, the sketch should contain the case number, the name of the individual who prepared the sketch, the date of the sketch, the type of crime, the location of the crime, and the name of the victim (or victims) depicted in the sketch (if any).

Sketches help document the exact position of the evidence something that cannot be done with the use of photographs or videos. With the measurements provided by the sketches, an investigator can re-create the crime scene in a controlled environment. This information helps the investigator determine which events took place at the crime scene and in what order. The investigator may also re-create the scene to make sense of the evidence found within it.

Photographs

Photographs are used to provide a visual depiction of the crime scene and supplement the evidence illustrated in an investigator's sketches and described in an investigator's notes. Photographs are taken to preserve the crime scene and evidence. Evidence exposed to the elements (e.g., rain) will eventually be destroyed. Evidence may also be contaminated by officers who inadvertently bring items into the crime scene (e.g., dirt from a police officer's shoes) and take items out of the crime scene (e.g., fibers attached to their shoes).

Additionally, depending on the location of a crime scene, cleanup may need to occur soon after a crime or incident. This is especially true when crimes are committed in public places and streets. For these reasons, a crime scene must be photographed as soon as possible after it has been discovered.

Photographs must be taken in such a way that they do not disturb the crime scene or the evidence in it. Some changes are unavoidable. Nevertheless, these changes must be reasonable and must be thoroughly documented (e.g., where the item was moved to, by whom, and for what reason was it moved). Failure to do so may render the evidence inadmissible in court. Ideally, a crime scene should be photographed before any changes are made. One notable exception to this guideline occurs when injured parties need assistance; responders may disturb the scene to treat them as necessary.

Photographs provide a much-needed visual representation of the crime for those not physically present at the scene (e.g., judge, jury, forensic scientists, and behavioral profilers). They can also bring to light information that was initially overlooked by an investigator when he or she was physically present at the crime scene. Photographs also are required for other reasons. Depending on the crime committed, law enforcement agencies may require assistance from other agencies—for example, state or federal investigators. Specifically, photographs are required for criminal profiling. Indeed, crime scene photographs may be sent to the Behavioral Analysis Unit of the Federal Bureau of Investigation (FBI) so that the personnel there can construct a profile of the perpetrator of the crime.

A **photograph log** of crime scene photos should be maintained. It will include the following types of photographs:

- Photos of the overall scene
- Medium-range photos of parts of the crime scene and any evidence included in it
- Close-range photographs of the evidence and/or victims (victim photos are needed if a homicide has taken place)

The photograph log should include the following information:

- Case number
- Name of the photographer
- Date and time the photograph was taken
- Location of the evidence
- Type of case (e.g., homicide, rape, stalking)
- Camera specifications (e.g., make and model of camera, lens used, shutter speed)

Videos

Today, **videos** are often used to document the overall crime scene; they complement the photographs of the crime scene and evidence. Similarly to photographs, they provide a visual depiction of the crime to those involved in the case (e.g., investigators, forensic scientists, prosecution, defense teams, relevant experts, judges and juries). Videos may reveal a vital clue that investigators initially missed while at the crime scene. Given that cases may take anywhere from several months to years to solve and reach trial, videos can refresh an investigator's memory of what happened during the event.

Videos also may provide an audio narrative of what is observed at the crime scene. With video and audio narration, investigators can transcribe what is recorded in the videos to their report at a later time. However, audio recordings are often discouraged, because the comments made by investigators may be considered unprofessional or inappropriate by a defense team or jury.

Finally, questions that may arise in court as to the proper execution of a warrant may be resolved by videotaping the actions of investigators at the crime scene.

Reports

Reports provide a narrative of what happened at the crime scene and how the investigation of the scene was conducted. All evidence should be included in the report. Moreover, the steps the investigator took to document, collect, package, label, transport, and preserve the evidence should be noted. The names of victims, witnesses, and suspects and their contact information should be included, as well as those persons' statements. Any observations made by the police officer concerning any of the parties involved must also be included in the report.

A report should include only the facts of the case. The personal opinions of the investigators should not be included in reports under any circumstances.

After Documentation

Once the crime scene and the evidence within it have been thoroughly documented, the investigator must properly collect the evidence (while, of course, wearing gloves), **label (tag)** it, and package (**bag**) it for later transport to a **forensic laboratory**. Once the evidence has been transported to the lab, all seized items need to be inventoried, recorded, analyzed, and subsequently secured in a locked room or container in a climate controlled environment. So that the evidence will be admissible in court, special care must be taken to ensure that the evidence is properly identified, retrieved, bagged, tagged, transported, analyzed, and preserved.

Special Considerations for Cybercrime Investigations

According to a manual written by the Technical Working Group for Electronic Crime Scene Investigation (hosted on the National Institute of Justice website), each investigator of a cybercrime is expected to carry a **forensic toolkit** that contains at least the following items: documentation tools (e.g., cable tags and stick-on labels), disassembly and removal tools (e.g., screwdrivers, pliers, and tweezers), package and transport supplies (e.g., evidence tape, antistatic bags, and packaging materials), and other essential items (e.g., flashlight, seizure disk, magnifying glass, and gloves).[6] It is always good practice when investigating cybercrime to have available extra cables, serial port connectors, extension cords, power strips, and batteries.

To acquire electronic evidence, an investigator must determine whether he or she will conduct an on-site or off-site search. Factors that influence this decision include the size and complexity of the computer system and related electronic devices, the technical demands of the search, and the specialized knowledge and skills required to successfully conduct the search.[7] Computer forensics investigations can be very time-consuming. Partly for this reason, searches of computers and electronic devices are frequently conducted off-site.

When the target of the search and seizure is a computer or information stored in the computer, the investigator needs to document how the computer was set up when it was found and what it was doing at the time of the seizure. It is important to note that computers can be accessed remotely. As such, computers should be isolated from networks (i.e., connections to other computers) and telephone lines to prevent the tampering with or destruction of data. Computers may also be connected to the Internet on a wireless basis. In all of these circumstances, an investigator should document these connections to computers and then disconnect them.

The status of the computer (on, off, or in sleep mode) must first be documented in the investigator's notes and photographed. To determine if the computer is on, the investigator will check whether the computer's light is on and whether the fan is running (which, of course, can be heard). If the computer is warm, that fact may indicate that the computer was on or that it was recently turned off. To secure the computer as potential evidence, if the computer is off, it should remain off. If the computer is on, the computer screen should remain on. If an investigator immediately shuts down the computer after arriving at the scene, potential evidence could be destroyed. Accordingly, if the computer is on, the investigator should photograph what is on the computer screen (e.g., documents or active programs).

Electronic evidence is fragile and could be changed if the investigator accidentally or even purposely hits a key on the keyboard or clicks on the mouse. In particular, data held in computer memory could be lost through such an action. As such, all volatile data should be immediately noted and photographed. The date and time of the computer system must also be documented. If the computer is in sleep mode, the investigator should move the mouse slightly, without touching any keys. Under no circumstance should the investigator click on the mouse or press any keys on the keyboard to display something on the computer screen; doing so may modify data. Moreover, the suspect may have created a kill switch. For instance, the suspect may have programmed the computer to write over the hard drive in such a manner as to render the data within it unusable if a particular key is pressed on the keyboard (e.g., "Enter").

The investigator should note how the computer is set up. Photographs of the computer from each side should be taken. The entire computer system configuration should also be photographed, including cable connections, electrical wires, and outlet configuration. The investigator should also thoroughly document the peripheral devices that are connected to the computer. The ports that the cables are connected to should be documented as well. Color-coordinated tags should be used to label the cables and their connections to ports of the computer. This will allow a computer forensics specialist to set up the computer at the forensic lab in exactly the same manner it was set up at the crime scene. The status of the peripheral devices—whether they were on or off must also be noted. If dealing with a desktop computer, the investigator should also note the position of the mouse in relation to the keyboard. Was it on the left side or the right side of the keyboard? This factor will help determine whether the computer user was left-handed or right-handed.

After the investigator has completed documenting the computer and its attached peripheral devices, the unit will be powered down if it was found on or in sleep mode. After the computer is powered down, its power cable should be disconnected from the wall socket, but not from the computer. This procedure, however, is not appropriate for each computer system; rather, the procedure used depends on the computer's operating system. The previously described procedure can be used for Windows operating systems. Other operating systems, such as UNIX/Linux and Macintosh, require different shutdown procedures. For example, to shut down computers with Macintosh operating systems, an investigator must click on the apple icon in the menu bar and then select "Shutdown."[8] When the screen indicates that it is safe to turn off the computer, the investigator should then pull the computer's power cord from the wall socket.[9]

The procedure followed for laptop computers requires a few extra steps. After removing the power cord from the computer, the battery pack should be removed from the laptop. Some laptops also contain a second battery.[10] As such, if investigators are unfamiliar with the laptop they encounter at the crime scene, they should check whether it contains a second battery.

The make, model, and serial numbers of all electronic devices should be recorded. Any damage to the computer or identifying marks should be noted and photographed, for both liability and evidentiary purposes.

Other electronic devices found on scene such as mobile phones, personal digital assistants (PDAs), and caller ID boxes should be handled with special care, as they contain volatile data. If such devices are on, then the data displayed must be documented as soon as possible. When seizing mobile devices or PDAs, special procedures are required for their packaging and analysis back at the forensic lab. The appropriate procedures for analysis of these devices must be followed to ensure the admissibility of data extracted from them in court. Investigations involving these devices are explored in further detail in Chapter 12.

Identifying Evidence

When identifying potential evidence in computer forensics investigations, an investigator should not overlook nondigital evidence. Specifically, the investigator should not forget about any physical evidence (e.g., trace and impression evidence, such as fibers, hair, dust, and fingerprints) that might be on or near the computer, keyboard, mouse, and other related electronic devices. Storage devices such as DVDs, CDs, and flash drives should be considered physical evidence as well. Because the chemicals used to process latent fingerprints may damage data and equipment, an investigator must first extract the electronic evidence from, for example, a CD before he or she dusts it for fingerprints so as not to damage it.

The investigator should photograph and collect any books, papers, notes, and hardware relating to his or her investigation. In addition, it may be necessary to seize any documentation that explains the hardware and software installed on the system.[11] Often, individuals keep passwords and decryption keys within their view. As such, notes, papers, Post-its, and other such items that are found on a desk, table, bookcase, computer, and related devices, as well as on walls or boards close to the computer, may hold such information and, therefore, should be collected. Recall the case of the suspected member of al-Qaeda, Ali Saleh Kahlah al-Marri, discussed in Chapter 2. When government agents seized his computer, they found nondigital evidence implicating him of credit card fraud in close proximity to his computer. Specifically, the agents found a folded two-page handwritten document, which included approximately 36 credit card numbers, the names of the account holders (none of which belonged to the suspect), the types of the cards (e.g., Visa or Mastercard), and their expiration dates.

As part of the search for evidence, an investigator should look under desks and tables, in areas of concealment near the computer, and inside manuals and books. Any trash bins should be checked for potential evidence as well. Basically, investigators should look everywhere they are authorized to by law.

Tagging, Bagging, and Transporting Evidence

All physical items that are collected as evidence must be labeled, packaged, and transported to a forensic laboratory. At a minimum, the label should include the case number, the initials of the investigator, the date when the evidence was found, a description of the evidence, and the location where the item was found. All evidence should be packed into antistatic packaging. **Faraday bags** are required to prevent messages from being sent or received by electronic devices (such as PDAs and mobile phones). Items should be wrapped in static-free bubble wrap and placed in separate containers to prevent shifting.[12] Each external hard drive, flash drive, and other electronic storage device must be placed in a separate paper envelope. Items should be disassembled for packaging and transport only as required.

Special factors that computer forensics investigators need to consider when packaging and transporting evidence are magnetic fields, static electricity, corrosive elements, and temperature. Caution should be taken to ensure that evidence is not altered, damaged, or destroyed by any of these factors during its packaging and transport.

Preservation of Evidence at the Forensic Lab

After its collection, evidence is sent to the laboratory for forensic analysis. At the forensic laboratory, all of the seized items need to be inventoried, recorded, and secured in a locked room. Access to this room must be restricted to essential personnel only. The room should be guarded, and access to it should be regulated and recorded in a log. In the lab, computer systems and related devices must be secured away from extreme temperatures, humidity, dust, and other possible contaminants. Thus, when electronic evidence has been transported to the forensic lab, it should be kept in a cool, dry place, away from magnetic fields or radio frequency interference sources and in a climate-controlled environment.

Analysis of Evidence

All of the actions of the computer forensics specialist must be documented. Investigators need to document which evidence was obtained, where the evidence was taken from, when the evidence was collected, how the evidence was acquired, and who retrieved the evidence. Investigators will need to provide proof in court that they preserved all of the data in a computer system without damaging or modifying it. If any modifications were made to the evidence, the investigators must be able to provide a reasonable explanation for why this change occurred. An exact copy of the data contained in the hard drive of a computer or electronic device is required for analysis. Security measures must be taken to ensure that computers and related electronic devices and the data within them are protected from potential damage or modification (with, for example, tamperproof storage devices and a write blocker). To minimize possible alternations, destruction, or damage of data, the computer forensics specialist should limit access to the data.

An investigator can analyze the copy of the hard drive in several ways. For example, he or she might search the copy for any files that the suspect may have purposely hid or deleted. An investigator could also conduct an analysis to find suspect application software—for instance, software such as Timestomp (discussed in Chapter 7), which seeks to modify or erase timestamp information of a file, a program, or the computer system itself. Data can be analyzed to determine the dates and times when files, e-mails, and programs were created, accessed, modified, and downloaded. The dates and times of the computer system can provide information on who accessed the system and, if shared computers are involved, which users were logged on to the computer. These data can further be analyzed to determine which individual accessed, created, or modified a file on the system in question—at least according to the computer. Note that an individual cannot be definitively linked to the computer through this kind of analysis, even if the individual owns the computer and has sole use of it; the evidence retrieved from these analyses is purely circumstantial.

The computer forensics specialist must write a report on the steps he or she took to extract evidence from the electronic devices and preserve the integrity of the evidence. The specific evidence and files that were found during the analysis must also be documented. These reports make it easier to demonstrate how the investigation was conducted and the evidence was obtained.

Once the analysis and report of the evidence have been completed, the evidence must be stored. Access to the area where evidence is stored must be regulated and limited to only few personnel.

How to Handle Evidence in an Investigation

The procedures for the handling and preservation of evidence must be thoroughly documented. A chain of custody log helps ensure the integrity and admissibility of the evidence that is collected. **Chain of custody** is the process by which investigators preserve the crime (or incident) scene and evidence throughout the life cycle of a case. It includes information about who collected the evidence, where and how the evidence was collected, which individuals took possession of the evidence, and when they took possession of it. The chain of custody log is normally kept to show that the evidence of a crime (or incident) has been properly handled and that it was never at a risk of being compromised. Chain of custody must be maintained at all times. Failure to maintain a chain of custody may result in the inadmissibility of evidence in court.

The famous case involving O. J. Simpson illustrates the importance of maintaining the integrity of the chain of custody. In that case, Simpson was tried for the murders of Nicole Brown Simpson (his ex-wife) and Ronald Goldman. The individuals tasked with the handling of the evidence in this case made numerous grievous mistakes that led to a suspect—whom the evidence overwhelmingly pointed to as having committed the crime—being found not guilty of the murders by a jury. Some of the mistakes made in the handling of the evidence are explored here.[13]

There were too many individuals at the crime scene who were involved in the investigation. Nonessential personnel were not removed from the scene. A limited number of individuals should have collected evidence at the crime scene to avoid contamination and to ensure that all evidence was thoroughly documented and properly labeled, packaged, and transported back to the forensic laboratory.

The names and contact information of all personnel at the crime scene were not recorded. As previously mentioned, all persons entering or exiting the crime scene must be documented either in an investigator's notes or in a separate log.

A clear search pattern of the residence was not established. Consequently, vital evidence was overlooked by the first officers and investigators on scene, most probably because of the sheer number of individuals searching for evidence. For instance, two law enforcement agents who were in the kitchen before other investigators arrived on scene (Officer Robert Riske and Investigator Mark Fuhrman) overlooked a butcher knife in that room as potential evidence. When other investigators arrived approximately two hours later, one of them (Phil Vannatter) observed the knife and documented it as evidence.

Individuals involved in the investigation did not communicate their observations and findings to one another. For example, Fuhrman observed a clear bloody fingerprint on the back gate of the Bundy residence and documented it in his notes. Fuhrman, however, did not have the photographer take pictures of the evidence. Fuhrman also did not inform other investigators and officers on scene of the presence of this fingerprint. Moreover, he did not notify his superiors of this piece of evidence. Thus a chain of custody for this evidence was never established to show that it existed. Ultimately, it was deemed inadmissible in court.

Photographs were not documented properly. A chronological record of the photographs taken at the crime scene did not exist. There was no indication of who directed the photographers to take the photos and when they were directed to do so. The photographs of the crime scene also did not include the name of the individual who took the photographs and the time the photographs were taken.

Evidence was improperly handled. One of the investigators on scene, Detective Burt Luper, "was upstairs in O. J.'s bedroom with a black leather glove in his hand that he took from O. J.'s closet. When someone called Luper he went downstairs with the glove and left it downstairs where [it was] captured . . . on videotape . . . [In court,] Luper admitted his distraction error."[14] Additionally, a DNA expert for the defense team testified that one of the individuals tasked with evidence collection, Andrea Mazzola, failed to change her gloves when collecting different evidence, which might have potentially contaminated the evidence.

Evidence was improperly logged. A blue plastic heart that was found at the crime scene was placed in a desk drawer of a law enforcement agent, instead of being labeled, packaged, and transported back to the forensic laboratory for analysis. Approximately two months after it was photographed, the blue heart was logged into evidence. This delay and irresponsible handling of the plastic heart rendered it inadmissible in court.

Evidence observed at the crime scene disappeared. Riske testified in court that he had observed a denim jacket at the crime scene. He discussed this finding with Fuhrman. Fuhrman, however, did not document this piece of evidence in his notes, nor did he inform anyone else about this item. The jacket was not photographed and eventually disappeared from the crime scene.

Blood evidence was improperly recorded. A male nurse, Thanos Peratis, drew O. J. Simpson's blood. Peratis did not log how much blood was drawn. The Los Angeles Police

Department had documented that 6.5 cubic centimeters (cc) of Simpson's blood was drawn. When Peratis was questioned during the trial, he claimed that he drew approximately 8 cc of blood. During the trial, the defense argued that the police had used the "missing" blood to plant evidence at the crime scene, thereby incriminating Simpson in the murders.

Vital blood evidence was carelessly handled. There was a significant delay in transferring the sample of Simpson's blood to the forensic laboratory for analysis. When it arrived at the lab, it had dried. When the sample was sent out of the lab, it was wet again. Ultimately, this evidence was ruled inadmissible in court.

Items with trace evidence were improperly packaged. Trace evidence consisting of hairs and fibers was found at the crime scene. Specifically, hair linking Simpson to the residence where the crime took place was found. Fiber evidence consistent with socks found in Simpson's residence and a Ford Bronco owned by Simpson was also found at the crime scene. This trace evidence was found on the shirt of one victim (Goldman) and on a glove and a cap found at the crime scene. The individual responsible for collecting this evidence did so haphazardly and against protocol by placing two distinct items into the same container (e.g., the knit cap and the victim's shirt). This rendered the evidence inadmissible, as one of the evidentiary items might have possibly contaminated the other. The individual should have placed each item in a separate container, as protocol dictates.

Evidence was improperly secured. For example, glasses were photographed and collected at the crime scene. These glasses had both lenses attached to them, one of which contained what appeared to be bloody fingerprint. The lens with the supposed bloody fingerprint went missing while the glasses were in storage. This occurred because the area where evidence was stored was inadequately secured.

Basically, all of the significant evidence that implicated Simpson in the crime was rendered inadmissible due to its improper handling by investigators and others involved in the case.

Hypothetical Criminal Investigation

Consider the following (fictitious) identity theft and credit card fraud case. Alice Stewart was considered the suspect in the crime. Investigators retrieved a search warrant and proceeded to her residence. Alice lived alone, a fact she confirmed when investigators handed her the search warrant. After securing the scene and establishing a single point of entry and exit, the scene was searched for evidence. An investigator located what appeared to be a home office, a fact later confirmed by the suspect. The evidence of the crime was believed to be located in the hard drive of Alice's computer.

The investigator photographed Alice's home office. First, overall pictures of the office were taken, showing the layout of the desk, items of potential evidentiary value, and the position of these items in relation to the computer and other electronic devices. Pictures of the monitor were then obtained to show the state in which the computer was found and the programs, if any, that were operating. In this case, the computer was in sleep mode, so the investigator moved the mouse slightly to reveal what was on the computer screen.

The next pictures taken showed the position of the mouse and the keyboard. The keyboard was then flipped over, which revealed a piece of paper attached to the keyboard with what appeared to be a password. A picture of this item was subsequently taken by the investigator. The investigator then took pictures of the computer tower and monitor (from various angles), including their make, model, and serial numbers. Photographs were also taken of various CDs (most of them were secured in cases) containing numerous software titles, a notebook on the desk, and several books on computers (including Windows 7 and computer programming books), website design, and hacking that were located on a shelf above the desk. The photographer also documented various items on the desk, including a printer, scanner, cordless phone, digital camera, and a magnetic strip reader/encoder. The make, model, serial numbers, and conditions of these electronic devices were documented as well.

A photo log was used to record the pictures taken of the crime scene.

Example of a Photo Log[15]

Case Number: 870-34-2317
Crime(s): Identity Theft and Credit
    Card Fraud
Date: 23 June 2010
Camera Make/Model: Nikon D80
Camera Lens Focal Length and
    Serial Number: 1:2 f=50mm No.600114

Name of Photographer: Gary P. Manning
Location: 123 West 59 St., Apt. # 5D,
    New York, NY 10019
Start Time: 08:00
Camera Serial Number: 4824543281
Camera Specifications: Shutter speed is 30
    to 1/4000 sec
Focus Modes: auto and manual
Lens Mount Type: F-mount
Built-in flash and external flash capabilities

Time	Photo Number	Type of Photo
0805	# 1	Overall picture of Alice's home office
0807	# 2	Overall picture of Alice's desk
0809	# 3	Right side of Alice's desk
0811	# 4	Left side of Alice's desk
0813	# 5	Picture of computer screen
0814	# 6	Picture depicting location of keyboard and mouse
0817	# 7	Bottom of keyboard with an attached paper with writing on it (possibly a password)
0820	# 8	Back of computer monitor
0823	# 9	Monitor identification number

(continues)

0827	# 10	Right side of computer monitor
0831	# 11	Left side of computer monitor
0835	# 12	Notebook
0840	# 13	Books
0845	# 14	CDs
0848	# 15	Printer
0852	# 16	Make, model, and serial number of printer
0857	# 17	Scanner
0900	# 18	Make, model, and serial number of scanner
0904	# 19	Digital camera
0906	# 20	Make, model, and serial number of digital camera
0910	# 21	Cordless phone with answering device and caller ID
0912	# 22	Make, model, and serial number of phone
0915	# 23	Magnetic strip reader/encoder
0917	# 24	Make, model, and serial number of magnetic strip reader/encoder

This is a very basic photograph log. The contents of photograph logs differ by agency. Some may also include, for example, distance and an area where an investigator can place any remarks.

After photographs were taken of the crime scene, the items of possible evidentiary value were seized. An **evidence log** was then created.

Example of an Evidence Log

Computer System (Monitor and Tower: All in One)

Manufacturer:	Hewlett-Packard
Model:	TouchSmart^{300} PC
Model Number:	HP 300-1379
Serial Number:	7MM7739779
Manufacture Date:	October 13, 2009
Labeled with:	Certificate of Authenticity
	Windows 7 Home Premium
	TIN92-4F79A-BD17R-L1EC3-3NX7R

Operating System
Microsoft Windows 7
Version 6.1

(continues)

Printer

Manufacturer:	Hewlett-Packard
Model:	Deskjet F4480
Model Number:	CB745A
Serial Number:	EQ28KDO2SJ06B5

Scanner

Manufacturer:	Hewlett-Packard
Model:	Scanjet 8270
Model Number:	L1975A
Serial Number:	FR31LGV60085CP

Digital Camera

Manufacturer:	Sigma
Model:	Digital SLR Camera
Model Number:	SD15
Serial Number:	1005759

Cordless Phone

Manufacturer:	AT&T
Model:	DECT 6.0 Cordless Phone with Digital Answering Device and Caller ID
Model Number:	CL82209
Serial Number:	FR31LGV60085CP

Magnetic Strip Reader/Encoder

Manufacturer:	Unitech
Model:	MSR206 Manual Swipe Magnetic Card Reader/Writer
Model Number:	N2907B
Serial Number:	GS42MHW51196DR

Home Office Area

Cables and wires
Keyboard with an attached paper with writing on it
14 CDs and 17 CD-Rs
Dark blue notebook, ruled paper with writing on it
Books:

1. *Hacking: The Art of Exploitation* (January 2008) by J. Erickson.
2. *Hacking Exposed: Network Security Secrets and Solutions* (6th Edition) (January 2009) by S. McClure, J. Scambrav, and G. Kurtz.
3. *Mastering Windows 7 Administration* (January 2010) by W. Panek and T. Wentworth.

(continues)

4. *Windows 7: The Missing Manual* (March 2010) by D. Pogue.
5. *C Programming: A Modern Approach* (April 2008) by K. N. King.
6. *Just Enough C/C++ Programming* (November 2007) by G. W. Lecky.
7. *Build Your Own website the Right Way Using HTML and CSS* (2nd Edition) (November 2008) by I. Lloyd.
8. *Building Findable Websites: Web Standards SEO and Beyond* (February 2008) by A. Walter.
9. *The Design of Sites: Patterns for Creating Winning websites* (December 2006) by D. K. van Duyne, J. A. Landay, and J. I. Hong.
10. *Stealing the Network: How to Own an Identity* (May 2005) by R. Russell, R. Eller, J. Beale, C. Hurley, T. Parker, B. Hatch, and T. Mullen.

Extracting Electronic Evidence

Recall the hypothetical scenario of identity theft and credit card fraud described in the previous section, in which Alice's computer and related items were seized. After the items were seized, they were labeled, packaged, and transported back to the forensic lab for examination. Back at the lab, the computer that was seized was subsequently hooked up in the same manner as it had been hooked up in Alice's home. A write blocker device was utilized before the computer was powered on. The write blocker (discussed in Chapter 7) was used to prevent the investigator from writing anything to the hard drive. Writing other data to the hard drive may destroy or damage electronic evidence; it may also alter data on the hard drive. Any change in the data on the hard drive would render the evidence unusable and inadmissible in court.

Once a write blocker has been applied to prevent any data from being written to a suspect's hard drive, the computer forensics investigator clones or copies the drive by writing each and every part of the drive exactly to a blank hard drive. That is, the investigator makes a **bitstream copy** of the hard drive. The process by which a duplicate copy of the entire hard drive is created is known as imaging. With imaging, all data on the hard drive are copied, including metadata, system-created files, and deleted files.

When investigating a computer system, the computer forensics technician must choose a computer forensics tool to image the hard drive whose validity for this purpose has been upheld in court. Otherwise, the validity of the tools used to create the image of the hard drive may be called into question and the evidence deemed inadmissible. The investigator must also have all of the appropriate tools with which to examine the evidence. For instance, most computers are password protected. As such, a **password cracking** tool would be required to gain access to the suspect's computer system. Consistent with applicable law, it is better to try to voluntarily obtain the user name, passwords, and decryption passphrase of the owner of the computer, if possible.[16] Password cracking software should be used cautiously, as an individual may have set the computer to recognize unauthorized access to the system (with repeated failed attempts to access the system) by setting a "booby trap" to delete the data that the investigator is trying to retrieve.

Once an exact copy of a suspect's hard drive has been made, the investigator must verify that it is an exact copy. As mentioned in Chapter 7, an investigator does so by computing an **MD5 hash algorithm** for the original hard drive and the copy. If they have the exact same MD5 values, then the copy of the suspect's hard drive is an exact copy of the original drive.

With this process in mind, let's go back to Alice's case. Suppose that the computer forensics investigator made an image of the hard drive after the computer was turned on. The investigator then verified that the acquired image was an exact copy of the source hard drive by comparing the hash values (MD5 hash algorithm); they were the same. Finally, before the source computer was disconnected and placed in evidence storage, photographs were taken of the computer to show the condition it was in during and at the end of the acquisition process. The copy was then searched.

A computer forensics investigator generally searches for both active data (data accessible to the computer user) and recovered data (data recovered after deletion). He or she also searches unallocated space (space available because it was never used or because the information in it was deleted) for evidence of a crime or incident. Furthermore, the investigator is likely to search for evidence of the criminal's attempts to conceal information in, among other things, the search history, shortcuts, and the application registry (list of all applications installed) of the computer.

The investigator in Alice's case searched all areas of the hard drive, including the unallocated space. The investigator's search of the hard drive revealed a list of credit card numbers. The relevant file contained the holder's name, account number, credit limit, current balance, and credit remaining. The file with this information was called "system23.dll". The investigator pulled the information out of this file and viewed it in its original format: an Excel spreadsheet. In other words, this file originally had an ".xls" extension. It was clear to the investigator that a conscious effort was made to rename this file to the name in which the investigator found it. Renaming files is one method that criminals use to conceal their illegal activities. The investigator also found a suspicious program on the computer. In particular, software was found on the computer to operate a magnetic strip reader/encoder, which was found at the crime scene.

Critical Thinking Exercise Concerning the Hypothetical Criminal Investigation

Consider the evidence seized by the investigators at Alice's residence. Which other evidence might potentially be obtained from the items seized and where can this evidence be found? Which analyses should the computer forensics specialist perform to find this evidence?

Which Items of Possible Evidentiary Value Would You Seize?

You are investigating Bob Terry, who is suspected of computer hacking and distribution of child pornography. You locate the suspect's home address. After obtaining a warrant, you go to the suspect's residence to search for evidence of the crimes. Upon entering the residence, you notice the following room (**Figures 8–1** to **8–6** are close-ups of the items in the room):

Figure 8–1

Figure 8–2

Figure 8–3

Figure 8–4

(continues)

Figure 8–5

Figure 8–6

Describe what you would collect as potential evidence of the crimes. That is, which of the items in these photographs might be useful for your investigation of the crimes?

Chapter Summary

When conducting a crime scene investigation, the investigator first needs to establish the legal authority to perform a search. Once at the crime scene, the boundaries of the scene need to be established, along with a single point of entry and exit. All nonessential personnel should be removed from the crime scene.

Once the crime scene is secure, the investigators must identify potential evidence and determine which evidence can be lawfully collected from the crime (or incident) scene. The evidence sought depends on the crime. Both the crime scene and the evidence to be seized should be thoroughly documented. Investigators should carry a forensic toolkit that contains gloves, pens, labels, tags, evidence bags, plastic ties, colored tape, and other items necessary for this purpose. When computers are to be seized, an investigator should photograph, document, and label all ports and cable connections and place evidence tapes over ports, power switches, and disk drives. The same steps should be taken with peripheral devices to be seized as well. After these items have been documented and labeled, they should be powered down and their cables disconnected.

Following the documentation of the evidence, an investigator must label, package, and transport the evidence to a forensic lab. At the lab, the items must be inventoried, recorded, and secured in a climate-controlled environment.

If there is a problem with how the evidence was collected, handled, or processed, the evidence may be rendered inadmissible in court. To make sure the evidence is considered admissible in court, the proper retrieval, identification, analysis, and preservation of evidence is required. The chain of custody documents the description of the evidence, the individuals who had possession of it, and the dates and times when they had possession of it. If the chain of custody of evidence is broken at any time, the evidence will be unusable in court. The O. J. Simpson case brought home the lesson that significant attention needs to be paid to chain of custody issues.

Key Terms

Bag	MD5 hash algorithm
Bitstream copy	Notes
Chain of custody	Password cracking
Documentation	Photograph
Evidence log	Photograph log
Faraday bag	Report
Forensic laboratory	Sketch
Forensic toolkit	Tag
Label	Video

Critical Thinking Questions

1. Which steps are necessary to preserve a crime scene and ensure that the evidence obtained (especially electronic evidence) is admissible in court?

2. Do you think that chain of custody places an undue burden on investigators of crimes?

Review Questions

1. What is the first thing that an investigator must do when he or she decides to conduct a search?

2. What is the first step when processing a crime scene?

3. Which other steps are involved when processing a crime scene?

4. What are the five ways to document a crime scene?

5. Why is proper crime scene documentation critical in investigations?

6. Why is it important for an investigator to take photographs of the crime scene and the evidence in it?

7. How should an investigator shut down a computer at the crime scene? Is the procedure similar for all computers?

8. What items should an investigator's toolkit contain?

9. Which evidence should not be overlooked when conducting a computer forensics investigation?

10. How should computers and electronic devices be packaged?

11. What must be done to preserve evidence in a forensic laboratory?

12. What is the chain of custody? Why is it important?

Footnotes

[1]See Chapter 4 for further information.

[2]No-knock warrants were covered in Chapter 4.

[3]Gardner, R. S. (2005). *Practical crime scene processing and investigation: Practical aspects of criminal and forensic investigations*. New York: CRC Press, p. 1.

[4]*Ibid.*, p. 1; Girard, J. E. (2008). *Criminalistics: Forensic science and crime*. Boston: Jones and Bartlett, p. 8; Saferstein, R. S. (2011). *Criminalistics: An introduction to forensic science* (10th edition). New York: Prentice Hall, p. 29.

[5]Girard, J. E. (2008). *Criminalistics: Forensic science and crime*. Boston: Jones and Bartlett, p. 8.

[6]Technical Working Group for Electronic Crime Scene Investigation. (2001, July). Electronic crime scene investigations: A guide for first responders. National Institute of Justice, US Department of Justice, pp. 23–24. Retrieved from http://www.ncjrs.gov/pdffiles1/nij/187736.pdf

[7]Brenner, S. W., & Frederiksen, B. A. (2002). Computer searches and seizures: Some unresolved issues. *Michigan Telecommunications and Technology Law Review, 8*, 62–63.

[8]US Department of Energy. (n.d.). First responder's manual. *Computer Forensic Laboratory*, p. B–3. Retrieved from http://www.linuxsecurity.com/resource_files/documentation/firstres.pdf

[9]*Ibid.*

[10]Technical Working Group for Electronic Crime Scene Investigation. (2001, July). Electronic crime scene investigations: A guide for first responders. National Institute of Justice, US Department of Justice, p. 31. Retrieved from http://www.ncjrs.gov/pdffiles1/nij/187736.pdf

[11]*Ibid.*

[12]US Department of Energy. (n.d.). First responder's manual. *Computer Forensic Laboratory*, p. D-1–D-4. Retrieved from http://www.linuxsecurity.com/resource_files/documentation/firstres.pdf

[13]Jones, T. L. (n.d.). O. J. Simpson. *Tru TV Crime Library*. Retrieved from http://www.trutv.com/library/crime/notorious_murders/famous/simpson/index_1.html; Walraven, J. (n.d.) The Simpson trial transcripts. Retrieved from http://walraven.org/simpson/; Chain of custody. (1998). Retrieved from http://www.smartfellowspress.com/chain_of_custody.htm

[14]Chain of custody. (1998). Retrieved from http://www.smartfellowspress.com/chain_of_custody.htm

[15]The sections here and the data included in them are only meant to serve as examples of what might be included in a photograph log.

[16]Technical Working Group for Electronic Crime Scene Investigation. (2001, July). Electronic crime scene investigations: A guide for first responders. National Institute of Justice, US Department of Justice, p. 26. Retrieved from http://www.ncjrs.gov/pdffiles1/nij/187736.pdf

Chapter 9

Corporate Crimes and Policy Violations Involving Computers: How to Conduct a Corporate Investigation

This chapter explores the circumstances in which corporate investigations are conducted. It examines the criminal, civil, and policy violations that may be committed in a corporate environment, and considers how a business prepares for and conducts an investigation of these violations. Moreover, it analyzes the steps that need to be taken to ensure that the evidence obtained through such an investigation is handled properly. Finally, this chapter illustrates the importance of clearly communicated policies in protecting the corporation from liability claims.

Corporate Investigations

Complaints or allegations of misconduct may be made by insiders within a company (e.g., the perpetrator's peers and superiors) or by outsiders to the company (e.g., suppliers, contractors, customers, private security, or police). Corporations and organizations are vulnerable to both internal and external threats. Internal threats are the result of employee, management, or board member misconduct. External threats are perpetrated by outsiders to the company or organization, including those who have an affiliation with the corporation (e.g., customers, vendors, and suppliers) and those with no affiliation whatsoever.

Contrary to popular belief, the greatest threat to a company is posed by insiders (e.g., employees)—who could be anyone under the corporate umbrella. These insiders have in-depth knowledge of the inner workings of the business and its security measures. In fact, temporary employees or employees who have been contracted to the corporation for a fee are the individuals who are most likely to engage in illicit activities, as they have access to the business and few, if any, ties to the corporation (i.e., they lack loyalty). According to the **American Society for Industrial Security** (ASIS), these individuals engage in unlawful activity by exploiting trust relationships with the corporation.[1]

Corporate investigations are conducted when allegations of misconduct are raised by any of the following parties:

- **Whistleblowers**[2]
- Senior management
- Board members
- Employees
- Vendors
- Contractors
- Suppliers
- Customers
- Media
- Watchdog groups
- Academicians
- External auditors
- Internal auditors
- Compliance officers
- Others (e.g., police, government agencies, and competitors)

Corporate Criminal Activities and Policy Violations

Corporate investigations are normally conducted for the following forms of misconduct:

- Corporate policy violations
- Labor violations
- Concerns for potential liability claims
- Board members, management, and employee misconduct
- Suspected criminal activity

Examples of noncriminal activities that may be investigated include computer use policy violations, such as e-mail and Internet abuse. Investigations are also conducted when civil lawsuits are brought against the corporation for any of the following issues:[3]

- **Negligent retention**[4]
- **Negligent hiring**[5]
- **Invasion of privacy**

- Harassment
- Sexual harassment
- Discrimination
- **Wrongful termination**
- Exposure to harm in the workplace
- Union and labor issues
- **Workplace injury**

In respect to criminal activity, the most common types of cybercrimes in the workplace are fraud, **intellectual property theft**, **data manipulation**, economic espionage, identity theft, hacking, **extortion**, and embezzlement (to name but a few possibilities). Within a corporation, money laundering, murder, burglary, robbery, vandalism, arson, and workplace violence may also occur.

Workplace violence may occur between employees on the job or when outsiders seek to harm an employee or employees within the company. In addition, threats of workplace violence have been communicated via e-mail. For instance, Gino Augustus Turrella had more than 100 firearms in his home and threatened, via e-mail, to inflict violence on various corporations. Turrella made these threats while posing as one of his former managers and coworkers. Specifically, his e-mails stated "he was going to bring a gun into a Boeing facility and 'shoot ever[y] employee [he] see[s],' and also that he would 'strap himself with explosives and detonate' if and when he was apprehended, in order to cause 'maximum death and destruction in the workplace'."[6] Turrella further threatened the Anacortes oil refinery and Chevron Oil by stating that he had placed a bomb "'at a strategic location at the oil refinery' and that he was 'going to set if off via remote control' so that it '[would] kill the most . . . employees and do the most destruction to [the] refinery'."[7]

Generally, corporate crimes can be classified into two categories:

- Crimes against the corporation
- Crimes for the corporation

Crimes Against the Corporation

Crime against the corporation can be divided into three main categories: input scams, output scams and throughput scams. Each of these categories is described in this section.[8]

Input Scams

In **input scams**, data within a computer database are altered, deleted, or fabricated. In computer programming, this process is known as data diddling. Data diddling occurs when data are changed before or during their input into a computer.[9]

Embezzlement is a form of input scam. Consider the case of Sujata Sachdeva, the former vice president of Koss Corporation. Sachdeva embezzled more than $31 million, which she used to pay off personal expenses and to purchase personal items.[10] To conceal her crime, she ordered employees of the corporation to make fraudulent entries into books and records so as to make the transfers and transactions she made appear legitimate.[11] Embezzlement can be detected from routine internal, external, or compliance audits, especially if these audits bring to light financial irregularities, questionable transactions, or asset losses.

Money laundering may also be uncovered during audits. **Money laundering** is defined as "the process by which criminals hide, disguise, and legitimize their ill-gotten gains."[12] A case in point involved Michael Daly, the president of Data Resource Group, a computer parts company.[13] Daly created fraudulent identities to receive computer networking equipment replacement parts from Cisco. Daly then engaged in money laundering by selling these parts to Cisco hardware resellers throughout the United States.

Another money laundering case involved E-Gold, an online money transmitting business. These types of businesses are required to comply with money laundering laws and regulations. Under the **Bank Secrecy Act of 1970** (31 U.S.C. § 1051 et seq.), institutions are required to report instances of money laundering. This Act also requires financial institutions to take an active role in detecting and preventing money laundering. E-Gold did not comply with this or other pertinent regulations. Among other things, this company hired employees with no relevant experience to monitor accounts for criminal activity.[14] E-Gold also allowed users to open accounts without providing proof of identification, even though it had been made aware that its service was being used by individuals to conduct criminal activities (such as identity theft, investment fraud, credit card fraud, and child exploitation).[15] Consequently, the Acting Assistant Attorney General, Matthew Friedrich, argued that "the E-Gold operation created an environment ripe for exploitation by criminals seeking anonymity in conducting online transactions."[16]

Earlier forms of money laundering regulation "focused on tracing criminal money—associated with narcotics or political corruption—after the crime."[17] After the war on terrorism, however, this perspective changed. Nowadays, data are collected and analyzed to identify suspicious activity concerning the movement of money; that is, financial data are collected and analyzed "to identify suspicious transactions that may indicate terrorist behavior."[18] Software is used to achieve this goal, by searching financial data for transactions that deviate from the norm.[19] Some of the patterns that are considered suspicious include individuals without regular income wire transferring money internationally and numerous wire transfers by the same individual of small amounts of money internationally.[20]

In another input scam, the criminal impersonates an authentic user to gain access to the system. Once access is gained, the data are manipulated for the user's personal gain or for the personal gains of others. Also, an individual who has access to the system may exceed his or her authorization by gaining access to parts of the system to which his or her authorization does not extend. Additionally, once access to a system is gained, the user may exceed his or her authorization by gaining access to other systems that he or she does not have permission to use.

Output Scams

In **output scams**, an individual seeks to use company-owned data (e.g., research and development plans, customer lists, marketing strategies, sensitive data of employees and customers) in the computer for personal gain. In particular, an individual may engage in the following activities:

- *Use the data to extort money from the company.* For example, Jeff Greer, a trucker assigned with transporting documents to a paper recycling mill, observed that some of these documents contained sensitive personal information of customers of gaming companies such as MGM Mirage and Harrah's Entertainment.[21] These companies had not first shredded these documents before they were shipped out for recycling. Greer sought to extort $250,000 and a 30-day consulting job contract from these companies. The contract, Greer argued, would provide these companies with information on how he was able to obtain the documents and identify which measures the companies could take to prevent this kind of breach from happening in the future. MGM Mirage and Harrah's Entertainment subsequently contacted the FBI and asked for the agency's assistance on this matter.

- *Resell the data to competitors.* Stolen trade secrets, files, programs, pricing structures, operational and strategic plans, and other forms of proprietary information are often offered to rival corporations for a price. One such example involved a former Director of Information Technology at Lightwave Microsystems, Brent Alan Woodard. Woodard stole trade secrets contained on a backup tape from his employer and tried to sell them to a competitor, JDS-Uniphase, via e-mail.[22]

- *Use the data in other ways for profit.* Sometimes data are stolen from a company so that the individual who stole the information can use it to secure employment with a competitor of the specific corporation. At other times, the experience and knowledge of an individual of the day-to-day operations of a corporation are enough to secure a position with a competitor. To ensure that their data remain private, some companies make their employees sign a **noncompete agreement**, which can restrict them from working with competitors within a particular geographical region for a reasonable period of time (no greater than two years).[23] This step is intended to prevent employees from transferring their knowledge of the corporation's operations and the skills they learned on the job to a competitor.

  Moreover, individuals may steal data from a corporation for their own personal use; that is, they may use the information to open up their own business. For instance, they may solicit former or current employees and clients from the corporation they were previously employed at. Corporations, however, may have all employees sign a **nonsolicitation agreement** that becomes effective when an employee resigns or is terminated from his or her position. This agreement prohibits an individual from trying to "steal" the employees or customers of the former employer.[24]

Policies should also be in effect that prohibit the misuse of the corporation's confidential business data, trade secrets, and other forms of intellectual property.

Throughput Scams

Throughput scams occur when individuals commit crimes against business computers. In one very popular throughput scam, known as *salami slicing*, an individual installs a program to steal "small amounts of money from a large number of sources by shaving a penny from each savings account during an interest calculation run or rounding off the mills (one-tenth of a cent) and accumulating them for transfer to a particular account."[25] In other words, this scam works by repeatedly stealing extremely small quantities of money.

Other forms of throughput scams involve hacking into corporate systems and manipulating the data within them. The same end could be achieved by using malicious software to the modify data on computer systems. "Logic bombs" have often been placed on computer systems by disgruntled employees (see Chapter 1). For example, after the name of the employee who has been fired has been deleted from payroll, a logic bomb might be set off that increases, decreases, or deletes payroll data.

Crimes for the Corporation

Crimes for the corporation include inflating sales, understating expenses, and overstating assets. In businesses that sell products, employees may cheat customers by substituting

---

Corporate Self-Protection from Crimes

There are several ways in which corporations can protect themselves from would-be criminals. For example, they can use effective security systems, such as intrusion detection systems (IDS; explored in Chapter 11), CCTV (closed-circuit television), and security personnel to deter and prevent employees from engaging in illicit activities. Procedural controls may also be implemented. Policies are an example of procedural controls. Other ways unlawful behavior can be deterred or prevented is by creating an environment of trust and cooperation within the organization, and especially between management and employees.

To protect their computer systems, corporations typically limit access to it and limit the functions that users can perform and the areas that they can access once in the system. Additionally, employees are required to periodically change their passwords. Most companies require these passwords to be of a specific length and to contain a variety of types of characters—a strategy that prevents the passwords from being easily cracked by unauthorized users seeking access to the system. Moreover, to protect their computer systems, corporations ensure that their antivirus and antimalware software is up-to-date. These systems are routinely scanned by security software and monitored by IT personnel to ensure that no viruses or malicious software has been downloaded and that the systems are working properly.

cheaper material for the expensive materials the customers actually paid for. Depending on the items sold by a corporation, an employee might sell the customer items of less weight or shorter measure than that which the customer really paid for. Customers may also be cheated by false advertising of products or services.

Investment fraud, such as an active role in a Ponzi or pyramid scheme (see Chapter 5), is another type of crime committed for the corporation. One now-notorious Ponzi scheme was implemented by Bernard Madoff. To maintain a Ponzi scheme, a constant flow of money is required. In 2008, investment contributions to Madoff's firm stagnated, and investors sought to withdraw a portion of their assets. In the past, Madoff had covered the withdrawals made by his clients with investments from new clients. However, in 2008, Madoff was unable to do so as the requests for withdrawals overwhelmed him—reaching $7 billion.[26] Owing to the lack of funds from new investors, he was unable to cover these requests. By the end of 2008, Madoff was forced to admit that he operated a $50 billion fraudulent investment business.[27]

Corporations may also engage in **tax evasion** (federal, state, or both). According to 26 U.S.C. § 7201,

> Any person who willfully attempts in any manner to evade or defeat any tax imposed by this title or the payment thereof shall, in addition to other penalties provided by law, be guilty of a felony and, upon conviction thereof, shall be fined not more than $100,000 ($500,000 in the case of a corporation), or imprisoned not more than 5 years, or both, together with the costs of prosecution.

Examples of tax evasion by U.S. corporate officials abound. Consider the case of Stanley Tollman, an executive of Tollman-Hundley Hotels in New York. Tollman intentionally failed to report approximately $18 million in income to the Internal Revenue Service (IRS).[28] Another such example involved Jeffrey Koger, the former Chief Financial Officer for Koger Management Group (KMG), which managed property for approximately 400 homeowners' associations. Koger embezzled approximately $3 million from these groups and failed to file his income tax returns from 2003 to 2006, thereby evading payment of more than $750,000 in federal income taxes.[29] A third case involved the former Chief Operating Officer of Holiday International Security (currently known as the USProtect Corporation), Richard S. Hudec. Hudec was found guilty of failing to pay more than $200,000 dollars in federal income taxes by falsely claiming that he made an annual income of only $18,000 and $21,000 in 2002 and 2003, respectively.[30]

Corporations may also engage in bribery to obtain highly coveted contracts or to evade detection by authorities for unlawful activities. **Bribery** is defined as "the offering or receiving of money, goods, services, information, or anything else of value for the purpose of influencing public officials to act in a particular way."[31] Under 18 U.S.C. § 201(b)(1), bribery of a public official occurs when a person

> directly or indirectly, corruptly gives, offers or promises anything of value to any public official or person who has been selected to be a public official, or offers or

promises any public official or any person who has been selected to be a public official to give anything of value to any other person or entity, with intent—

(A) to influence any official act; or

(B) to influence such public official or person who has been selected to be a public official to commit or aid in committing, or collude in, or allow, any fraud, or make opportunity for the commission of any fraud, on the United States; or

(C) to induce such public official or such person who has been selected to be a public official to do or omit to do any act in violation of the lawful duty of such official or person.

There have been numerous cases of public official bribery—both foreign and domestic. The **Foreign Corrupt Practices Act of 1977** (15 U.S.C. § 78dd-1, et seq.) prohibits individuals from paying foreign government officials in an attempt to "obtain or retain business" within these countries.[32] One such case involved KBR, a U.S. engineering and construction company; in 2009, it was charged with bribing Nigerian officials to secure contracts.[33] In another international case, the U.S. Department of Justice accused Siemens of paying more than $1 billion in bribes in numerous countries around the globe to secure infrastructure contracts.[34]

A domestic example of bribery involved a Washington, D.C., government official. In this instance, Yusuf Acar, the Acting Chief Security Officer of the Washington, D.C., Office of the Chief Technology Officer, was accused of accepting bribes (for contracts) from the president and chief executive of Advanced Integrated Technologies Corporation, Sushil Bansal.[35] Another domestic bribery case involved a developer, Anthony Spalliero.[36] Spalliero had bribed the mayor of Marlboro Township, New Jersey, Matthew Scannapieco, with approximately $100,000 to support his projects. In 2005, the former mayor, Scannapieco, pled guilty to bribery and federal income tax evasion (for hiding the proceeds of the bribe from the IRS). In yet another case, a former employee of the purchasing department of the New York Power Authority (NYPA) received approximately $167,000 from vendors in kickbacks and bribes.[37]

Corporations may also engage in unfair trade practices to inhibit fair competition in the market. Free markets and fair competition ensure that businesses provide the best products and services at the best prices; under these conditions, consumers will continue to buy these products and services, inflation will remain low, and the economy will stay healthy. Unfair trade practices, which upset this balance, include price fixing and bid rigging.

Price fixing occurs when two or more parties act together to fix prices of particular goods or services. Price fixing has been broadly condemned by the courts.[38] Price fixing also violates the **Sherman Antitrust Act of 1890**. Specifically, Section 1 of the Act states: "Every contract, combination in the form of trust or otherwise, or conspiracy, in restraint of trade or commerce among the several States, or with foreign nations, is declared to be illegal." A recent example of price fixing involved three leading electronic manufacturers—LG Display Company, Sharp Corporation, and Chunghwa Picture Tubes.[39] These manufacturers conspired to fix the sales prices of liquid crystal display (LCD) panels.

By contrast, if two potential purchasers are set to bid on a house and one simply agreed to stand down, letting the other win the bid, the two parties would be involved in criminal **bid rigging**.[40] One such case highlighting this practice was *New York v. Feldman*.[41] *Feldman* involved three states—New York, Maryland, and California—which "sued eight individual stamp dealers and two corporations in a federal civil court action, alleging a bid rigging conspiracy involving auctions for collectible postage stamps over a twenty-year period."[42]

In another type of crime intended by persons to benefit the company, employees may purposely violate administrative regulations (e.g., regulations established by the Occupational Safety and Health Administration) under the direction of management (see Chapter 2). These administrative regulations may cover compensation, termination and hiring practices, safety, and working conditions (as discussed in the nearby box). For instance, senior management may have explicitly directed middle management to ignore safety hazards because the costs of tending to them would be prohibitive. Such an action violates the **Occupational Safety and Health Act of 1970**, which mandates that companies ensure safe working conditions for their employees.

Legislation Governing the Relationship Between an Employer and an Employee

Fair Labor Standards Act of 1938: established minimum wages and the 40-hour work week.

Equal Pay Act of 1963: mandated equal pay for equal work, regardless of gender.

Title VII of the Civil Rights Act of 1964: prohibits employers from discriminating against a potential employee on the basis of race, gender, religion, or national origin.

Age Discrimination in Employment Act of 1967: prohibits an employer from discriminating against an individual 40 years or older on the basis of his or her age.

Occupational Safety and Health Act of 1970: requires employers to adopt certain practices to ensure safe working conditions for their employees.

Pregnancy Discrimination Act of 1978: amended *Title VII of the Civil Rights Act of 1964* and prohibits employers from discriminating against pregnant women.

Americans with Disabilities Act of 1990: prohibits employers from discriminating against an individual based on his or her disability.

Family and Medical Leave Act of 1993: provides employees with unpaid leave (maximum of 12 work weeks) to tend to family or medical emergencies.

Uniformed Services Employment and Reemployment Rights Act of 1994: protects individuals from workplace discrimination because of their military service.

Other regulations that may be ignored or violated cover **discrimination**. According to the **Equal Employment Opportunity Commission** (EEOC), discrimination based on age, sex, race, national origin, religion, disability, and pregnancy is prohibited in all aspects of employment, including "hiring, firing, pay, job assignments, promotions, layoff, training, fringe benefits and any other term or condition of employment."[43] A recent example of sex discrimination involved a Swiss pharmaceutical company, Novartis. In a lawsuit, jurors found that Novartis engaged in "sexual discrimination, denying women promotions, pay raises and derailing the careers of pregnant employees."[44] The jury awarded the female employees who filed the lawsuit more than $3 million.

Corporations with government contracts may overcharge the government for products and/or services. A case in point involves Fluor Daniel, which is a subsidiary of one of the largest engineering and construction companies in the United States—Fluor Corporation.[45] This case was initiated by a whistleblower named Cosby Coleman. The Sarbanes-Oxley Act of 2002 protects whistleblowers from being fired from the business after reporting its unlawful activities. Under the **False Claims Act of 1863** (31 U.S.C. § 3729-3733), an individual (whistleblower) can file a civil lawsuit against a federal contractor on behalf of the government. Lawsuits are filed when an individual accuses the

---

Insider Trading

Some corporate crimes do not fit into the categories identified earlier in this chapter. One such crime is insider trading. **Insider trading** "occurs when a director, officer, or shareholder who holds more than 10 percent of the stock of a corporation listed on a national exchange buys and sells corporate shares" based on insider information, which is "information known generally by security officers before it is made available to the general public."[47]

According to the Securities and Exchange Commission, examples of insider trading include the following:[48]

- Directors, managers, and employees trade securities after learning of confidential corporate developments.

- The above mentioned individuals inform family members, friends, associates, and others of these developments. In turn, these individuals trade the company's securities based on that information.

- Outsiders to the company who were made privy to this information (e.g., lawyers, government employees) use it to trade the securities of the corporation.

One famous case of insider trading involved media mogul Martha Stewart. Stewart was informed that the share price of a biotech company, ImClone, was going to drop. Consequently, she sold all 3928 of her ImClone shares and in so doing, avoided $45,673 in losses.[49] Stewart was sentenced to serve five months in prison. She was also ordered to pay $195,000 in fines and penalties.[50] As one can clearly see, this amount is significantly more than what she would have lost had she not engaged in insider trading.

contractor of committing claims fraud against the government. For example, in the *Fluor* case, among other things, it was claimed that "expenses for corporate properties, such as a $20,000 antique Chippendale chair and a $410,000 condo in Palm Springs, [were] charged to government contracts."[46] In 2005, the lawsuit was settled and the Fluor Corporation agreed to pay the government $12.5 million in compensation.

Preparing for the Investigation

The first step in preparing for a corporate computer forensics investigation is identifying the purpose of the investigation. An investigation may be initiated to determine if authorization is exceeded in a computer system, harassing e-mails were sent, workplace fraud was committed, images of child pornography were found, money laundering has been discovered, and so on.

After the purpose of a computer forensics investigation is established, the resources required to conduct the investigation are identified. These resources include the investigator and the tools he or she requires to conduct the investigation. The jurisprudence of the courts must have accepted the tools the investigator uses as valid and reliable to ensure that the evidence that is retrieved by them is not rendered inadmissible in court.

To handle the investigation, a corporation may use either personnel from its own IT department, a computer forensics investigator on the company's staff, or an outside computer forensics expert it hires specifically for the investigation. As a general rule, corporate IT professionals should not be used to conduct the investigation. Computer forensics requires specialized knowledge, and knowledge of computer and information technology alone will not suffice to properly conduct the investigation. As such, corporate IT professionals are not the best suited to conduct the investigation. Caution should be exercised if corporate IT professionals are charged with managing the investigation, as they may improperly handle electronic evidence, thereby rendering it inadmissible in court. Larger organizations can afford to keep computer forensics investigators on staff. The same cannot be said about small and medium-size corporations, which often hire external computer forensics experts when an investigation is deemed necessary.

The use of external computer forensics investigators has certain advantages. For instance, external investigators can preserve the secrecy of the investigation. In fact, they are most often used for sensitive enquiries that must be made unbeknownst to employees. Additionally, external computer forensics investigators are often better trained to conduct the investigation, especially when compared to employees from the company's IT department. In a corporate environment, there is an increased risk that those corporate personnel engaged in computer forensics investigations may not be the most qualified to do so. Pressure to conduct the investigation surreptitiously may result in the retrieval of electronic evidence in a less than sound manner. Moreover, external computer forensics investigators are not influenced by internal politics or other biases. Of course, there are also some dangers in providing outsiders with access to the proprietary information in the corporation's computer systems.

Computer forensics investigators from law enforcement agencies are normally used if a crime has been committed. Their involvement depends on whether the corporation reported the crime to the police. Nevertheless, there are certain exceptions to this practice. In particular, law enforcement agencies may initiate an investigation of possible unlawful activities perpetrated by the corporation and/or a specific individual (or individuals) within the corporation. Corporations are required by law to contact the police if certain crimes are suspected to have been committed (e.g., child pornography and money laundering).

In contrast, corporations often do not officially report other crimes such as theft and hacking. There are many reasons why a corporation might not report a crime to law enforcement agencies. They may choose to do so to avoid any negative publicity that might result from making the incident known. The publicity surrounding the event might also result in a loss of consumer confidence, new and existing customers taking their business elsewhere, damage to corporation's name brand, and loss of new revenue (to name a few negative consequences), especially if proprietary information (including the sensitive data of customers) was the target of the crime. Indeed, a National Survey on Data Security Breach Notification conducted by the Ponemon Institute on 51,000 adult consumers found that consumers had an extremely negative reaction to notification that their confidential data were mishandled or lost by companies; in fact, 60% of the consumers terminated or seriously considered terminating their relationship with the company following such notification.[51] Similarly, a study conducted by Javelin Strategy and Research found that 40% of consumers reported that a report of data loss adversely affected their relationship with the business.[52] Additionally, this study showed that consumer confidence in the business significantly decreased after data breaches. In particular, 55% of consumers stated that they trusted the business less after the breach and 30% explicitly stated that they would never buy goods or services from that business again.[53]

Conducting the Investigation

Before an investigation is conducted, the source of the allegations of misconduct must be qualified; that is, the credibility and reliability of the source must be verified and corroborated by other sources or documents. The response of investigators to corporate criminal activities or policy violations depends on the alleged crime or violation in question and the circumstances of the case. Once a determination is made on how to respond to the alleged crime or violation, an investigation will be conducted. If a criminal act is discovered while investigating a violation of corporate policy, it becomes a law enforcement investigation and should be referred to the appropriate authorities. However, as mentioned earlier, this transfer of investigational responsibilities may not always occur. The corporation may deem that the incident itself is not important enough to notify the police and risk jeopardizing the reputation of the business.

Computer forensics investigations occur once an incident or offense has been identified. The purpose of this investigation is to find the perpetrator and to build a case against the individual so as to administer disciplinary action or take the suspect to court. Violations of computer usage policy usually result in disciplinary action or dismissal (depending on the severity of the violation). For example, Xerox fired 40 employees for visiting pornographic websites and shopping online while at work.[54] Additionally, in 2000, Dow Chemical Company fired approximately 100 employees and disciplined more than 200 employees for distributing pornographic and violent images via the company's e-mail system.[55] Other companies, such as Merck and Company (a pharmaceutical corporation), have also disciplined and dismissed employees and contractors for improper e-mail and Internet use.[56] In either case (disciplinary action or dismissal), evidence of contravention of corporate policy is required.

Computer misuse can involve anything from playing Solitaire on the computer to surfing pornographic websites. Computers, e-mail, and Internet privileges are misused when they are used for reasons other than conducting business. Instead of working, individuals may use the Internet for any of the following purposes:[57]

- Send personal e-mails
- Use social networking websites
- Read online newspapers, magazines, or other materials
- Trade stocks online
- Shop online
- Use online chat rooms
- Watch television shows, movies, sports, and other live media
- Hunt for job opportunities
- View pornographic websites

The misuse of these privileges may result in financial loss, reduction in productivity, wasted resources, and unnecessary business interruptions. This misuse has come to be known as **cyberslacking**.

Accessing particular websites during work hours may also make the corporate network susceptible to computer viruses and malicious software. This outcome is especially likely if employees are visiting pornographic websites, which are notorious for containing such software that infects users' computers. To curtail this practice, companies should create policies explicitly prohibiting the viewing of sexually oriented websites. Likewise, the sending and/or soliciting of sexually oriented images and e-mail or instant messages via computer-owned technology should be strictly prohibited.

Corporations investigate cyberslacking to determine more than just whether an employee is wasting resources. Specifically, monitoring primarily occurs because corporations can be held liable for employees' misuse of computer, e-mail, and Internet resources, especially when such misuse leads to workplace harassment and discrimination lawsuits.

To respond to these issues, corporations have implemented filtering software programs, which block access to certain websites.[58] However, these programs are easily circumvented. Other corporations use computer and Internet monitoring software, such as WebSense.[59] With this software, if the computer user accesses a pornographic website, the program will issue an alert. Keylogger programs can also be used to record every keystroke made by the company's employees.

The corporation should have policies that expressly prohibit the use of its computer, e-mail, or Internet services to violate state, federal, or international law. If a violation of criminal law is found, the corporation may choose to pursue legal action against the offender. Consider the following example: An employee in the IT department observed that an image of child pornography had been downloaded to a corporate computer. Under no circumstance should that employee delete the image of child pornography. If the IT employee does so, he or she may be criminally prosecuted for deleting the evidence. In addition, the employee must not make a copy of the image for evidentiary purposes. If a copy is made, the employee may be charged with the possession (and possibly for distribution) of child pornography. Nothing should be done to the image. Instead, the police should be immediately notified as to its existence on the corporate computer or server.

The investigator should begin by documenting the corporate incident scene. Items of potential evidentiary value on scene should then be identified, documented, and collected. It is always possible that evidence may be contaminated after its identification because an employee who detects an incident or crime will probably first contact his or her supervisor to show that person the evidence. Subsequently, the supervisor will contact the IT department and show its members the evidence. After the IT department has reviewed the evidence, the corporate security staff and police may be contacted.

Corporations need to conduct such investigations carefully so as not to alter the computer data, especially when civil and criminal law violations are suspected. Failure to do so can render the data retrieved from the investigation inadmissible in court. Even if an incident with an employee is handled internally, if the employee subsequently retaliates against the corporation, the latter case may end up in court. Accordingly, the data that are gathered during a corporate computer forensics investigation may end up being used in a civil, criminal, or administrative proceeding. As such, the methods used to obtain such information must be rigorous and unassailable.

The corporate investigation of the suspected policy or criminal violation should be meticulously conducted. Failure to do so may expose the company to liability claims. Consider the following example: Legal action was taken against Edward for numerous harassing e-mails that were sent from his account to his colleagues. Edward filed a lawsuit against his employer for wrongful termination. During the court proceedings, the plaintiff's attorney asked the respondent if a virus check had been performed on the computer to determine whether malicious software might have been responsible for the dissemination of the e-mail. The respondent answered that such a check had not been performed. The failure of the respondent to conduct a thorough investigation of the incident (including a check for malicious software on the system) may result in a judgment in favor of the plaintiff. History has shown that malicious software can be used to send e-mails unbeknownst to the user. Recall the Melissa computer virus discussed in Chapter 1. When an individual opened an e-mail infected by this virus, the virus would spread by sending a copy of itself to every name in the individual's address book if the computer ran the Microsoft Outlook program.

Evidence can be acquired from both computer systems and personnel. In addition, it can be acquired from the company's servers and network (explored in further detail in Chapter 11). When a computer suspected to hold evidence is connected to the network, an individual trained on the seizure of networked computers should be contacted. If possible, technical assistance from the IT department should be obtained (wherever required). The acquisition of the evidence would require specialized computer forensics tools, such as Encase and FTK (described in Chapter 7).

If items are to be seized from the employee's workspace, care should be taken to seize only those items that belong to the corporation. The personal belongings of the employee cannot be seized unless the investigation is being conducted by law enforcement agents who have obtained a search warrant for the items in question or one of the exceptions that authorize the seizure of the items without a warrant applies. The computers and the networks to which they are attached are corporate property. Desks, cabinets, and other furniture within an office are also corporate property. However, if an employee has taken measures to ensure the privacy of items within, for example, a desk or cabinet by securing them with a lock that the employee owns, then these items may not be searched by law enforcement agencies without a warrant (unless, of course, they are permitted to do so under one of the legal exceptions to a search warrant, such as with the suspect's consent).

Confidentiality must be maintained during an investigation. One way to ensure that this requirement is met is by limiting the number of individuals involved in the investigation. Time is also of the essence. The longer the investigation takes, the more likely that employees will realize that the company and/or its employees are under investigation. Secrecy is vital when conducting the investigation, because in the majority of corporate cases, the targets of the investigation are active employees. The simple knowledge that an investigation is occurring may result in crucial evidence being destroyed by the suspect (or suspects). In one highly publicized case, Enron employees started shredding documents

---

Trade Secret Theft

Imagine that trade secrets were stolen from EFG Corporation and sent to its competitor, HIJ Corporation.

1. Who would be the most likely suspect? And why do you think this is the case?

2. Describe the steps that an investigator would take to conduct this investigation.

3. Where would an investigator search to find evidence of this crime?

and deleting files and e-mails after they were informed that their company was under investigation.

Sometimes an investigation may require the covert surveillance of employees when using computers and the network of the corporation. A business should have sound computer and network use policies. These policies should help deter and minimize computer policy violations (e.g., using the computer for personal reasons), illegal behavior (e.g., fraud and harassment), and company liability. Computer use (including e-mail and Internet use) could be monitored to ensure that a computer is used only for legitimate business purposes. It can also be monitored to determine the extent of an individual's misuse of the corporation's computer and network privileges.

Corporations must have unambiguous policies in place that alert employees to the possibility that their computing and networking privileges may be monitored in the workplace. These policies should be clearly communicated to employees, as failure to do so may expose a corporation to a lawsuit. All new hires should be made to read this policy and acknowledge their understanding of it. This acknowledgment usually takes the form of the employee signing the computer and network use policy statements. For existing employees, this policy should be distributed to them, and they should be required to read and sign it to indicate that they have read and understood what is required of them. This signed policy should then be placed in each employee's human resources folder. Also, a message should appear each time a user logs on to his or her work computer, which states the policy. Unless the computer user clicks on the box that says "accept" relative to the policy, the user should not be able to log on to the computer. In this way, an electronic record of the user's acceptance of the corporate computer and network use policy may be kept each time an individual uses a corporate computer.

The policy should also contain the applicable disciplinary measures for incidents of computer and network misuse. Examples of such consequences include the following:

- Violations of the computer and network use policy will be documented in the employee's record.
- Criminal acts will be prosecuted in a court of law.
- Lawsuits will be filed against civil law violators.
- Serious company policy violations will result in employee termination.

The company policy should clearly state that computers and related electronic and peripheral devices are the property of the business and that the activities and data on them are subject to search. These devices should be used for business purposes only. Moreover, the policy should stipulate that the company reserves the right to search an employee's workspace. It also reserves the right to seize items that belong to the corporation—even if they are used exclusively by the employee—pursuant to a criminal investigation, civil investigation, or investigation of noncompliance.

The importance of a computer and network monitoring policy is illustrated in *United States v. Simons*.[60] The Foreign Bureau of Information Services (FBIS), which is a division of the Central Intelligence Agency (CIA), contracted Science Applications International Corporation (SAIC) to monitor its computer network. The FBIS had a policy that stated employee computer and network use were subject to monitoring. While monitoring the network, a manager of SAIC observed that the defendant, Mark Simons, had visited child pornography websites. The manager subsequently remotely obtained evidence from Simons' computer and provided this evidence in court. Simons objected to the admission of this evidence, claiming that this search violated his Fourth Amendment rights. The U.S. Supreme Court disagreed with his contention, stating that Simons did not have a reasonable expectation of privacy in his computer and network use because a policy was in place that warned him that his activities were subject to monitoring.

In many instances, phone calls made from and to a corporation can be monitored. Calls between employees and customers can be monitored for quality assurance, although some states require that such monitoring be announced. For example, a pre-recorded message might explicitly state: "This call may be monitored for quality assurance purposes." The Electronic Communications Privacy Act of 1986 allows the monitoring of business calls without notification. State laws, however, may differ (for example, in California). According to the ruling made in *Watkins v. L. M. Berry & Co.*[61] if the employer recognizes that a call from the business is personal, the monitoring of the call must stop pursuant to federal law.

In summary, the location of the corporation will dictate whether monitoring is lawful. The corporation must ensure that its monitoring practices are in accordance with state and federal laws.

Investigating Sexual Harassment

Clear policies can help protect corporations from liability. The availability of grievance (i.e., complaint) procedures also helps minimize their liability, especially in regard to **sexual harassment** cases. According to the Equal Employment Opportunity Commission (EEOC):

> [U]nwelcome sexual advances, requests for sexual favors and other verbal or physical conduct of a sexual nature constitute sexual harassment when submission to or rejection of [this] conduct explicitly or implicitly affects an individual's employment, unreasonably interferes with an individual's work performance or creates an intimidating, hostile, or offensive work environment.[62]

The statutory basis for sexual harassment claims is Title VII of the Civil Rights Act of 1964. Case law has explicitly stated the conditions under which an employer may be held liable for sexual harassment occurring in the workplace. Notably, the employer can be held liable for sexual harassment in the workplace that occurs via e-mails and instant messaging. Additionally, posts on electronic bulletin boards are sufficient to establish employer liability for sexual harassment.[63]

(continues)

Determining Employer Liability *(continued)*

The Supreme Court has held that if an "employer exercised reasonable care to prevent and correct promptly any sexually harassing behavior" and the "employee unreasonably failed to take advantage of any preventive or corrective opportunities provided by the employer,"[64] then the corporation may not be held liable for the harassment. As such, liability is determined by the absence or presence of a policy against sexual harassment and the availability of grievance procedures should sexual harassment occur in the workplace.

If the offense being investigated is sexual harassment, an investigator must review the harassment policy:

- What does it state?
- Which behaviors does it prohibit?
- Are employees aware of this policy? If so, to what extent?
- What disciplinary actions for this behavior are included in the policy?

The investigator must also determine whether the following services are available in the corporation:

- Sexual harassment training: Is it provided? If so, how often?
- Grievance procedures: Are these procedures available to employees? Are they adequate?

The computer forensics investigation should be conducted in such a way as to ensure minimum disruption of day-to-day operations of the corporation. Corporate investigations of criminal (or civil) activities or policy violations normally involve complex configurations of multiple computers networked to each other, to a common server, to network devices, or a combination of these connections.[65] These linkages pose significant challenges to investigators. The clandestine nature of this investigation further complicates matters, especially when the IT department is not involved in the investigation (even in a consulting role).

Under no circumstances should an employee be informed of the alleged policy violation or illegal behavior in front of a third party. Any matters that relate to the investigation should be dealt with in private. Ideally, items should be seized after work hours to avoid causing a scene. Discussing the investigation in front of third parties or seizing items during work hours in front of the plain view of the coworkers of the employee being investigated may expose the corporation to a defamation lawsuit. **Defamation** occurs when a person either as an individual or as a representative acting on behalf of a corporation damages an individual's reputation "by making public statements that are both false and malicious."[66]

To avoid claims of defamation, the investigator should also not seize any property that belongs to the employee. Additionally, if the employee confronts the investigator, the investigator should not accuse the employee of wrongdoing. Moreover, the investigator

---

Hypothetical Corporate Computer Forensics Investigation

A memo (both in electronic and hard copy version) was sent within the PQR Corporation that disparaged an employee, George Smith. Smith informed his manager of the incident. Rather than receiving the response that he expected, the manager belittled Smith and told him to lighten up. Two more memos were subsequently sent, which contained vicious and scathing remarks against Smith, including comments informing him that he was not wanted in the corporation and that the senders of the memos would not stop until Smith had quit. Alarmed by the memos and his manager's indifference, Smith went to senior management with his concerns, only to encounter a similar response to the one he had received from his manager. Smith informed senior management that he would file a formal grievance. His employment was terminated the following day. Smith then filed a lawsuit against the company for harassment and wrongful termination. The only evidence Smith had with him consisted of the hard copies of the memos.

1. Is the evidence Smith has enough to prove his case in court?
2. Which other evidence does Smith need?
3. How would the investigation of Smith's case be conducted to find this evidence?
4. Where would the investigator look for evidence?
5. Which hurdles might the investigator encounter in his or her investigation?

---

should not alert the employee as to the nature of the investigation. If confronted, the investigator should state only that he or she is authorized to conduct the investigation.

Chapter Summary

There are many crimes committed within and for businesses. The majority of these crimes are committed by insiders. The type of corporate investigation conducted will depend on whether a policy, civil, or criminal violation has occurred. If a violation of criminal law is brought to the attention of the corporation, then the police must be notified. Often companies do not wish to prosecute violations because of the damage that publicizing the incident could cause to customer confidence in the company and the corporation's reputation.

Corporations may use IT personnel, their own computer forensics investigators, or externally hired computer forensics experts to conduct the investigation of a suspected violation. Investigations must be conducted in secret so as not to bring the investigation to the attention of the potential suspect or suspects. Care must be taken when conducting the investigation. A cursory inspection may be authorized by the corporation, but the investigator must be aware that property within a workspace may include both the suspect's personal items and items that belong to the corporation. Law enforcement agencies must ensure that they have legal authority to conduct a search if they are brought into the case. The surreptitious nature of these investigations makes them especially challenging. However, failure to conduct a clandestine investigation may result in the destruction of evidence or bringing of a defamation lawsuit. Corporations must have clear policies in place, which unambiguously state the disciplinary actions that will be taken should these policies be violated.

Key Terms

Age Discrimination in Employment Act
   of 1967
American Society for Industrial Security
Americans with Disabilities Act of 1990
Bank Secrecy Act of 1970
Bid rigging
Bribery
Cyberslacking
Data manipulation
Defamation
Discrimination
Equal Employment Opportunity
   Commission
Equal Pay Act of 1963
Extortion
Fair Labor Standards Act of 1938
False Claims Act of 1863
Family and Medical Leave Act of 1993
Foreign Corrupt Practices Act of 1977
Input scam
Insider trading
Intellectual property theft
Invasion of privacy

Money laundering
Negligent hiring
Negligent retention
Noncompete agreement
Nonsolicitation agreement
Occupational Safety and Health Act
   of 1970
Output scam
Pregnancy Discrimination Act of 1978
Price fixing
Sexual harassment
Sherman Antitrust Act of 1890
Tax evasion
Throughput scam
Title VII of the Civil Rights Act of 1964
Trade secret theft
Uniformed Services Employment and
   Reemployment Rights Act of 1994
Whistleblower
Workplace injury
Workplace violence
Wrongful termination

Review Questions

1. Who can report misconduct in the workplace?

2. For which forms of misconduct is an investigation normally conducted in the workplace?

3. What are the two categories of corporate crime?

4. What are input, output, and throughput scams?

5. Which crimes are corporations required to report to the police?

6. What are some reasons why certain crimes committed against or by employees are not reported to the police?

7. When is a business liable for sexual harassment in the workplace?

8. What is the first step in preparing for a corporate computer forensics investigation?

9. Who can conduct a corporate investigation? What are the pros and cons of using those individuals?

10. What is cyberslacking? How does it affect a business?

11. How can a corporate investigation be conducted to avoid raising privacy and defamation claims by an employee under investigation?

12. If child pornography is found on a corporate computer, what should an employee avoid doing?

13. What should a corporate computer, e-mail, and Internet use policy contain?

Footnotes

[1]American Society for Industrial Security. (2007, August). Trends in proprietary information loss. *Survey Report*, p. 3. Retrieved from http://www.asisonline.org/newsroom/surveys/spi2.pdf

[2]A whistleblower is an individual who exposes a person, group, or business for engaging in illegal activities.

[3]This list is by no means exhaustive.

[4]Failure to take action against an employee after being made aware that the employee in question poses a risk to the company and the employees within the company (e.g., an employee had a violent outburst, but no investigation was undertaken and no disciplinary action was taken against the employee).

[5]Failure to conduct a diligent inquiry into the background of a prospective employee and his or her suitability to the position in question before deciding to hire that person.

[6]United States Attorney's Office, Western District of Washington. (2008, August 28). Des Moines man indicted for Internet threats of violence and identity theft. US Department of Justice, Computer Crime & Intellectual Property Section. Retrieved from http://www.justice.gov/criminal/cybercrime/turrellaIndict.htm

[7]United States Attorney's Office, Western District of Washington. (2010, February 12). Des Moines man sentenced to prison for Internet threats of violence and possessing a firearm during threats of violence. Federal Bureau of Investigation, Seattle. Retrieved from http://seattle.fbi.gov/dojpressrel/pressrel10/se021210.htm

[8]McLellan, V. (1984, June 1). Of Trojan horses, data diddling and logic bombs: How computer thieves are exploiting companies' hidden vulnerabilities, p. 3. Retrieved from http://www.inc.com/magazine/19840601/2515_pagen_3.html

[9]Bologna, J. (1993). *Handbook on corporate fraud*. London: Butterworth-Heinemann, p. 91.

[10]United States Attorney's Office, Eastern District of Wisconsin. (2010, January 20). Former Koss Corporation executive charged in $31 million dollar fraud. Federal Bureau of Investigation, Milwaukee. Retrieved from http://milwaukee.fbi.gov/dojpressrel/pressrel10/mw012010.htm

[11]*Ibid.*

[12]Bachus, A. S. (2004). From drugs to terrorism: The focus shifts in the international fight against money laundering after September 11, 2001. *Arizona Journal of International and Comparative Law, 21*, p. 835.

[13]United States District Attorney's Office, Northern District of California. (2009, April 10). Owner and operator of Massachusetts computer parts company pleads guilty to wire fraud and money laundering in connection with $15.4 million dollar Cisco networking equipment fraud. US Department of Justice, Computer Crime & Intellectual Property Section. Retrieved from http://www.justice.gov/criminal/cybercrime/dalyPlea.pdf

[14]US Department of Justice. (2008, July 21). Digital currency business E-Gold pleads guilty to money laundering and illegal money transmitting charges. Retrieved from http://www.justice.gov/criminal/pr/2008/07/07-21-08egold-plea.pdf

[15]*Ibid.*

[16]*Ibid.*

[17]de Goede, M. (2004, September 9–11). *The risk of terrorist financing: politics and prediction in the war on terrorist finance.* Paper presented at Constructing World Orders Conference, Standing Group on International Relations, Transnational Politics of Risk Panel, Den Haag [The Hague]. Retrieved from http://www.sgir.org/conference2004/papers/de%20Goede%20-%20The%20Risk%20of%20Terrorist%20Financing.pdf

[18]Amoore, L., & deGoede, M. (2005). Governance, risk, and dataveillance in the war on terror. *Crime, Law and Social Change, 43*, 151–152.

[19]*Ibid.*, p. 154.

[20]*Ibid.*

[21]United States Attorney's Office, District of Nevada. (2008, October 24). Man convicted of attempted extortion of gaming companies. Federal Bureau of Investigation, Las Vegas. Retrieved from http://lasvegas.fbi.gov/dojpressrel/pressrel08/lv102408.htm

[22]United States Attorney's Office, Northern District of California. (2005, August 1). Former IT director of Silicon Valley company pleads guilty to theft of trade secrets. US Department of Justice, Computer Crime & Intellectual Property Section. Retrieved from http://www.justice.gov/criminal/cybercrime/woodwardPlea.htm

[23]Ortemeier, P. J. (2009). *Introduction to security: Operations and management* (3rd ed.). Upper Saddle River, NJ: Pearson/Prentice Hall, p. 70.

[24]*Ibid.*

[25]Bologna, J. (1993). *Handbook on corporate fraud.* London: Butterworth-Heinemann, p. 91.

[26]Nunziato, M. C. (2010). Aiding and abetting, a Madoff family affair: Why secondary actors should be held accountable for securities fraud through the restoration of the private right of action for aiding and abetting liability under the federal securities laws. *Albany Law Review, 73*, 609.

[27]Stempel, J., & Plumb, C. (2008, December 11). Investors scramble after Madoff is charged with $50B fraud. *USA Today* [online]. Retrieved from http://www.usatoday.com/money/markets/2008-12-11-madoff_N.htm

[28]US Department of Justice. (2008, November 21). Hotel executive pleads guilty and agrees to pay restitution and forfeiture in connection with tax evasion scheme. Federal Bureau of Investigation, New York. Retrieved from http://newyork.fbi.gov/dojpressrel/pressrel08/nyfo112108a.htm

[29]United States Attorney's Office, Eastern District of Virginia. (2008, November 10). Herndon man pleads guilty to $3m embezzlement and tax evasion. Federal Bureau of Investigation, Washington. Retrieved from http://washingtondc.fbi.gov/dojpressrel/pressrel08/wf111008.htm

[30]US Department of Justice. (2008, June 13). Former USProtect chairman sentenced for concealing prior fraud judgments to obtain over $150 million in federal security contracts and tax evasion. Federal Bureau of Investigation, Baltimore. Retrieved from http://baltimore.fbi.gov/dojpressrel/pressrel08/ba061308.htm

[31]Reid, S. T. (2009). *Crime and criminology* (12th ed.). New York: Oxford University Press, p. 553.

[32]Center for Corporate Policy. (2004). Crackdown on corporate bribery: Fix the Foreign Corrupt Practices Act. *Corporate Crime and Abuse*. Retrieved from http://www.corporatepolicy.org/issues/FCPA.htm

[33]Frieden, T. (2009, February 6). KBR charged with bribing Nigerian officials for contracts. *CNN* [online]. Retrieved from http://www.cnn.com/2009/US/02/06/KBR.bribery/

[34]Searcey, D. (2009, May 26). U.S. cracks down on corporate bribes. *Wall Street Journal* [online]. Retrieved from http://online.wsj.com/article/SB124329477230952689.html

[35]Wilber, D. Q., & Stewart, N. (2009, March 13). D.C. tech official is accused of bribery. *Washington Post* [online]. Retrieved from http://www.washingtonpost.com/wp-dyn/content/article/2009/03/12/AR2009031201426.html?hpid=moreheadlines

[36]United States Attorney's Office, District of New Jersey. (2010, January 29). Marlboro Township developer sentenced to probation with home confinement for bribery and tax evasion conspiracy. Federal Bureau of Investigation, Newark. Retrieved from http://newark.fbi.gov/dojpressrel/pressrel10/nk012910.htm

[37]US Department of Justice. (2008, August 26). Former New York Power Authority employee pleads guilty to fraud and tax charges. Federal Bureau of Investigation, New York. Retrieved from http://newyork.fbi.gov/dojpressrel/pressrel08/taxcharges082608.htm

[38]*Arizona v. Maricopa County Medical Society*, 102 S. Ct. 2466, 2473-74 (1982); *Kiefer-Stewart Co. v. Joseph E. Seagram & Sons, Inc.*, 340 U.S. 211, 213 (1951); *United States v. Socony-Vacuum Oil Co.*, 310 U.S. 150 (1940); See also Harrison, J. L. (1982). Price fixing, the professions, and ancillary restraints: Coping with Maricopa County. *University of Illinois Law Review*, p. 925.

[39]United States Attorney's Office, Northern District of California. (2008, November 12). LG, Sharp, Chunghwa agree to plead guilty, pay total of $585 million in fines for participating in LCD price-fixing conspiracies. Federal Bureau of Investigation, San Francisco. Retrieved from http://sanfrancisco.fbi.gov/dojpressrel/pressrel08/sf111208.htm

[40]Hovenkamp, H. (2003). Antitrust violations in securities markets. *Journal of Corporation Law*, 28, 619. Cited in Piraino, T. A. Jr. (2008, September). The antitrust implications of "going private" and other changes of corporate control. *Boston College Law Review*, 49, 1028.

[41]210 F. Supp. 2d 294, 298 (S.D.N.Y. 2002).

[42]Conners, P. A. (2003). Current trends and issues in state antitrust enforcement. *Loyola Consumer Law Review*, 16, 52.

[43]See the Equal Opportunity Employment Commission website (http://www.eeoc.gov/).

[44]Martinez, J. (2010, May 17). Drug giant Novartis Loses $3.3M sex-discrimination suit against 12 women. *New York Daily News* [online]. Retrieved from http://www.nydailynews.com/money/2010/05/17/2010-05-17_jury_drug_giant_novartis_discriminated_against_women_.html?r=news

[45]Rosenzweig, D. (1999, November 5). U.S. will prosecute Fluor Daniel complaint. *Los Angeles Times* [online]. Retrieved from http://articles.latimes.com/1999/nov/05/business/fi-30147

[46]Pae, P. (2005, November 2). Fluor settles lawsuit for $12.5 million. *Los Angeles Times* [online]. Retrieved from http://articles.latimes.com/2005/nov/02/business/fi-fluor2

[47]Reid, S. T. (2007). *Criminal law* (8th ed.). Oxford, UK: Oxford University Press, p. 228.

[48]US Securities and Exchange Commission. (2001). Insider trading. Retrieved from http://www.sec.gov/answers/insider.htm

[49]Hoffman, D. (2007). Current development 2006–2007: Martha Stewart's insider trading case: A practical application of Rule 2.1. *Georgetown Journal of Legal Ethics, 20,* 708.

[50]Thomas, L. Jr. (2006, August 7). Martha Stewart settles civil insider-trading case. *New York Times* [online]. Retrieved from http://www.nytimes.com/2006/08/07/business/07cnd-martha .html?ex=1312603200&en=457785a2ba7ccd4e&ei=5088

[51]Ponemon Institute. (2006). 2006 annual study: Cost of a data breach: Understanding financial impact, customer turnover, and preventative solutions, p. 10. Retrieved from http://www.com-puterworld.com/pdfs/PGP_Annual_Study_PDF.pdf. See also Ponemon Institute. (2007). 2007 annual study: U.S. cost of a data breach: Understanding financial impact, customer turnover, and preventative solutions. Retrieved from http://www.bomgar.com/research/external/ Ponemon-Cost-of-a-Data-Breach-2007.pdf

[52]Javelin Strategy and Research. (2008, June). Consumer survey on data breach notification, p. 2. Retrieved from http://tawpi.org/uploadDocs/Data_Breach_survey.pdf

[53]Javelin Strategy and Research. (2008, June). Consumer survey on data breach notification, p. 2. Retrieved from http://tawpi.org/uploadDocs/Data_Breach_survey.pdf

[54]Masterson, M. (2000, January 4). Cyberveillance at work. *CNN Money* [online]. Retrieved from http://money.cnn.com/2000/01/04/technology/webspy/

[55]Trombly, M. (2000, August 24). Dow to fire up to 40 employees over sexually explicit e-mails. *CNN* [online]. Retrieved from http://archives.cnn.com/2000/TECH/computing/08/24/dow .sex.firing.idg/; Institute for Global Ethics. (2000, August 7). Dow Chemical fires 50 workers after searching email records for pornography. *Ethics Newsline.* Retrieved from http://www .globalethics.org/newsline/2000/08/07/dow-chemical-fires-50-workers-after-searching-email-records-for-pornography/

[56]Disabatino, J. (2000, July 11). E-mail probe triggers firings. *CNN* [online]. Retrieved from http://archives.cnn.com/2000/TECH/computing/07/11/email.firing.idg/index.html

[57]Miles, J. E., Hu, B., Beldona, S., & Clay, J. (2001, October/November). Cyberslacking! A liability issue for wired workplaces: The Internet has brought more than distraction to the workplace employee. *Administration Quarterly.* Retrieved from http://www.entrepreneur.com/tradejournals/ article/82671387.html

[58]Greengard, S. (2000, December). The high cost of cyberslacking: Employees waste time online. *Workforce,* p. 2. Retrieved from http://findarticles.com/p/articles/mi_m0FXS/is_12_79/ai_ 68325779/pg_2/?tag=content;col1

[59]Powell, A., & Vincent, C. (2006, July 20). Content filtering: Sifting through the mess. *NASA SEWP Security Center,* p. 9. Retrieved from http://www.sewpsc.sewp.nasa.gov/documents/content_ filtering.doc

[60]206 F.3d 392 (4th Cir. 2000).

[61]704 F. 2d 577, 583 (11th Cir. 1983).

[62]See US Equal Employment Opportunity Commission. (n.d.). Facts about sexual harassment. Retrieved from http://www.eeoc.gov/eeoc/publications/fs-sex.cfm

[63]*Blakey v. Continental Airlines,* 751 A. 2d 538, 551 - 552 (N. J. 2000).

[64]*Faragher v. City of Boca Raton,* 524 U.S. 775, 807 (1998); *Burlington Industries v. Ellerth,* 524 U.S. 742, 765 (1998).

[65]Network forensics is explored in further detail in Chapter 11.

[66]Ortemeier, P. J. (2009). *Introduction to security: Operations and management* (3rd ed.). Upper Saddle River, NJ: Pearson/Prentice Hall, p. 65.

Chapter 10

E-mail Forensics

This chapter examines the basics of e-mail investigations and the various sources that can be utilized to assist computer forensics specialists, law enforcement agents, and other relevant practitioners in the field in conducting these investigations. It also explores the primary tasks involved in e-mail investigations, including identifying the e-mail, retrieving it, and documenting the examination process. It further considers the techniques required for reviewing e-mail headers and tracking the transmission path and origin of the e-mail. Finally, it covers the problems encountered by computer forensics investigators during e-mail investigations.

The Importance of E-mail Investigations

E-mail investigations are included as a separate chapter in this book because many cyber-crimes have involved the use of e-mails, either as the means with which to commit the crime or as evidence of the crime.

In respect to the former, e-mails may be sent, for example, to extort money from the recipient. A case in point is the crime committed by Oleg Zezev.[1] Zezev sent an e-mail to Michael Bloomberg, the president and chief executive officer of Bloomberg, Inc. (a multinational financial data company), in an attempt to extort $200,000 from him. Specifically, in the e-mail to Bloomberg, Zezev stated that he would inform the media and Bloomberg's customers that Zezev had gained unauthorized access to Bloomberg's computer system, where confidential customer information was retained, if Bloomberg did not pay him the money.

Another similar case involved Ethan Mikeal Avalos.[2] Avalos used e-mails to extort money from a medical professional. In particular, Avalos threatened to expose that the victim had several paid sexual encounters with a woman the victim contacted on Craigslist on several occasions if he did not pay Avalos.

In respect to the use of e-mails as evidence, they may contain evidence of many different types of crimes:[3]

- Death
- Domestic violence
- Cyberbullying
- Cyberharassment
- Cyberstalking
- Extortion
- Embezzlement
- Fraud
- Identity theft
- Computer intrusions
- Child exploitation and abuse
- Terrorism
- Organized crime
- Cyberprostitution
- Drug dealing
- Gambling
- Intellectual property theft
- Electronic espionage

One well-known case involving the use of evidence from e-mails was the Enron scandal. According to the FBI, during the Enron investigation, agents "collected over four terabytes ... of data, including e-mail from over 600 employees."[4] Within these e-mails, FBI agents found numerous remarks and jokes from employees about their shredding of thousands of crucial auditing materials under the direction of Arthur Andersen.

Another case involved Rodney King, an African American male who was brutally beaten by police officers. In this case, an e-mail surfaced in which one of the police officers involved in beating, Lawrence Powell, bragged that he had not beaten up anyone that bad in a long time (when referring to Rodney King).[5]

A third case involved suspected al-Qaeda terrorist member, Ali Saleh Kahlah al-Marri (discussed in Chapter 2), who had opened up several Yahoo and Hotmail e-mail accounts. Using these accounts, he then sent messages to an account identified as belonging to the mastermind of the terrorist attacks on the United States on September 11, 2001, Khalid Shaykh Muhammed.[6]

E-mails are currently the most commonly used form of communication in the United States and hold a wealth of data. The types of information that e-mails contain will be explored in further detail next.

E-mail: The Basics

Two types of e-mail systems exist:

- **Client/server e-mail:** The client is the computer that sends or receives messages; the server stores any messages received until they are retrieved by the user. In client/server systems, e-mails are downloaded to a user's computer.

- **Web-based e-mail:** With a Web-based system, e-mail accounts are accessed through a Web browser and e-mails are stored in the e-mail service provider's server. A few well-known examples of Web-based systems are Yahoo, Gmail, and Hotmail.

To communicate with one another, e-mail systems use a variety of protocols:

- *Simple Mail Transfer Protocol* (SMTP). **SMTP** is part of the Transmission Control Protocol/Internet Protocol (TCP/IP) suite, which is the primary protocol for transmitting messages on the Internet. SMTP is the protocol that is used to send e-mail across the Internet or across a network.

- *Post Office Protocol 3* (POP3). **POP3** is used to read the e-mail. It is designed to store e-mails in a single mailbox until they are downloaded by the user. Specifically, POP3 is designed in such a way as to delete e-mails on the server immediately after they have been downloaded. However, a user or an administrator may opt to save e-mails on a server for a set period of time. As such, these servers should not be overlooked by investigators when they are seeking e-mails that have already been downloaded by the user.

- *Internet Message Access Protocol* (IMAP). Similarly to POP3, **IMAP** is a protocol used to retrieve and read e-mails. It is a more powerful protocol than POP3 but is not used as often as POP3 is. E-mails can be received and stored on the server. Unlike POP3, IMAP affords users with the opportunity to create and manage multiple folders within which to store e-mails on the server.

In summary, an e-mail is sent with SMTP, an e-mail handler receives this message, and it is then read using POP3 or IMAP.

E-mail Address

An e-mail address is used to identify the e-mail box that the message should be sent to. It is made up of two parts: the username and the **domain name**. Consider the following e-mail address: marika.filipopoulou@yahoo.com. The characters on the left side of the @ symbol comprise the username—that is, the mailbox where the message should be sent. The characters on the right hand side of the @ symbol are the domain. The domain name is an identifying label by which computers are familiarly known on the Internet (e.g., yahoo).

Parts of an E-mail

An e-mail primarily has two parts: the body and the header. The body contains the actual message. There are two versions of an **e-mail header**—the condensed and full versions. The data retrieved by the investigator will depend on the version of the e-mail header used. The data included in the e-mail header also largely depend on the e-mail system. With the condensed version of the e-mail header, four basic fields of header information are provided:[7]

- *From*: This field consists of the sender's address. The name of the sender may also be included (whether real or fake). The investigator should keep in mind that the e-mail address of the sender may be faked. SMTP does not verify e-mail headers, so the suspect may disguise (or spoof) his or her address to make it look like another individual sent the e-mail.

- *To*: This field consists of the recipient's address. The name of the recipient may also be included. This address can also be faked or spoofed.

- *Subject*: Sometimes this field may be left blank. The investigator must also keep in mind that this field may contain misleading information.

- *Date*: This field includes the date, day of the week, time and time zone, such as GMT (Greenwich Mean Time), UTC (Coordinated Universal Time), ADT (Atlantic Daylight Time), and CST (Central Standard Time), to name a few. This field is recorded by the computer where the message was sent. It may not be accurate if the sender's clock was set incorrectly (either intentionally or accidentally).

As indicated above, header information cannot be trusted. To confirm header information, you need to expand the header. When expanded, the header reveals much more information. When the header is expanded, the header fields described next are revealed.

X-Originating-IP

The **X-Originating-IP field** reveals the IP address of the computer from which the e-mail was originally sent. The IP address is a unique identifier that is assigned to a computer by service providers when it connects to the Internet.[8] An IP address is composed of four groups of numbers known as **octets** (consisting of 8 bits), separated by periods in between them. Each IP address consists of 32 bits. Each octet must contain a number between 0 and 255. Accordingly, there are 256 possible numbers in each octet, and approximately 4 billion possible IP addresses (4,294,967,296—to be exact).[9] As such, the IP addresses that are assigned to computers are finite. Due to their finite nature, IP addresses are assigned more than once to computers.

IP addresses are assigned to computers by Internet service providers (ISPs). ISPs assign a limited number of **static IP addresses**—that is, permanent addresses of specific computers on the Internet. More commonly, they assign **dynamic IP addresses**. Specifically, computers are randomly assigned temporary IP addresses from a pool of available IP addresses. The user will continue to use this IP address until he or she logs off the Internet. When the user logs off, his or her assigned IP address is placed back into the pool of available IP addresses. When the user logs on to the Internet, the process starts again, and a temporary IP address is randomly assigned to that individual.

The *X-Originating IP* header field provides an investigator with the real IP address of the sender.

Received

To determine the original address from which a message was sent, a user should also pay close attention to the ***Received* field** in the full header. An example of what a user would see in the *Received* field is shown here:

Received Field

Received: by mail-gy0-f180.google.com with SMTP id 13so4018650gyg.25 for <marika.filipopoulou@yahoo.com>; Wed, 09 Jun 2010 07:40:46 -0700 (PDT).

What information does this field reveal to the user? It indicates that the server "mail-gy0-f180.google.com" received the message on June 9, 2010, at 7:40:46 A.M. Pacific Daylight Time (PDT) via SMTP.

Depending on the e-mail system, the *Received* field may also include the IP address of the sender. The format in this case may be:

Received: from [IP address] by [server name] with [Internet protocol], day of the week (first three letters), date [format: day month (first three letters) year], at [time (format is hour: minute: seconds)] time zone.

For example:[10]

Received: from 70.19.46.22 ([70.19.46.22]) by vms170003.mailsrvcs.net (Verizon Webmail) with HTTP; Thu, 03 Jun 2010 22:06:45 -0500 (CDT).

Accordingly, the *Received* field provides the domain name and/or the IP address of the sender of the e-mail. Both can be used to trace the e-mail back to the offender. If an IP address is not available, a request can be made to the domain to retrieve further information on the user.

Sometimes e-mails contain more than one *Received* fields, as shown in the following example.

Multiple *Received* Fields[11]

Received: from 127.0.0.1 (EHLO vms173013.mailsrvcs.net) (206.46.173.13) by mta109.mail.ac4.yahoo.com with SMTP; Thu, 03 Jun 2010 20:06:58 -0700

Received: from vms170003.mailsrvcs.net ([unknown] [172.18.12.133]) by vms173013.mail-srvcs.net (Sun Java(tm) System Messaging Server 7u2-7.02 32bit (built Apr 16 2009)) with ESMTPA id <0L3G00G2DZB94Y5A@vms173013.mailsrvcs.net> for marika.filipopoulou @yahoo.com; Thu, 03 Jun 2010 22:06:47 -0500 (CDT)

Received: from 70.19.46.22 ([70.19.46.22]) by vms170003.mailsrvcs.net (Verizon Webmail) with HTTP; Thu, 03 Jun 2010 22:06:45 -0500 (CDT)

The *Received* field allows the computer forensics investigator to trace the e-mail from the user's mailbox back to mailbox it originated from. As the message travels across the Internet, it passes through a series of routers on the way to the recipient. Each router adds code with the IP addresses and timestamp to the header, as follows: When an e-mail is sent, timestamp information is recorded along with the IP address of the sender. When the e-mail arrives at the e-mail server of the sender's ISP, the timestamp information and the IP address of the e-mail server are recorded. When the e-mail subsequently arrives at the e-mail server of the receiver's ISP, the timestamp information and an IP address of the e-mail server are recorded. Finally, when the e-mail reaches its destination, the recipient's e-mail account, the timestamp information, and the receiver's IP address are recorded.

To determine where the e-mail was sent from, an investigator must start at the bottom. If the e-mail traveled across several servers, then each mail server will be added on top of the existing *Received* header field.

Multiple *Received* field headers can also reveal whether a sender tried to send the e-mail with a false IP address. To make this determination, the investigator compares the sending location listed next to the word "by" in each *Received* field header and the receiving location listed next to the word "from" in the *Received* field header below it. If they do not match, the sender has used a fake IP address.

Return-Path

The **Return-Path field** indicates where the e-mail should be sent back to if it fails to reach the recipient. If the e-mail address provided in the *From* field does not match the e-mail address listed in the *Return-Path*, then the user should suspect that the address provided in the *From* field has been faked or spoofed.

Message ID

The **Message ID field** consists of the name of the server and a unique string that the sending e-mail server assigned to the message. This number can be used to track the message (along with timestamp information) on the originating e-mail server in the logs held by ISPs.

Received-SPF

Consider the examples of the **Received-SPF**[12] **field** shown here.

Excerpts from the Full E-mail Header

Example 1

Received-SPF: pass (mta1077.mail.re4.yahoo.com: domain of smtpreturnpath@skysiteonline .com designates 208.184.87.85 as permitted sender)

Example 2

Received-SPF: neutral (google.com: 92.247.202.229 is neither permitted nor denied by domain of uufas1247@spectrumnet.bg)

Example 3

Received-SPF: fail (mta1019.mail.re4.yahoo.com: domain of bestmeds1@hatfuel.ru does not designate 198.185.97.95 as permitted sender)

The *Received-SPF* field is intended for spam filtering (although it is by no means perfect at this task). The function of this header field is as follows: The receiver of an e-mail makes an SPF query, which checks to see whether the sending server is allowed to send an e-mail for the sender's domain. If the result of the query is "fail," as it is in Example 3, the e-mail should be rejected and not delivered. However, SPF specification does not mandate that rejection occur, so many times receivers will see such messages delivered to their mailboxes (albeit typically to their spam folders). If the result of the SPF query is "neutral" or "pass," another spam filter will take this header information into consideration when deciding whether the e-mail in question constitutes spam.

Authentication-Results

Authentication asserts the "validity of a piece of data about a message (such as the sender's identity) or the message in its entirety."[13] The **Authentication-Results field** decides and/or makes a recommendation to the user as to the validity of the origin of the message and the integrity of its content (if possible).[14] In this sense, this field is similar to the *Received-SPF* field in that it is intended for spam filtering. An example of an *Authentication-Results* field follows.

Authentication-Results Header

Authentication-Results: mta1032.mail.ac4.yahoo.com from=gmail.com; domainkeys=pass (ok); from=gmail.com; dkim=pass (ok)

MIME-Version

The **MIME-Version field** indicates that the message that was sent was composed in compliance with RFC 1341, **Multipurpose Internet Mail Extensions (MIME)**.[15] The MIME protocol standardizes headers and allows e-mails to contain, for example, a textual e-mail header and body in character sets other than **American Standard Code for Information Interchange (ASCII)** format; text with unlimited line and overall length; numerous objects in a single message; messages with multiple fonts; and images, video, audio, and multimedia messages.[16] MIME uses headers to inform the system which type of processing is required to re-create the message. Non-MIME messages may not always be recognized, so the system may not be able to interpret the message. If there is no *MIME-Version* header, the message is not in MIME format and the *Content-type* header (explored later in this section) should not be considered to be in conformance with the MIME format.

Another Excerpt from the Full E-mail Header

MIME-version: 1.0
Content-type: multipart/mixed;
boundary="――=_Part_834179_1894254106.1275620805664"
X-Mailer: Verizon Webmail

Content-type

The **Content-type field** indicates the type of data included in the message, such as text, image, audio, video and multipart (i.e., different types of data in a single message). It also includes the subtypes of these data.[17] For instance, in the preceding example, the *Content-type* header field indicates that the type of data is multipart and the subtype is mixed. This field thus specifies that the body of the message is made up of multiple independent parts that should be bundled in a particular order to be viewed by the recipient.[18]

X-Mailer

The **X-Mailer field** specifies the e-mail system used to send the message. For example, if an e-mail was sent from Microsoft Outlook, an *X-Mailer* header field will be attached to the outgoing message that indicates Microsoft Outlook sent the message. In the preceding example, the e-mail client that sent the message was Verizon Webmail.

Depending on the type of e-mail system from which the message was sent, there may be more or fewer header fields than are depicted in the example.

How to Conduct an E-mail Investigation

The primary tasks of a computer forensics investigator in e-mail investigations involve identifying the e-mail, retrieving it, and documenting the examination process.

Obtaining the E-mail

The first step in an e-mail investigation is to make an evidentiary copy of the digital evidence. The copy of the e-mail must include the header information and any attachments. A computer forensics investigator should keep in mind that if an individual sent an incriminating document via e-mail, even if the e-mail on the receiving end has been deleted, a copy of it will still reside in the sent folder of the suspect's e-mail program. Assume that the suspect deletes the copy of the e-mail that is in his or her sent folder. Even in this scenario, a computer forensics investigator can still find a copy of the e-mail attachment in the computer's hard drive or the backup tape of a network server. Other places where e-mails can be found include temporary files and in the unallocated space if the temporary files have been erased.

Searching the E-mail for Evidence

For e-mail investigations, a computer forensics investigator can search the body and headers of the e-mail and e-mail server log files for potential evidence. Once a copy of the e-mail is made, the two main parts of an e-mail message—the header and the body—must be reviewed. The information in both parts can be of value to an investigation. An investigator should also check, if applicable, attachments, individuals who have received copies of the e-mail as secondary recipients (i.e., carbon copies [CC] or blind carbon copies [BCC]), individuals to whom the message was forwarded, and the original message to which the e-mail under investigation represents a response.

The investigator should examine both the condensed and full versions of the header. E-mail headers can assist investigators in numerous ways, as they include a record of the entire transmission of the e-mail. Within these headers, the most reliable and most important information to an investigation is the IP address.

The **PING command** can be used to validate IP addresses found in the e-mail header. It checks the accessibility of the IP address. The IP address can also be obtained by using the PING command, as long as the domain name is known. **Figure 10–1** depicts how the PING command can provide this information. The first PING command is directed at retrieving the IP address of a domain name (www.google.com). The second PING command is used to determine whether the computer that the IP address is allocated to is accessible.

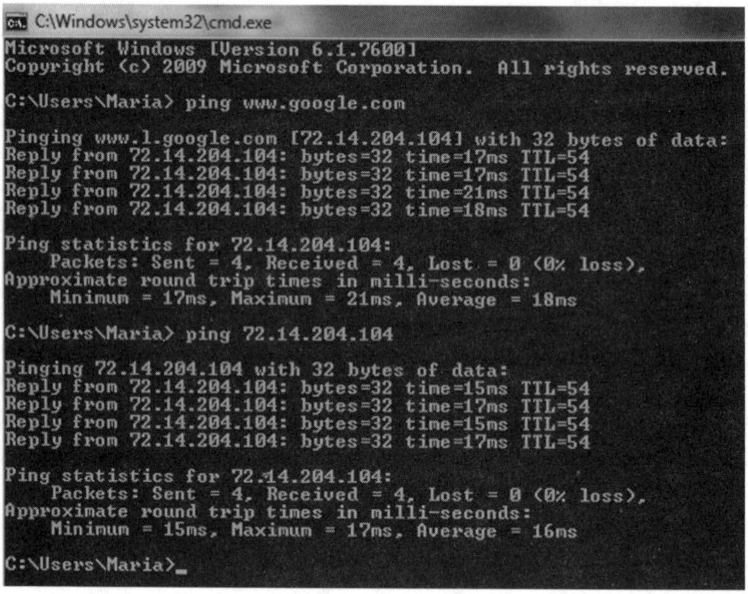

Figure 10–1

Once the IP address has been retrieved, the owner of the IP address needs to be verified.

Verifying the Owner of an IP Address

The **Internet Assigned Numbers Authority (IANA)** is responsible for coordinating the general pool of IP addresses and providing them to **Regional Internet Registries (RIRs)**. RIRs administer and register IP addresses within a defined region. Several RIRs exist:

- **American Registry for Internet Numbers (ARIN).** ARIN is responsible for assigning and registering IP addresses in the North American region.

- **African Network Information Center (AfriNIC).** AfriNIC serves as the registry of Internet number resources for Africa.

- **Asia Pacific Network Information Center (APNIC).** APNIC is responsible for IP address allocation and registration in the Asia-Pacific region.

- **Latin American and Caribbean Internet Addresses Registry (LACNIC).**
  LACNIC serves as the regional Internet registry for this part of the Southern
  Hemisphere.

- **Regional Internet Registry for Europe (RIPE).** This registry assigns and registers IP addresses from Europe, the Middle East, and parts of Central Asia.

Of particular importance to U.S. investigators is the North American registry, ARIN.
ARIN's database holds a number of resource records (e.g., IP addresses), organizational
data, and point of contact information (e.g., the address of the organization). According
to the ARIN website, it is the responsibility of the organization associated with the records
in the database to ensure the validity of the information contained in these records. A
query tool on ARIN's website, known as **WHOIS**, allows a user to find out the contact and
location information of the owner of an IP address. To use WHOIS, the computer forensics investigator types the IP address retrieved from the e-mail into this query tool, and the
tool then retrieves information about the ISP.

Assume that the IP address retrieved from the suspect's e-mail was 72.14.204.104 (the
same IP address used in the PING command in Figure 10–1). When the investigator types
this IP address into the WHOIS query box and presses the Enter key, the following information is displayed:

WHOIS Query for IP Address

OrgName: Google Inc.
OrgID: GOGL
Address: 1600 Amphitheatre Parkway
City: Mountain View
StateProv: CA
PostalCode: 94043
Country: US

NetRange: 72.14.192.0 - 72.14.255.255
CIDR: 72.14.192.0/18
NetName: GOOGLE
NetHandle: NET-72-14-192-0-1
Parent: NET-72-0-0-0
NetType: Direct Allocation
NameServer: NS1.GOOGLE.COM
NameServer: NS2.GOOGLE.COM
NameServer: NS3.GOOGLE.COM
NameServer: NS4.GOOGLE.COM

(continues)

Comment:
RegDate: 2004-11-10
Updated: 2007-04-10

RTechHandle: ZG39-ARIN
RTechName: Google Inc
RTechPhone: +1-650-318-0200
RTechEmail: arin-contact@google.com

OrgTechHandle: ZG39-ARIN
OrgTechName: Google Inc
OrgTechPhone: +1-650-318-0200
OrgTechEmail: arin-contact@google.com

Domain names can also be used to conduct a WHOIS search. Consider once again the example of www.google.com. The user types Google into the WHOIS query box and presses Enter. Numerous hits are returned, some of which are shown here:

Part 1: WHOIS Query for Google

Google Apps. (GOOGL-2)
Google Inc. (GOGL)
Google Incorporated (GOOGL-1)
Google Apps (GOOGL-ARIN) apps-arin-contact@google.com +1-650-253-0000
Google Inc (ZG39-ARIN) arin-contact@google.com +1-650-318-0200
Google Inc. (AS15169) GOOGLE 15169
Google Inc. (AS36039) GOOGLE 36039
Google Inc. (AS36040) GOOGLE 36040
Google Inc. (AS15169) GOOGLE 15169
Google Inc. (AS22859) ON2-TECH-DUCK 22859
Google Inc. (AS36039) GOOGLE 36039
Google Inc. (AS36040) GOOGLE 36040
Google Incorporated (AS36384) GOOGLE-IT 36384
Google Incorporated (AS36385) GOOGLE-IT 36385
Google Apps. GOOGLE-APPS (NET-173-194-240-0-1) 173.194.240.0 - 173.194.247.255
Google Apps. SPRINTLINK (NET-206-160-135-240-1) 206.160.135.240 - 206.160.135.255
Google Apps. SPRINTLINK (NET-208-21-209-0-1) 208.21.209.0 - 208.21.209.15
Google Apps. GOOGLE-APPS (NET-173-194-248-0-1) 173.194.248.0 - 173.194.255.255
Google Apps. GOOGLE-APPS-1 (NET-173-255-112-0-1) 173.255.112.0 - 173.255.127.255
Google Inc. GOOGLE (NET-216-239-32-0-1) 216.239.32.0 - 216.239.63.255
Google Inc. GOOGLE (NET-64-233-160-0-1) 64.233.160.0 - 64.233.191.255
Google Inc. GOOGLE (NET-66-249-64-0-1) 66.249.64.0 - 66.249.95.255

(continues)

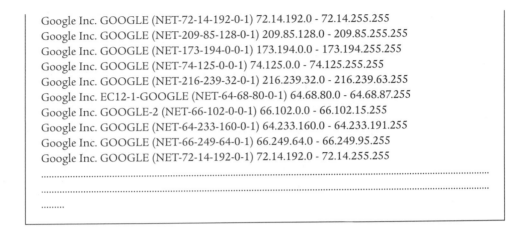

```
Google Inc. GOOGLE (NET-72-14-192-0-1) 72.14.192.0 - 72.14.255.255
Google Inc. GOOGLE (NET-209-85-128-0-1) 209.85.128.0 - 209.85.255.255
Google Inc. GOOGLE (NET-173-194-0-0-1) 173.194.0.0 - 173.194.255.255
Google Inc. GOOGLE (NET-74-125-0-0-1) 74.125.0.0 - 74.125.255.255
Google Inc. GOOGLE (NET-216-239-32-0-1) 216.239.32.0 - 216.239.63.255
Google Inc. EC12-1-GOOGLE (NET-64-68-80-0-1) 64.68.80.0 - 64.68.87.255
Google Inc. GOOGLE-2 (NET-66-102-0-0-1) 66.102.0.0 - 66.102.15.255
Google Inc. GOOGLE (NET-64-233-160-0-1) 64.233.160.0 - 64.233.191.255
Google Inc. GOOGLE (NET-66-249-64-0-1) 66.249.64.0 - 66.249.95.255
Google Inc. GOOGLE (NET-72-14-192-0-1) 72.14.192.0 - 72.14.255.255
```

To be effective, an investigator would have to know exactly which Google from the options in this list he or she is looking for. If the investigator did know this information, he or she could identify the one that was pertinent to the investigation, and the WHOIS search might provide the following results:

Part 2: WHOIS Query for Google

```
OrgName: Google Inc.
OrgID: GOGL
Address: 1600 Amphitheatre Parkway
City: Mountain View
StateProv: CA
PostalCode: 94043
Country: US
Comment:
RegDate: 2000-03-30
Updated: 2009-08-07

AdminHandle: ZG39-ARIN
AdminName: Google Inc
AdminPhone: +1-650-318-0200
AdminEmail: arin-contact@google.com

TechHandle: ZG39-ARIN
TechName: Google Inc
TechPhone: +1-650-318-0200
TechEmail: arin-contact@google.com
```

The Murder of Bobbie Jo Stinnett

E-mails have often helped an investigator find the perpetrator of a crime. A notable example involves the death of Bobbie Jo Stinnett. Stinnett was strangled and her eight-month-old fetus was cut from her womb and kidnapped. An investigator searched Stinnett's computer for evidence related to the crime and found that just before her murder, the young woman had received a message from someone requesting to buy a dog from Stinnett. The e-mail was supposedly sent by someone named Darlene Fisher from the e-mail address, fischer4kids@hotmail.com.[19] An investigator traced the IP address back to a computer at Lisa Montgomery's house. In light of further evidence that the investigator found linking her to the crime, Montgomery was subsequently convicted of murdering pregnant Bobbie Jo Stinnett and kidnapping her fetus.

Domain-related information is best located at the following address: whois .internic.net. ARIN's WHOIS query tool shows that domain names and IP addresses can allow an investigator to track an e-mail to a particular ISP and its location. It is important to note that IP addresses can be traced back to the user. Static IP addresses are easy to trace back to the computer. To trace a dynamic IP address back to a specific user, however, the investigator must provide the ISP with the IP address and a specific date and time that a suspect was online or a crime was committed. With this information, the ISP can look up the individual who was temporarily assigned the specific IP address during the particular date and time in question.

In summary, after determining the originating IP address, the investigator should use a query tool to retrieve the name and contact information of the ISP that assigned the IP address in question. Once this ISP is located, the investigator should draft a subpoena or search warrant—whatever is required for the particular situation—for ISP records pertaining to the e-mail message and its sender. Generally, a search warrant is needed to view the contents of a computer. By contrast, data on ISP servers requires only a subpoena.

As the final step in the e-mail investigation process, the investigator secures and documents the evidence. The chain of custody must also be maintained at all times.

Hypothetical Corporate E-mail Investigation

An employee was fired from XYZ Corporation. The disgruntled employee sought revenge against his manager, whom he believed was responsible for his termination. To exact revenge, the employee spoofed his e-mail address to match that of his manager and sent an e-mail to a competitor of XYZ, STU Corporation. This e-mail informed STU Corporation that the manager had stolen trade secrets from XYZ Corporation that he was willing to sell for $500,000. Instead of replying to the sender of the e-mail, STU Corporation forwarded the

(continues)

e-mail to its legal department. The legal department, after reviewing the e-mail, contacted the legal department of XYZ Corporation. Senior executives were subsequently notified of the incident. Without conducting a formal investigation of the matter, the manager, whom they believed had sent the e-mail, was terminated.

1. Which issues may arise as a result of the employee's termination?

2. What would you have done differently?

3. Provide a step-by-step breakdown of how the investigation of the incident should have been conducted.

Problems Encountered by Computer Forensics Investigators

Criminals use different techniques with which to communicate undetected online. They may also seek to avoid detection by changing their IP or e-mail addresses. In other cases, criminals may use **proxy servers** to hide or mask their IP addresses. If an individual uses a proxy server (i.e., a server that acts as an intermediary for client requests for resources from other servers; see Chapter 11) when accessing a website, the user's identity is not revealed because the proxy server provides its own identity when it retrieves the website for the user. Frank Tribble was one offender who used this technique: He pleaded guilty to, among other things, fraud and violating the **CAN-SPAM Act of 2003** by using proxy servers and fake e-mail headers to send spam touting Chinese penny stocks.[20]

Criminals may also use anonymous remailers to communicate undetected. These services allow an individual to send an e-mail without revealing the sender's identity to the receiver. Moreover, a criminal may intentionally bounce (or route) his or her communication through numerous intermediate computers all over the world before arriving to the target computer. To find the criminal, the investigator will have to identify each router or bounce point through which the message traveled to eventually find the e-mail's point of origin. This is likely to be a slow process, as investigators may have to retrieve data from each point pursuant to a subpoena, court order, or search warrant (depending on which location in the world the message bounced to) to trace the message back to the computer from which it originated.[21]

Criminals may attempt to evade detection by setting up a foreign POP3 or IMAP e-mail account and then accessing this account by using certain Web-based systems such mail2web.com.[22] With this approach, the mail is stored in and accessed from a foreign account,[23] so that the criminal never uses publicly available electronic communications services or public communications networks. Investigators may run into significant obstacles when attempting to obtain e-mails retained on foreign servers.

Moreover, criminals have tried to avoid detection by accessing the Internet from Internet cafés. For instance, terrorists may be able to avoid detection in this venue by using passports and IDs—either obtained by sympathizers or altered or counterfeited documents—if required to provide identification to use these services.[24]

Criminals have also been known to use library computers to commit crimes, believing that the anonymity afforded to them in these venues will shield them from detection by authorities. This, however, is not always the case. Jeff Henry Williamson managed to evade capture by the FBI after he posted a message threatening to kill FBI agents on the FBI's tip Web page from the public access computers at the Washoe County Law Library in Nevada.[25] When he sent a threatening e-mail to several government officials a few weeks later, he was not as lucky. Specifically, Williamson sent an e-mail to the United States Attorney's Office in downtown Houston, the Department of Justice, the Inspector General, and the House Select Intelligence Committee from the public access computers at the Houston Public Library. This message stated:

> Please advise FBI Director Mueller I will take justice into my hands and blow the front of the J. Edger Hoover building off to get everyone's attention—then the CIA HQ and DOJ.[26]

The FBI was able to trace the e-mail back to Williamson's Yahoo account. In March 2010, he was found guilty of sending a threatening communication in interstate commerce.[27]

Marx identified several behavioral techniques used by individuals to subvert surveillance.[28] One such technique is **avoidance**. Avoidance moves are passive and do not attempt to directly engage or tamper with surveillance.[29] With this technique, individuals' actions are displaced to "times, places and means in which the identified surveillance is presumed to be absent."[30] Consider al-Qaeda's use of the Internet to spread propaganda. Videos messages from Abu Musab al-Zarqawi (a known al-Qaeda leader who died in June 2006) were spreading like a computer virus on the Internet in 2005. websites and message boards were used to distribute these videos. While different websites simultaneously uploaded these videos, the videos were subsequently removed after a short period of time and placed on other websites for further distribution. These websites were frequently updated and propaganda removed and switched to different online locations to avoid detection. This practice made it extremely difficult for authorities to track both the individual who was uploading these videos and the locations from which these videos were being uploaded.

Another technique described by Marx is **piggybacking** moves. In this approach, when surveillance is being performed, information that needs to be passed on to a third party undetected can be attached to a legitimate object. One way this goal can be accomplished is through the use of steganography (discussed in Chapter 7). Al-Qaeda operatives and supporters, for example, have used steganography to send messages to one another online. This technique was reportedly used by the al-Qaeda operatives who plotted to blow up the U.S. embassy in Paris in 2001. With steganography, only those individuals with the appropriate software can see the hidden messages.

As mentioned in Chapter 7, terrorists may also hide messages in spam. Several tools are available online that allow terrorists and other criminals to do so. For instance, spammimic is a program that turns e-mail messages into spam. This program is designed to encode the e-mail to appear as spam. The spam is then decoded by the recipient to reveal the original message.

Yet another technique that is used to avoid surveillance is **blocking moves**. In this case, individuals "physically block access to the communication or, if unable or unwilling to do that, to render it (or aspects of it such as the identity, appearance or location of the communicator) unusable."[31] An example provided by Marx is encryption, which is frequently used by individuals to protect their privacy (see Chapter 7). For example, individuals may encrypt messages by transforming the **plaintext** message (or parts of it, such as any personal information of the sender) into a **ciphertext** message. The ciphertext message is gibberish; thus, if eavesdroppers get their hands on this ciphertext, they will not be able to determine what the message means. The intended recipient of the message should have the **decryption** key—the means for transforming the ciphertext back into the plaintext message.

Explicit instructions exist regarding how members of al-Qaeda should communicate with one another to avoid detection. Indeed, terrorists are taught how to use the Internet to send messages and how to encrypt those communications to avoid detection by authorities and make sure that their messages are read only by the intended recipients. With easily available spoofing (electronic concealment of Internet addresses) techniques, terrorists can send encrypted e-mails from hidden locations on the Internet to and from one another undetected.[32] The extent to which terrorists have used these advanced techniques to avoid surveillance remains unclear. Some analysts argue that terrorist tactics do not need to be so sophisticated in their use of computer technology, when coded language sent in plaintext messages or placing advertisements on websites would suffice.[33]

Perhaps terrorists will mimic the Sicilian Mafia and resort to using **pizzini** (small slips of paper—either tiny handwritten or typewritten notes—used for high-level communications to evade fax or phone taps) as a means of communication to avoid the surveillance of their telecommunications and electronic communications. Future technology might also introduce other means for communicating undetected. These criminals may even resort to primarily communicating through ordinary mail. Numerous online and printed manuals have been published by al-Qaeda on how to mail letters and items in an effort to avoid detection by law enforcement agencies.[34]

Criminals find creative ways to inhibit law enforcement efforts by using privacy-enhancing technology. An example of such technology is **Tor**—an anonymous Internet communication system that, among other things, provides individuals (and organizations) with the ability to share information and communicate over public networks without compromising their privacy. Tor works by distributing a user's transactions

over several places on the Internet, so no single point can link [a user] to [his or her] destination . . . Instead of taking a direct route from source to destination, data packets on the Tor network take a random pathway through several relays that cover your tracks so no observer at any single point can tell where the data came from or where it [is] going.[35]

Nevertheless, criminals are not the only persons who seek to use privacy-enhancing technologies, such as Tor. Some individuals use these technologies to retain **anonymity**, wherein an individual seeks and finds freedom from identification and surveillance in public or while engaging in public activities (this can include Internet activities).[36] Anonymity concerns the ability of individuals to conduct their lives without making their activities known to others. It allows individuals to avoid justifying themselves and their personal preferences (religious, political, and sexual) in the face of scrutiny and allows them to arrange their lives in ways that may differ from those exercising disciplinary power or, in the case of sexual preferences (maybe even political views), allows them to differ from social expectations (e.g., conservative views frowning on bisexuality, transexuality, or homosexuality). Thus there may exist perfectly legitimate reasons for wanting to remain anonymous. Perhaps individuals want to present some idea publicly but do not want everyone to identify them, even authorities, especially if they are expressing unpopular views. As Gandy explained, as long as the purpose for remaining anonymous (or even changing one's identity) is nonfraudulent and not an attempt to escape legitimate debts and responsibilities, a person should have the right to remain anonymous (or change his or her identity).[37]

Respect for anonymity is also important because it provides a vital diversity of speech and behavior by allowing individuals to make autonomous and unmonitored choices.[38] Without anonymity, unpopular or different views concerning constitutionally protected decisions about speech, beliefs, religious, political, and intellectual association, which anonymity seeks to shelter, may be stifled.[39] Yet, while the objectives of anonymity are desirable, there also exist arguments against its maintenance. In online forums, Lessig argues that while anonymity may encourage freedom of speech, it may also facilitate anti-social behaviors (e.g., hate speech, incitement to violence, and verbal attacks on other users); conversely, while forbidding anonymity may restrict these antisocial behaviors, it may also inhibit protected rights such as the freedom of speech.[40]

The social costs of the loss of anonymity may include individuals' disengaging from the behavior altogether (not expressing their beliefs, views, and so on, especially if they are unpopular) to avoid privacy disutility (when an individual experiences a loss of privacy caused by surveillance) and taking measures to protect their privacy.[41] Measures to protect an individual's privacy include the use of privacy-enhancing technologies (for example, encryption). These technologies help protect personal privacy by providing users with anonymity and with the ability to choose if, when, to whom, and, more generally, under which circumstances their personal information is disclosed. In practice, these techniques hide some or all of the user's characteristics, thereby making the individual's identification much more difficult.

Chapter Summary

This chapter focused on how a computer forensics investigator conducts e-mail investigations. E-mails can be the tool used to commit a crime or can contain evidence of a crime.

To find evidence, the investigator examines the body and header information of the e-mail. The most useful piece of information in the e-mail header fields is the originating IP address. The IP address can be used to track down the person who sent the e-mail—although criminals have discovered many different ways to make the tracing of an e-mail back to the source quite difficult (e.g., by using proxy servers, anonymizers, or privacy-enhancing technologies). Tools available on the Internet, such as the ARIN WHOIS database, allow a user to look up the ISP that assigned a specific IP address. If investigators have been given the legal authority to request data from the ISP—that is, if they have obtained a subpoena, court order, or search warrant (depending on the location of the ISP)—the provider will make the requested information about the computer from which the e-mail was sent and the contact information of the owner of the computer available to the investigator.

Key Terms

African Network Information Center (AfriNIC)

American Registry for Internet Numbers (ARIN)

American Standard Code for Information Interchange (ASCII)

Anonymity

Asia Pacific Network Information Center (APNIC)

Authentication-Results field

Avoidance

Blocking moves

CAN-SPAM Act of 2003

Ciphertext

Client/server e-mail

Content-type field

Decryption

Domain name

Dynamic IP address

EHLO

E-mail header

ESMTPA

HTTP

IMAP

Internet Assigned Numbers Authority (IANA)

Latin American and Caribbean Internet Addresses Registry (LACNIC)

Message ID field

MIME-Version field

Multipurpose Internet Mail Extensions (MIME)

Octet

Piggybacking

PING command

Pizzini

Plaintext

POP3

Proxy server

Received field

Received-SPF field

Regional Internet Registries (RIRs)

Regional Internet Registry for Europe (RIPE)

Return-Path field

SMTP

Static IP address

Tor

Web-based e-mail

WHOIS

X-Mailer field

X-Originating-IP field

Practical Exercise

Find a spam or phishing e-mail in your e-mail account. Analyze the condensed and full e-mail header fields.

1. What did you observe?
2. Which header fields were included in the chosen e-mail that were not covered in this chapter?
3. What forensic value do they hold, if any?

Critical Thinking Question

Given what you know about anonymity, would you argue for its maintenance online or against it? Support your answer with the material in this chapter and independent research.

Review Questions

1. What are the two types of e-mail systems?
2. What does an e-mail address consist of?
3. Can the condensed header fields be trusted? Why or why not?
4. Which information can e-mail headers provide?
5. What is an IP address?
6. Are IP addresses finite or infinite? Why?
7. How do you verify an IP address?
8. What is the difference between a static IP address and a dynamic IP address?
9. What is the purpose of the PING command?
10. How do you conduct an e-mail investigation?
11. What is anonymity?
12. What are some ways in which criminals evade detection of their communications?

Footnotes

[1]United States Attorney's Office, Southern District of New York. (2003, July 1). Kazakhstan hacker sentenced to four years prison for breaking into Bloomberg systems and attempting extortion. US Department of Justice, Computer Crime & Intellectual Property Section. Retrieved from http://www.justice.gov/criminal/cybercrime/zezevSent.htm

[2]United States Attorney's Office, Eastern District of Missouri. (2010, April 29). St. Louis city man indicted on extortion charges. Federal Bureau of Investigation, St. Louis. Retrieved from http://stlouis.fbi.gov/dojpressrel/pressrel10/sl042910.htm

[3]This list is by no means exhaustive.

[4]Federal Bureau of Investigation. (2009). Crime in the suites: Enron. *The FBI: A centennial history, 1908–2008.* Retrieved from http://www.fbi.gov/fbihistorybook.htm

[5]Axelrod, D. F., Phillips, J. E., & Conners, K. R. (2001, August). Hard times with hard drives: Paperless evidence issues that can't be papered over. *The Champion, 25,* 18.

[6]See Jeffery Rapp's report about al-Marri for further information: http://www.washingtonpost .com/wp-srv/nation/documents/jeffreyrapp_document.pdf

[7]For details on e-mail headers, see document RFC 2076 (Common Internet Message Headers) at http://www.ietf.org/rfc/rfc2076.txt

[8]See Chapters 2 and 5 for more information on IP addresses.

[9]Given that four octets make up an IP address and there are 256 possible numbers in each octet, this number was calculated as follows: $256 \times 256 \times 256 \times 256$.

[10]In this example, CDT stands for Central Daylight Time. **HTTP** stands for Hypertext Transfer Protocol. HTTP is a protocol that is used to transmit files, including Web pages, across the Internet or a network.

[11]One of these *Received* fields included the Internet protocol **ESMTPA** (Extended SMTP with SMTP Authentication). ESMTPA is a protocol used to transmit e-mail. Another *Received* field included **EHLO** (Extended Hello). The receiving server logged EHLO because the sending server identified itself when it initiated the transfer of the e-mail to the receiving server.

[12]SPF stands for Sender Policy Framework.

[13]Kucherawy, M. (2009, April). Message header field for indicating message authentication status. In Network Working Group, *Request for Comments: 5451 (RFC 5451).* Retrieved from http://www.ietf .org/rfc/rfc5451.txt

[14]*Ibid.*

[15]Borenstein, N., & Freed, N. (1992, June). MIME (Multipurpose Internet Mail Extensions): Mechanisms for specifying and describing the format of Internet message bodies. In Network Working Group, *Request for Comments: 1341 (RFC 1341).* Retrieved from http://www.faqs.org/rfcs/ rfc1341.html

[16]The specifications of MIME are defined in RFC 2045, RFC 2046, RFC 2047, RFC 2049, RFC 4288, and RFC 4289.

[17]For descriptions of each type and subtype in the *Content-type* header field, see the following source: Borenstein, N., & Freed, N. (1993, September). MIME (Multipurpose Internet Mail Extensions) Part One: Mechanisms for specifying and describing the format of Internet message bodies. In Network Working Group, *Request for Comments: 1521 (RFC 1521).* Retrieved from http://www.faqs .org/rfcs/rfc1521.html

[18]Borenstein, N., & Freed, N. (1993, September). MIME (Multipurpose Internet Mail Extensions) Part One: Mechanisms for specifying and describing the format of Internet message bodies. In Network Working Group, *Request for Comments: 1521 (RFC 1521).* Retrieved from http://www.faqs .org/rfcs/rfc1521.html

[19]Becker, M., & Bode, N. (2004, December 26). *Daily News* special report on the lives of the two women whose destinies met in one shocking crime. *NY Daily News* [online]. Retrieved from http://www.nydailynews.com/archives/news/2004/12/26/2004-12-26_daily_news_special_ report_on.html

[20]United States Attorney's Office, Eastern District of Michigan. (2008, October 17). Second guilty plea entered in Ralsky spam conspiracy. US Department of Justice, Computer Crime & Intellectual Property Section. Retrieved from http://www.cybercrime.gov/tribblePlea.pdf

[21]The USA Patriot Act 2001 alleviated some of the burdens placed on law enforcement agents seeking access to data nationwide (see Chapter 3). However, the speed with which the tracing of

an offender's message can occur largely depends on the country with which law enforcement agencies are dealing, its laws, and any existing treaties, conventions, and agreements with that country.

[22]Walker, C., & Akdeniz, Y. (2003). Anti-terrorism laws and data retention: War is over? *Northern Ireland Legal Quarterly, 54*(2), 171.

[23]*Ibid.*

[24]For more information on how such tactics are being used by terrorists, see Rudner, M. (2008). Misuse of passports: Identity fraud, the propensity to travel, and international terrorism. *Studies in Conflict and Terrorism, 31*(2), 95–110.

[25]Williamson, J. H. (2010, June 2). Threat to blow up FBI building in D.C. results in prison term. Federal Bureau of Investigation, Houston. Retrieved from http://houston.fbi.gov/dojpressrel/pressrel10/ho060210.htm

[26]*Ibid.*

[27]*Ibid.*

[28]Marx, G. T. (2003, July). A tack in the shoe: Neutralizing and resisting the new surveillance. *Journal of Social Issues, 59*(2), 369–390.

[29]*Ibid.*, p. 375.

[30]*Ibid.*

[31]*Ibid.*, p. 379.

[32]See Carter, P. (2003, March 12). Al Qaeda and the advent of multinational terrorism: Why material support prosecutions are key in the war on terrorism. Retrieved from http://writ.news .findlaw.com/student/20030312_carter.html

[33]Costigan, S. S. (2007). Terrorists and the Internet: Crashing or cashing in? In S. S. Costigan & D. Gold, (Eds.), *Terrornomics*. Aldershot, UK: Ashgate, p. 119.

[34]Gunaratna, R. (2002). *Inside Al-Qaeda: Global network of terror*. London: Hurst and Company, p. 81.

[35]The Tor Project. (2010). Tor: The solution. Retrieved from http://www.torproject.org/overview .html.en#whyweneedtor

[36]Westin, A. F. (1966, June). Science, privacy and freedom: Issues and proposals for the 1970s: Part I—The current impact of surveillance on privacy. *Columbia Law Review, 66*(6), 1021.

[37]Gandy, O. H. Jr. (1993). *The panoptic sort*. Oxford, UK: Westview Press, p. 285.

[38]Cohen, J. E. (2000, May). Examined lives: Informational privacy and the subject as object. *Stanford Law Review, 52*, 1424–1425.

[39]*Ibid.*

[40]Lessig, L. (1999). *Code and other laws of cyberspace*. New York: Basic Books. Cited in Jones, R. (2005). Surveillance. In C. Hale et al. (Eds.), *Criminology*. Oxford, UK: Oxford University Press, p. 489.

[41]For the social costs of privacy disutility, see Song, A. (2003). *Technology, terrorism, and the fishbowl effect: an economic analysis of surveillance and searches* (Research Publication No. 2003-045/2003). Cambridge, MA: Berkman Center for Internet and Society, Harvard Law School.

Chapter 11

Network Forensics: An Introduction

This chapter begins by exploring the differences between stand-alone and networked devices. It then looks at the different types of computer networks. Next, it briefly introduces network forensics, examining the components of a network that are critical to an investigation. Subsequently, it analyzes the types of evidence that can be retrieved from network forensics investigations and the tools that can be used to capture, inspect, and preserve that evidence. Finally, this chapter identifies the procedures used to conduct network forensics investigations.

Stand-alone Versus Networked Devices

A **stand-alone computer** is not connected to another computer or network. These computers are more secure than those connected to a network, because they are less prone to attacks by malicious software. By contrast, a **networked computer** is connected to one or more computers in a manner that allows them to share data, software, and hardware. Networked computers are more vulnerable to attacks by hackers and malicious software.

Computer Networks

Several types of computer networks exist. The simplest computer network is the **local area network** (LAN). LANs are used to connect computers within a small area and provide these systems with the ability to share resources (e.g., hardware, software, and information). LANs are normally used to connect computer systems within a building.

When network access is sought beyond a building, **wide area networks** (WANs), **metropolitan area networks** (MANs), or **campus area networks** (CANs) may be used. WANs and MANs seek to connect computers in different locations with one important difference: MANs are restricted to a particular city or town. CANs are used to connect computer systems in a particular area. Examples of where CANs are used include college campuses, university campuses, and military bases.

Network Configuration

Network configuration depends on the size of a corporation, organization, or agency and its needs. Two types of network configurations are described here:

- **Peer-to-peer networking configuration.** Each peer (or computer) manages authentication and access to its own resources. Therefore, each computer must be individually configured to attach devices to it (e.g., printer), allow access to it, and share its data. This type of configuration is normally implemented when a limited number of computer users need to be connected.

- **Server-based network configuration.** This configuration is designed for a larger group of users than the peer-to-peer network. Unlike with the peer-to-peer network configuration, individual configuration of each computer is not required for the server-based network. Instead, different servers manage authentication and access to resources from a database often referred to as a directory. The individual responsible for managing a server-based network is the **network administrator**. A network administrator controls all users' access to the network and resources by implementing **access control systems**, which are used to restrict access to protected network resources. The **access control list** (ACL) consists of rules that define network security policies and govern the rights and privileges of users of a specific system. Specifically, ACLs "can restrict access by specific user, time of day, IP address, function (department, management level, etc.) or specific system from which a logon or access attempt is made."[1]

Network Forensics: Defined

Several definitions of network forensics exist. The most widely used definition among computer security and network forensics professionals is the one provided in *The Report of the First Digital Forensic Research Workshop* (DFRWS). According to the DFRWS, **network forensics** is defined as follows:

> [The] use of scientifically proven techniques to collect, fuse, identify, examine, correlate, analyze, and document digital evidence from multiple, actively processing and transmitting digital sources for the purpose of uncovering facts related to the planned intent, or measured success of unauthorized activities meant to disrupt, corrupt, and or compromise system components as well as providing information to assist in response to or recovery from these activities.[2]

Network forensics seeks to capture, analyze, and preserve **network traffic** (i.e., data in a network). This traffic consists of **packets**, units of data transmitted over the network. Each packet contains a *header* and a *body*. The header is located at the beginning of the packet and includes information on the source address, destination address, total number of packets, and the specific packet's position within a sequence of packets (e.g., packet 1 of 3). The body includes the content (i.e., data) that the packet is delivering.

To put it briefly, network forensics is about monitoring. It checks the network for any anomalies (or malicious activity). If such anomalies are detected, it tries to determine the nature of the attack or intrusion within legal and/or corporate policy constraints. Network forensics also seeks to reconstruct events that have occurred and retrieve any potential evidence for later use in a criminal, civil, or administrative proceeding.

Network forensics investigations are conducted when attacks, intrusions, or network misuse is observed. These investigations help answer the following questions:

- What incident was observed?
- When was it observed?
- Where did it take place?
- Why did the incident happen?
- How did the incident occur?
- Who was responsible for the incident?

The responses to these questions are not meant solely for incident response purposes; they are also used for prevention purposes. As with the Roman deity Janus (who was conceptualized as having two faces, where one face looked into the past and the other looked into the future), one notices the retrospective and prospective capabilities of network forensics. Retrospectively, this discipline tries to make sense of the incident and respond to it by answering the previously mentioned questions. As such, this capability of network forensics seeks to deal with the incident after the fact. Prospectively, network forensics seeks to prevent something (such as malicious activity) from occurring in the future. This is accomplished by implementing measures that will help thwart future instances of the undesirable activity.

Network Components

A computer network is created by connecting computers with cables or wireless adapters and using the network interface of each computer (a special jack located in the back of the tower of most computer systems). A network is used to exchange data (i.e., files, programs, and resources) between two or more computers. The network interface allows these computers to share information and resources. A media access control (MAC) address is a unique identifier that is used to connect to the network. A **MAC address** is associated with a particular **network interface card** (NIC) or network adapter. It is used to identify the NIC or adapter on LANs. A MAC address consists of a 12-digit hexadecimal number.[3]

How Do You Find the MAC Address of a Computer?

The following set of steps is tailored toward a Windows operating system:

1. Click on the Start menu.

2. Type "cmd" in the Run text box.

3. Once in the command prompt window, type "ipconfig/all".

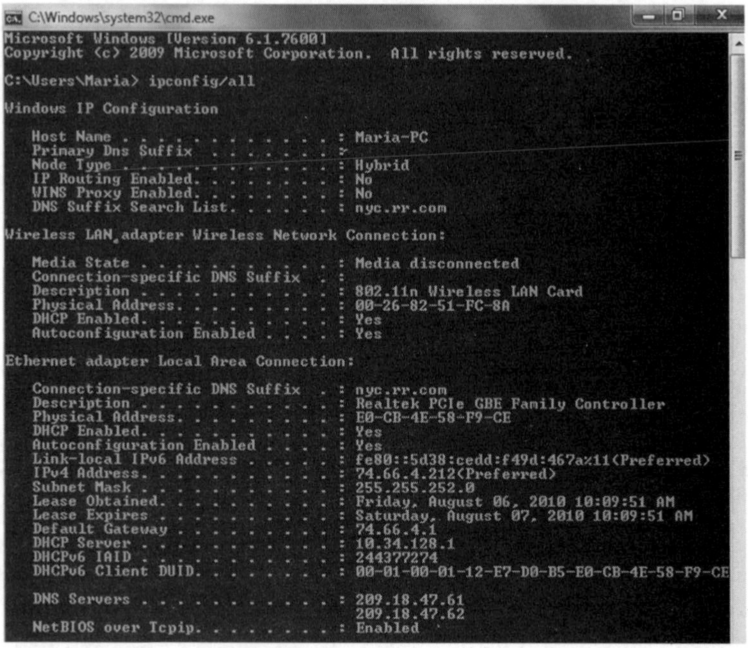

Figure 11–1   An example of what will be seen after the command is typed and executed.

The MAC address is the same as the physical address in Figure 11–1 (E0-CB-4E-58-F9-CE). Other information that can be retrieved from this utility includes the IP address of the computer user and the **Domain Name System (DNS)** server that the particular computer is using.

It is important to note that while MAC addresses can be retrieved from the **Address Resolution Protocol (ARP)** table, they can easily be changed by software. The NIC can also be easily removed and replaced.

The device that is used to connect computers is known as a **hub**. Hubs send data (packets) between networked computers. A copy of each packet is sent out from all of the ports of the hub, which creates a lot of unnecessary traffic.

A more efficient and intelligent version of a hub is a **switch**. A switch functions in a similar manner to the hub, except that it sends the packet only to the port that it is supposed to travel through.

Another device that is used to connect computers is known as a **router**. A router connects two networks (e.g., a LAN to an ISP network) and routes data between them. Both routers and switches read the header of the packet to determine where to send it.

Attacks on Network Components

Several attacks can be conducted on network components. A few of them are explored in further detail here.

MAC Address Spoofing

When an offender spoofs a MAC address, he or she can pose as an approved network device. This deception could trick other networked devices into sending network traffic to the offender's computer or receiving traffic from it. Not all of the devices mentioned earlier are affected by **MAC address spoofing**. Specifically, routers deal exclusively with IP addresses and, as such, are unaffected by MAC address spoofing. The same cannot be said about switches, as they deal solely with MAC addresses.

ARP Poisoning

ARP poisoning (also known as ARP spoofing) occurs as follows: An offender sends a poisoned ARP packet,[4] which contains an IP address and a request for a MAC address for the corresponding IP address. If the network device has the IP address assigned to it, it sends its MAC address in response to the request. The poisoned ARP request causes the network devices to update their ARP tables with the false information, which "poisons" them. Once the ARP tables are poisoned, the offender could redirect traffic and launch attacks.

An example of this kind of attack is the denial of service attack (see also Chapters 1 and 5). Denial of service attacks flood the intended target with illegitimate network traffic in an attempt to overwhelm its resources and prevent legitimate traffic from getting through.

MAC Flooding

MAC flooding occurs when numerous ARP requests are sent requesting multiple MAC addresses, which eventually overwhelms the resources of network switches. As a result, switches enter **fail-open mode**, which causes them to function as hubs and send the incoming traffic through all of their ports (rather than choosing the correct one as a switch is originally designed to do).

Man-in-the-Middle Attack

A **man-in-the-middle attack** hijacks a Transmission Control Protocol (TCP) connection between a client and a server. This connection is then split in two new connections: one between the offender and the client, and the other between the offender and the server. The offender relays the messages (i.e., communications) between them. Unbeknownst to those engaged in communication, the offender is eavesdropping on the conversation.

Where Can Network-Related Evidence Be Found?

The type of evidence that can be retrieved from networks includes full content data (i.e., the entire contents of packets) and session data (i.e., traffic data). File servers can also contain data of interest in an investigation. A **file server** is a computer that handles requests from other computers on the network for data that are stored on one or more of the server's hard drives. File servers hold the data that all of the computers on the network can use. In addition, they may contain logs of e-mails, instant messages, and Internet activities. Accordingly, these logs should be examined for potential evidence of a crime or policy violation. Such logs may not always be available, however: Some corporations, organizations, and agencies periodically clear them out. Additionally, data may not always be fully stored in these servers. Indeed, log files of computers may be stored on proxy servers. As such, the data on these proxy servers must be reviewed during a computer forensics investigation.

The **Dynamic Host Configuration Protocol (DHCP)** logs may also contain evidence. When a computer on the network starts up, it requests an IP address from the DHCP server. DHCP server logs may, therefore, link an IP address to a particular computer at a specific date and time. If these logs are not available, this information can be retrieved from the event logs (discussed in Chapter 7).

Peripheral devices, such as network printers, scanners, and copiers, may also contain data vital to an investigation.[5] Furthermore, information pertinent to an investigation may be retrieved from routers, firewalls, intrusion detection systems, sniffers, and keyloggers (if used). Each of these sources is explored in further detail in this section.

Routers

Routers are often the targets of attacks by hackers and other criminals, who know that gaining full access and control of a router often leads to full access and control of the network. Indeed, if a router is hacked, an offender can bypass existing firewalls and intrusion detection systems, attack the network and router, disable the router, redirect traffic to any destination desired, and record all outgoing and incoming traffic.

Access control lists can be used to prevent attackers from gaining access to the network through this means. With these lists, routers can be configured to allow or block certain IP addresses from traversing them. In addition, tools such as **tracert** (also known as

---

Live Analysis

Live analysis retrieves data from a running system, which will be lost once a device is powered down. This type of investigation is performed in lieu of traditional forensic analysis when the circumstances of the case warrant the live collection of data. However, live analysis always involves a tradeoff. When conducting live analysis, an investigator must choose between maintaining the integrity of the contents of the hard drive (i.e., ensuring that the contents remain unaltered) and acquiring live volatile data, which will be lost if the device is rebooted or shut down. This choice is necessary because live data analysis interacts with the device while it is operating and, in so doing, alters the stored contents of the hard drive. Accordingly, an investigator must determine which evidence is more important to the case: the live volatile data or the stored data. This choice will dictate whether live analysis is performed.

---

traceroute) can determine the route a packet used to travel across the network—that is, the routers the packet traversed to reach its destination.

Evidence within routers is found in the configuration files.[6] In addition to the nonvolatile data (**NVRAM**) stored within them, routers contain volatile data (**DRAM/SRAM**) that may be vital to an investigation.[7] If the device is powered down, such data will be lost. Examples of volatile information that will be lost if the system is powered down include the following items:

- NIC configuration settings
- Running processes and services
- Open sources, ports, and connections
- List of users currently logged on
- Active sessions
- Shared drives
- Files opened remotely

Live analysis must be performed on a router if volatile data are sought as evidence. To minimize the alteration to the contents of the hard drive, the software that is used to perform the live analysis should already be installed in the target system in anticipation of the event and the metadata should not be changed.[8] To perform live analysis, the router should be accessed through its console port. An investigator should record the entire session and all volatile data. Both the time on the router and the actual time should be documented as well.

Firewalls

A **firewall** is used to block incoming network traffic based on certain predetermined criteria. It "allows or disallows traffic to or from specific networks, machine addresses, and port numbers."[9] The firewall guards the ports and allows traffic only through those ports that have been specified for use by the network administrator. With firewalls, traffic cannot enter and exit from an unguarded port (i.e., one that has not been specified for use by a network administrator).

Firewalls contain detailed logs that hold a wealth of information pertinent to network forensics investigations. For example, these logs retain data about the network traffic that was blocked and allowed in, any attempted intrusions, and attacks that the firewall recognized. Firewall logs also include information about hardware failures, successful and unsuccessful connection attempts, users added to the system, and any permissions changed. In respect to permissions, a user may try to give himself or herself access to certain resources and/or create a backdoor for reentry at a later date and time.

Backdoors

A **backdoor** can be created to hide evidence of the offender's unauthorized access to the system. One tool widely used by hackers for this purpose is a **rootkit**. Rootkits are installed after an offender has gained full (root) access to a system. They help hide the attacker's intrusion into the system and any access or modification to data that may have occurred during this incursion. Backdoors are normally revealed when an offline analysis of the system occurs. Evidence of the rootkit may be found once the system is brought offline, as this software needs a live system to run its program and conceal its presence.

Intrusion Detection Systems

An **intrusion detection system** (IDS) is designed and used to detect attacks on systems and the unauthorized use of such systems, networks, and related resources.[10] With IDS, intrusions and attacks are determined by an existing set of rules that describe patterns of interests in network activity (i.e., anomalous activity). As a consequence, an IDS cannot identify all threats—just those that it has been set to discover. When new threats are discovered, the IDS should be adjusted so that it will detect such threats in the future.

Generally, two types of IDS are used. The first type is based on signature detection. This form of IDS contains a signature database; if a match is found between network activity and a signature in the database, the intrusion or attack will be documented in a log. Unfortunately, IDS signatures have been proven to be practically useless against new malicious software. The second form of IDS is based on pattern detection. This type of IDS performs statistical modeling of the network to determine normal activity. If abnormal activity is detected, an alarm is issued to notify the appropriate individuals of the breach.

Although both types of IDS are effective in detecting most forms of network attacks and intrusions, human analysis of network traffic is also required to identify incursions. In particular, network administrators should review encrypted traffic and traffic from entrusted sources. The latter is of particular importance because traffic from entrusted sources may have been compromised. This traffic is not checked, so allowing it in may put the system at risk. An offender may specifically target incoming traffic that is needed by the system (e.g., e-mail traffic) as means to surreptitiously entering the system. An investigator or network administrator must observe network traffic for unusual patterns, investigate them, and update the IDS to detect similar attacks in the future.

If an IDS alert is followed by high packet rates, this outcome indicates that the system has been compromised. In particular, high packet rates may indicate that a worm is present or that port scanning is occurring. Port scanning is an action taken to learn something about the target network. For example, the PING command (covered in Chapter 10) probes the network to determine if the destination address actually exists. After determining which hosts and associated ports are active, different types of probes are initiated on these active ports. **Port scanners** "send out successive, sequential connection requests to a target system's ports to see which one responds or is open to the request."[11] The targets are usually ports that have been left unguarded. To evade detection, an offender may use a port scanner "to slow the rate of port scanning—sending connection requests over a long period of time—so the intrusion attempt is less likely to be noticed."[12]

An example of an IDS is **Snort**,[13] which is the most popular of these tools in use today. Snort is designed to identify anomalies by inspecting network traffic. This program uses its analytical abilities to identify suspicious activity and alert the appropriate authorities (e.g., network administrator) of the observation of such activity. This tool "can be implemented into various locations across your network to provide early detection of unauthorized traffic so that administrators can curtail it before it gets out of control."[14] Snort can be configured to function as both an intrusion detection system and an **intrusion prevention system** (IPS). An IPS analyzes network traffic in real time to avert a potential attack or intrusion. It takes action by dropping a packet if it observes suspicious activity.

Another example of an IDS tool is the **Bro Intrusion Detection System**. This software passively monitors network traffic for suspicious activity. According to its website, it analyzes network traffic to detect "specific attacks (including those defined by signatures, but also those defined in terms of events) and unusual activities (e.g., certain hosts connecting to certain services, or patterns of failed connection attempts)."[15]

Intrusion detection systems can capture data from a variety of electronic sources and store these data in a centralized repository. An IDS normally "has a daily audit file to record important activity, including the times of user login and logout, website connection and document access and even a record of users' commands."[16] An IDS may also log the originating IP addresses of the intruders or attackers, recognized threats (e.g., worms and viruses), attempts to enter the network, and traffic coming in on strange ports or protocols.

An IDS produces one of four results:

- **True positive.** An actual attack or intrusion occurs, and authorities are alerted as to its presence.
- **True negative.** Network activity is normal. No attack or intrusion occurs, and no alarm is raised.

- **False positive.** In a false alarm, the IDS reads an activity as an attack or intrusion when, in fact, it is not (it is either a legitimate activity or no activity at all).
- **False negative.** In this case, the IDS misses an actual intrusion or attack.

Many IDS tools produce an unacceptably high level of false positives and false negatives, making it extremely difficult to identify true threats. Moreover, it is becoming increasingly difficult for IDSs to keep up with new attacks and intrusions, because offenders' goals, abilities, tactics, and tools are continuously evolving.

Honeypots

Honeypots are a form of intrusion detection and prevention system. These decoy mechanisms are implemented to lure offenders away from valuable network resources and can help capture new and unknown attacks. A **honeynet** consists of a collection of honeypots and is used to mimic a more complex environment. Both honeypots and honeynets seek to catch a crime as it occurs.

Honeypots are a "security resource whose value lies in being probed, attacked, or compromised."[17] Monitoring of these mechanisms can provide detailed insight as to how a specific exploitation occurred. To be effective, honeypots must be isolated from the network. If they are not, honeypots that have been compromised might be used to launch attacks on the real network. Additionally, honeypots should not be used for legitimate services or network traffic. Given that there is no legitimate reason for interacting with a honeypot, all activities that target it are considered suspicious. Furthermore, it is imperative that the honeypots created be able to record and monitor all activities.

Sniffers

A sniffer is a device that is used to capture network traffic.[23] Depending on the size and activity of the network, sniffers may collect an enormous amount of data.[24] For this reason, investigators should narrow the scope of their search of these logs. Network traffic can reveal data about the source, destination, and content of communications.

Sniffers have been used for both legitimate and illegal purposes. Notably, criminals have used sniffers to steal passwords and personal information. For instance, Jesus Oquendo, a former computer security specialist, installed a sniffer program on the network of one of his employer's investors, Five Partners Asset Management. This program intercepted and stored network traffic that contained passwords, which Oquendo later used to hack into another network.[25] Sniffers also have been used legitimately, to capture data and inspect it for any potential attacks or intrusions. Indeed, sniffers can provide a computer forensics investigator with information that can identify the suspect (or suspects) in an attack, the modus operandi of the suspect, and the full content of data of incoming and outgoing network traffic.[26]

Wireshark is an example of a sniffer program that captures network traffic in real time and records it.[27] This tool is not designed to alert authorities of any suspicious activity, so it is not considered a form of intrusion detection. Instead, it is designed to troubleshoot

Honeypots for Terrorists and Child Predators

Honeypots have been created to try to detect terrorists. For example, both the U.S. and Saudi Arabian intelligence agencies have been known to set up such traps in an attempt to ensnare terrorists. The intelligence obtained from such sites has proved invaluable, especially since the terrorist attacks on U.S. soil on September 11, 2001.

In the last decade, the war on terrorism has helped wipe out more than 70% of al-Qaeda's leadership. As an unfortunate side effect, this success has resulted in al-Qaeda's decentralization. In fact, intelligence reports indicate that today "Al-Qaeda has the most widely dispersed network in the history of modern terrorism. It still has global reach, with a presence in an estimated 65 countries. Its decentralized network makes it particularly hard to suppress: it is a true hydra."[18] Indeed, because al-Qaeda consists of nomadic and amorphous networks characterized by the diffusion of its groups, it challenges the abilities of governments to infiltrate the networks and to track these terrorists.[19] Accordingly, the identities of the terrorists and their whereabouts today are largely unknown. The terrorist attacks on Madrid and London (in 2004 and 2005, respectively) also brought to light a new threat: individuals engaging in terrorist attacks who were inspired by al-Qaeda's cause rather than directed by its leaders.[20] It is thus becoming increasingly difficult for authorities to find these terrorists. Honeypots help shed some light as to terrorists' identities and their plans.

Honeypots have also been used to catch child predators and child pornographers. For instance, in 2008, the FBI set up a honeypot for child pornographers. Specifically, agents created a website with fake links that claimed to contain videos of minors engaging in sexual activity.[21] Individuals who clicked on these links were subsequently arrested by the FBI.

Numerous joint sting operations have involved the use of honeypots to lure pedophiles and child pornography distributors. Operation Pin was one such example. This joint operation involved law enforcement agencies in the United States, United Kingdom, and Australia. For Operation Pin, several fake websites were created that held child pornography. When individuals tried to download the material on these sites, they were redirected to a law enforcement website that informed them that their information had been recorded and that they had committed a crime.[22]

Critical Thinking Questions

1. Which legal issues might be raised with the use of honeypots or honeynets?
2. Do you think honeypots should be used to a greater or lesser extent by law enforcement agencies? Why do you think so?

network problems and is used to perform both live and offline analysis of captured network data.

Keyloggers

Keyloggers can be installed either in person or remotely (e.g., via Trojan horse) on a computer and have the ability to record every keystroke of a user (see Chapter 1 for further information). Keyloggers are extremely difficult to detect on computer systems.

These programs have been used for illegitimate reasons to steal the target's personal information and passwords. However, like sniffers, keyloggers have also been used for legitimate purposes. For example, they have been used by corporations, organizations, and agencies to monitor employees and identify misuse of computer resources. Given that keyloggers can capture all of the data that a user inputs into a computer, a wealth of information can be retrieved from them if they have been installed on a system.

Network Forensics Analysis Tools

Network forensics is applicable in environments where security tools, such as intrusion detection/protection systems and firewalls, have been deployed at various strategic points on the network. **Network forensics analysis tools** may work synergistically with intrusion detection systems and firewalls in two ways: They may "preserve . . . a long-term record of network traffic [and they may allow] quick analysis of trouble spots identified by the other two tools."[28]

According to Garfinkel, two types of network forensics analysis tools may be distinguished:[29]

- **"Catch it as you can" tools.** Within the network, packets that traverse certain points are captured and stored. Analysis of this recorded traffic is subsequently done in **batch mode** (i.e., all at once). Such a system requires a significant amount of storage space for the data.

- **"Stop, look and listen" tools.** Unlike the "catch it as you can" tools, large amounts of storage are not required because packets are analyzed in a rudimentary manner and only certain types of data are stored for further analysis.

Numerous network forensics analysis tools are used to capture and record network traffic, including the following products:

- **NetWitness** (http://www.netwitness.com/). NetWitness detects threats and performs network forensics. Specifically, it proactively detects threats and automatically alerts authorities as to their presence. This tool also captures network traffic data and provides real-time analysis of it.

- **NetIntercept** (http://www.sandstorm.net/). This tool captures and archives network data for analysis at a later date and time. In addition, this tool is used to detect spoofing. According to its website, its hashing capabilities can demonstrate that the integrity of the data has been maintained during analyses. Moreover, NetIntercept can generate reports from the analyses of packets. This tool also has **data mining** capabilities; that is, it can sort and sift through vast quantities of data to find valuable information.

- **NetDetector** (http://www.niksun.com/). Similarly to NetWitness and NetIntercept, NetDetector engages in continuous, real-time surveillance of the network. According to its website, NetDetector helps provide a comprehensive defense by complementing existing routers, firewalls, and intrusion detection/protection systems. This tool is specifically designed to detect network intrusions based on signatures and anomalies. It also has advanced capabilities to reconstruct raw data.

- **OmniPeek** (http://www.wildpackets.com/). As the name implies, this tool allows an investigator or network administrator to see every part of the network in real time. It is also used to capture and store network traffic. This capability is particularly important for later analyses and submissions of information retrieved from these analyses as evidence in legal and administrative proceedings. OmniPeek allows users to search and mine traffic data as well.

- **SilentRunner** (http://www.accessdata.com/silentrunner.html). According to the AccessData website, this tool can capture, analyze, and visualize network activity by uncovering misuse, intrusion attempts, and abnormal activity before during and after an incident has occurred. It aids investigators by allowing authorities to play back events in the sequence they occurred. This tool is also used to monitor network traffic for compliance with existing regulations and policies.

- **NetworkMiner** (http://networkminer.sourceforge.net/). NetworkMiner is a sniffer tool that is used to capture network traffic. According to its website, the main purpose of this network forensics tool is to collect data "about hosts on the network rather than to collect data regarding the traffic on the network. The main view is host centric (information grouped per host) rather than packet centric (information showed as a list of packets/frames)."

- **PyFlag** (http://www.pyflag.net). PyFlag captures and analyzes network traffic. This advanced network forensics tool can be used to analyze disk images and large volumes of log files. Additionally, it enables the reconstruction of Web pages that emulate how the user viewed those pages when they were captured by the tool.[30]

Other forensics tools can be used in a networked computer environment. In particular, certain tools can be used to remotely acquire drive images in a live environment and securely investigate and extract network evidence from running systems. Two examples of such tools are the following:

- Encase Enterprise Edition (http://www.encaseenterprise.com/products/ ee_index.asp), from Guidance Software

- **ProDiscover** (http://www.techpathways.com/DesktopDefault.aspx?tabindex= 4&tabid=12), from Technology Pathway

The forensics tools used in a specific investigation will depend on the needs of the investigator or network administrator. The tool chosen must ensure that the integrity of the data is maintained. If it fails on this count, then the evidence retrieved via the tool will be inadmissible in a criminal, civil, or administrative proceeding.

Special Issues When Conducting Investigations in a Networked Environment

Upon arriving at a crime or incident scene, the procedures one will follow to investigate incidents, crimes, or policy violations will depend on whether the computer (or computers) involved is a stand-alone device or a networked machine. One obvious indication that computers are part of a network is the presence of multiple computers connected (either wirelessly or with cables) to hubs or switches. An investigator may also be alerted to the presence of networked computers by individuals at the scene. An investigator should use a wireless frequency scanner to determine the existence of any wireless storage devices at the scene. Securing and processing networked computers in a corporate environment poses serious problems to an investigator "as improper shutdown may destroy data [and] result in [the] loss of evidence and potential severe civil liability."[31] The same concerns apply to computer networks located within a home.

The primary purpose of a network forensics investigation is to find evidence of a crime, incident, or policy violation. During the investigation, an investigator specifically looks for the following:

- Network access around the time of the incident

- Access to the network at unusual times or from unusual locations

- Repeated failed attempts to access the network

- Evidence of port scanning or probing the network that preceded an incident

---

The Pros and Cons of Conducting Investigations on Networked Computers

Benefits

- Real-time surveillance of network activity

- Stored network traffic data from devices is usually available

- High probability that backups of data are conducted regularly (so that data required for an investigation are likely to be available)

- Centralized data storage on servers

Drawbacks

- Jurisdiction issues may arise for large networks

- Data can be remotely erased and easily hidden

- It is difficult to secure multiple computers at various sites

- It is difficult to isolate and secure servers at multiple sites

---

- Data transfers that occurred after the incident (e.g., a large volume of outgoing traffic after the incident may indicate theft of data)
- Detection of malicious software or exploitation methods

An investigator must not forget to look for or at encrypted data and hidden network traffic. In a corporate environment, the daily audit logs of corporations should be reviewed as part of the investigation, as they could reveal information about user logon and logout, websites accessed, and documents viewed. Other server logs, temporary files, and deleted files must be reviewed as well. Of particular note is that for the most part, certain computer-generated hard copies (e.g., printouts of logs) and electronic records (e.g., IDS alerts) are considered admissible in court under the hearsay exception.[32]

Preliminary Analysis

A preliminary analysis must first be conducted to determine whether an actual incident occurred. If an incident has not occurred, the investigator must determine if the activity under investigation was the result of employee (or user) negligence or mistake. If an incident has, in fact, occurred, all relevant evidence must be collected. The organization's security policies (e.g., ACLs) and security settings (e.g., access control and signatures) of network devices (e.g., IDS, firewalls, routers and switches) should be evaluated. These policies and settings should also be reviewed to determine if any changes were made to them as part of the offender's attempt to gain access to the network or to execute an attack. Indeed, the router, firewall, and IDS logs (described earlier in this chapter) can reveal any attempted or realized attack and intrusion. Moreover, these policies and settings should be evaluated to determine whether an offender made any additional changes to secure future access to the system (e.g., by providing himself or herself with permissions to access the network). If needed, a network administrator (or other relevant authority) should assist the investigator in determining whether any security settings or policies have been modified in any way.

Intrusion detection systems provide critical information to an investigation. When IDS alerts occur, the network administrator will analyze the anomaly to determine whether a policy violation, intrusion, or attack has occurred. If any of these events are observed, the investigation will continue. If not, the alert will be ignored and it will be considered a false alarm.

The response to the breach, crime, or policy violation detected depends on the type of incident identified. A decision must be made about whether to continue the investigation and gather more information. This decision is always guided by policy, legal, and economic constraints.

Documentation and Collection

The entire incident scene and all related devices must be photographed. In a networked computer environment, photographs of all network and phone cables connected to the computer (or computers) of interest in the investigation should be taken. The investigator should also photograph both ends of the cable (or cables) to prove that a computer was connected in a specific manner (to a certain network and phone line) when he or she first arrived at the scene.

Depending on the nature of the intrusion or the attack, certain parts of the system may need to be isolated from the network. Disconnecting systems is usually not a feasible option for corporations. Generally, investigators can disconnect networked computers and seize them only if by doing so they do not significantly disrupt normal business operations. However, there are certain exceptions to this rule. For example, an investigator can disconnect and seize such devices when permission to do so has been obtained from corporate authorities. Additionally, the investigator can disconnect networked computers if he or she has a search warrant specifying the seizure of such devices. Accordingly, an investigator can disconnect computers from a networked environment and collect them based on the circumstances of the case and the investigator's legal authority to do so.

The evidence and related devices that can be legally seized should be thoroughly documented and collected. The investigator must ensure that he or she logs the location of evidence and the network devices of interest to the investigation, including the serial numbers, model numbers, MAC addresses, and any IP addresses of these devices. Moreover, the contact information (e.g., telephone number, work number, and home address) of each individual who collected or handled the evidence in any way during an investigation must be noted. The exact date and time when the evidence was found or handled by the individuals should be documented. Given that logs may be stored on servers in multiple time zones, the respective time zone for each log must be documented as well.

Every individual who worked on incident-related tasks (other than handling evidence) must be noted. A description of the activities performed and the duration spent performing these tasks must also be documented. In addition, all of the data, accounts, services, systems, and networks that have been affected by the incident, and the manner in which they were affected, must be identified and logged. Furthermore, any damage that occurred as a result of the incident must be documented. This element of the investigation is critical, especially if a corporation is seeking to recover the costs from the damages incurred by submitting an insurance claim or suing the offender in civil court.

The scope of the investigation must be clearly established. This step, in turn, facilitates a comprehensive and methodical approach to conducting a network forensics investigation. Initial network traffic should be obtained (at the very least, photographed and documented in notes) as soon as possible (after an incident has occurred) to prevent the loss or change of volatile data. For each network device seized, an investigator must determine the order of volatile data and collect it accordingly.

If the investigation is conducted in a corporate environment, any passwords to the corporate computer accounts that cannot be obtained from the suspect (either because he or she refuses to provide them or is unavailable to do so) can be obtained from the system or

Collecting Volatile Data

RFC3227 outlines the order of volatility collection procedures, from the most volatile to the least volatile data. According to RFC3227, an example of the order in which evidence should be collected is as follows:[33]

1. Registers
2. Cache
3. Routing table
4. ARP cache
5. Memory
6. Temporary file systems
7. Disk
8. Remote logging and monitoring data that is relevant to the system in question
9. Physical configuration
10. Network topology
11. Archival media

network administrator. Specifically, the administrator can gain access to the account by canceling the password or bypassing it with another password. The investigator should also try to obtain all pertinent information related to a suspect's account, passwords, and programs used. If the suspect is not present or his or her whereabouts are unknown, the information contained in the account can potentially help investigators or administrators locate the suspect. Nevertheless, the offender may have taken significant steps to hide his or her location.

Special care must be taken when retrieving evidence. An investigator must be cognizant of the network he or she is working with and the effects of network imaging on daily operations. Network bitstream imaging can transfer an enormous volume of data across a network, rendering it unusable for normal operations. Investigators must streamline their search and seizure procedures when conducting network bitstream analysis so as to avoid unnecessary and unreasonable disruptions in the organization's operations. The liability associated with downtime (i.e., taking systems offline) has made live system imaging almost a requirement for investigators.

Analysis and Preservation

From the data obtained during the investigation, the investigator must develop a timeline for the incident. The source of the incident and the methods the offender used to carry it out must also be determined. Indeed, investigations of networked devices and traffic data can help determine which IP address (or IP addresses) successfully compromised the system on a specific date and time. If systems have been compromised, investigators can look at the **http sessions**, which include data about the requests and transfers of files across the network. Specifically, these sessions can reveal information about any files downloaded. For instance, if a network administrator of a company found a very long http session, intellectual property theft may be suspected. The main goal of network traffic analysis is to trace an intrusion or attack back to an offender, including through any intermediate systems that may have been used in the process. This process can also provide insight into the weaknesses that exist in current network security. The information retrieved from this analysis is then subsequently used to prevent future attacks.

Network traffic data can provide proof that a suspect performed a specific action (e.g., misused company resources by viewing pornographic websites on company time). Traffic data captured by network forensics analysis tools must be correlated with other information (e.g., data retrieved from computer events logs, browser history, and so on). The captured network traffic data can then be used as evidence in legal and administrative proceedings. All of the relevant network traffic data analyses must be included in a report for evidence to be considered valid.

As with other forms of evidence, the original captured network traffic data must be kept intact. An investigator must ensure that any programs that are run to obtain evidence do not modify the data on the system. Any modifications to the networked devices while performing live analysis must be documented. If an image is acquired while users are accessing the system, the introduction of evidence retrieved from this image may be challenged because the cryptographic hash of the device before and after acquisition will be different; the integrity of evidence collected may then be considered in doubt. Analyses should be performed on copies of the evidence so as to avoid any modifications, thereby preserving the validity and reliability of the data.

The electronic evidence obtained must be cryptographically hashed using forensically sound procedures. The appropriate network forensics analysis tools should be employed to ensure that evidence is not altered during the acquisition of data. The tool chosen must be able to deal with different log data formats and handle a vast amount of data that requires significant memory space.

The acquired evidence is methodically searched to reveal and collect specific indicators of the incident. The indicators found during such analyses are then classified and correlated in an attempt to match them with preexisting known intrusion and attack patterns. Data mining is used to search the data and match potential intrusion or attack patterns. The observed intrusion or attack patterns are combined and reconstructed to enable an investigator to visualize the incident by playing back an event (or events) in the sequence that it (they) occurred. This process assists an investigator in understanding the intent of an offender. Specific questions that should be answered when trying to ascertain the offender's intent from electronic evidence include the following:

- Did the offender target a specific system(s) or area(s) of the system(s)?
- Was any confidential information compromised?
- Did the offender access or download any files?
- Were any modifications made to the system by the offender? If so, which ones?

Finally, as with the other types of investigations covered in this book, a strict chain of custody must be maintained when conducting network forensics investigations. The chain of custody must include, among other things:[34]

- The location of the evidence when found
- The time and date when the evidence was documented and collected
- Identifying information of the individual who discovered the evidence (e.g., full name and contact information of the employee)
- Identifying information of the individual who secured the evidence (e.g., full name and contact information of the network administrator or investigator)
- Identifying information of all individuals who handled the evidence, including the date, time, duration, and reason for handling it
- Identifying information of all individuals at the scene

Access to obtained evidence should be restricted to avoid both unauthorized access and modifications to it.

Hypothetical Network Forensics Investigation

An employee was fired from MNO Corporation. The disgruntled employee sought revenge against the corporation, whom he believed wrongfully terminated his contract. To exact revenge, the prior employee hacked into MNO Corporation's network and gained access to its computer systems. Once in the system, the former employee deleted customer lists and other proprietary information within the system.

The next day, an employee in the IT department noticed the unauthorized access to the computer system and the modifications that were made to the database. Senior management and executives were subsequently notified of the incident.

1. Describe the step-by-step process that would be used to investigate this type of incident.
2. Where would the evidence for this crime be found?
3. Which problems might an investigator encounter?

Chapter Summary

Network forensics requires individuals to monitor network traffic, find any anomalies in that traffic, and determine whether this anomaly indicates an intrusion or an attack. With network forensics, an investigator or network administrator seeks to trace an incident back to an offender. By analyzing network traffic data, an investigator tries to determine how an incident or attack occurred, including the specific steps an offender took to execute it. Network forensics investigations can reveal, among other things, user logon IDs, the last date and time a user accessed the system, the means through which the user logged on to the network, and the location at which login took place.

The main purposes of network forensics are to uncover malicious activity, reconstruct incidents that have occurred, and extract evidence from the network that is pertinent to the case at hand. Network security tools include firewalls, intrusion detection systems, and intrusion protection systems. These tools (or a combination of them) have both retrospective and prospective capabilities. To be effective, network security tools must be deployed at various strategic points on the network. These tools provide alerts when they detect a security breach or a policy violation. The logs created by these tools assist investigators and network administrators in determining whether the threat that set off the alert was real or false. The network forensics tools that are used to examine logs and network traffic data must try to preserve the validity and reliability of the data. Special procedures should be followed to identify, collect, analyze, and preserve network traffic data.

Key Terms

Access control list

Access control systems

Address Resolution Protocol (ARP)

ARP poisoning

Backdoor

Batch mode

Bro Intrusion Detection System

Campus area network

"Catch it as you can" tools

Data mining

Domain Name System (DNS)

DRAM

Dynamic Host Configuration Protocol (DHCP)

Fail-open mode

False negative

False positive

File server

Firewall

Honeynet

Honeypot

http sessions

Hub

Intrusion detection system

Intrusion protection system

Live analysis

Local area network

MAC address

MAC address spoofing

MAC flooding

Man-in-the-middle attack

Metropolitan area network

NetDetector

NetIntercept

NetWitness
Network administrator
Network forensics
Network forensics analysis tool
Network interface card
Network traffic
Networked computer
NetworkMiner
NVRAM
OmniPeek
Packet
Peer-to-peer networking configuration
Port scanner
ProDiscover
PyFlag

Rootkit
Router
Server-based network configuration
SilentRunner
Snort
SRAM
Stand-alone computer
"Stop, look and listen" tools
Switch
Tracert
True negative
True positive
Wide area network
Wireshark

Review Questions

1. What is the difference between stand-alone and networked computers?

2. What are the different types of computer networks?

3. What are the two types of network configurations?

4. What are packets? What do they consist of?

5. Which questions do network forensics investigations seek to answer?

6. What is a MAC address? How can it be found?

7. What are the similarities and differences between a hub, a switch, and a router?

8. Describe the different types of attacks that can be conducted on network components.

9. What are firewalls?

10. What are intrusion detection systems?

11. What are honeypots?

12. What are sniffers?

13. What are keyloggers?

14. What are two types of network forensics analysis tools?

15. Which types of evidence do investigators look for during network forensics investigations?

Footnotes

[1]West, M. (2010). Preventing system intrusions. In J. A. Vacca (Ed.), *Network and system security*. New York: Syngress, p. 76.

[2]Palmer, G. (2001). A road map for digital forensic research. In *The report of the First Digital Forensic Research Workshop* (DTR-T001-01FINAL), p. 27. Retrieved from http://www.dfrws.org/2001/ dfrws-rm-final.pdf

[3]A 12-digit hexadecimal number is equivalent to 6 bytes or 48 bits in length.

[4]A network can be "poisoned" by providing false information.

[5]For information that can be retrieved from these objects, see Chapter 7.

[6]Technical Working Group for Electronic Crime Scene Investigation. (2001, July). Electronic crime scene investigations: A guide for first responders. National Institute of Justice, US Department of Justice, p. 17. Retrieved from http://www.ncjrs.gov/pdffiles1/nij/187736.pdf

[7]NVRAM, SRAM, and DRAM stand for nonvolatile random access memory, static random access memory, and dynamic random access memory, respectively.

[8]Kenneally, E. E. (2005). Confluence of digital evidence and the law: On the forensic soundness of live-remote digital evidence collection. *UCLA Journal of Law and Technology*, p. 5.

[9]Corey, V., Peterman, C., Shearin, S., Greenberg, M. S., & Van Bokkelen, J. (2002, November/ December). Network forensic analysis. *IEEE Internet Computing*, p. 60. Retrieved from http://www .sandstorm.net/support/netintercept/downloads/ni-ieee.pdf

[10]Alder, R., Carter, E. F. Jr., Foster, J. C., Jonkman, M., Marty, R., & Poor, M. (2007). *Snort: IDS and IPS toolkit*. Canada: Syngress, p. 3.

[11]West, M. (2010). Preventing system intrusions. In J. A. Vacca (Ed.), *Network and system security*. New York: Syngress, p. 64.

[12]*Ibid.*

[13]http://www.snort.org

[14]Street, J. E., Nabors, K., & Baskin, B. (2010). *Dissecting the hack: The forbidden network*. New York: Syngress, pp. 158–159.

[15]http://www.bro-ids.org

[16]Wang, S-J. (2007). Measures of retaining digital evidence to prosecute computer-based cyber-crimes. *Computer Standards & Interfaces*, 29, 221.

[17]Spitzner, L. (2003). *Honeypots: Tracking hackers*. New York: Addison-Wesley, p. 40.

[18]Wilkinson, P. (2005, July 9). We can beat the al-Qaeda hydra. *The Times* [online]. Retrieved from http://www.timesonline.co.uk/article/0,,1072-1686511,00.html

[19]Whine, M. *The new terrorism*. Tel Aviv: Stephen Roth Institute, Tel Aviv University, p. 8. Retrieved from http://www.tau.ac.il/Anti-Semitism/asw2000-1/whine.htm

[20]For further information on the Madrid and London bombings, see Chapter 12.

[21]McCullagh, D. (2008, March 20). FBI posts fake hyperlinks to snare child porn suspects. *CNET News* [online]. Retrieved from http://news.cnet.com/8301-13578_3-9899151-38.html

[22]Schrage, M. (2004, January 11). We can trap crooks with a net full of honey. *Washington Post* [online]. Retrieved from http://www.washingtonpost.com/ac2/wp-dyn/A5056-2004Jan9?language=printer

[23]Sniffers were also briefly discussed in Chapter 1.

[24]National Institute of Justice. (2007, October). Investigative uses of technology: Devices, l. US Department of Justice, p. 85. Retrieved from http://www.ncjrs.gov/pdffiles1/nij/213030.pdf

[25]United States Attorney, Southern District of New York. (2001, June 13). Computer security expert sentenced to 27 months' imprisonment for computer hacking and electronic eavesdropping. US

Department of Justice, Computer Crime & Intellectual Property Section. Retrieved from http://www.justice.gov/criminal/cybercrime/OquendoSent.htm

[26]National Institute of Justice. (2007, October). Investigative uses of technology: Devices, tools and techniques. US Department of Justice, p. 84. Retrieved from http://www.ncjrs.gov/pdffiles1/nij/213030.pdf

[27]http://www.wireshark.org

[28]Corey, V., Peterman, C., Shearin, S., Greenberg, M. S., & Van Bokkelen, J. (2002, November/December). Network forensic analysis. *IEEE Internet Computing*, p. 60. Retrieved from http://www.sandstorm.net/support/netintercept/downloads/ni-ieee.pdf

[29]Garfinkel, S. (2002, April 26). Network forensics: Tapping the Internet. Retrieved from http://www.oreillynet.com/pub/a/network/2002/04/26/nettap.html

[30]For further information see Cohen, M. I. (2008, September). PyFlag: An advanced network forensic platform. *Digital Investigation*, 5(1), 112–120.

[31]Technical Working Group for Electronic Crime Scene Investigation. (2001, July). Electronic crime scene investigations: A guide for first responders. National Institute of Justice, US Department of Justice, p. 32. Retrieved from http://www.ncjrs.gov/pdffiles1/nij/187736.pdf

[32]See Chapter 2 for more information on this topic.

[33]Brezinski, D., & Killalea, T. (2002, February). Guidelines for evidence collection and archiving. In Network Working Group, *Request for Comments: 3227 (RFC 3227)*. Retrieved from http://www.faqs.org/rfcs/rfc3227.html

[34]See Chapter 8 for further information on chain of custody issues.

Chapter 12

Mobile Phones and PDAs in Computer Forensics Investigations

This chapter explores the role of mobile phones and personal data assistants (PDAs) in criminal or civil investigations, as well as the important evidence that may be contained within these devices. Additionally, it considers the similarities and differences between mobile phones, PDAs, and other electronic devices. This chapter also examines the processes and techniques for identifying and collecting evidence from mobile devices without altering or tainting it. Furthermore, it analyzes the tools and techniques used on mobile phones and PDAs in computer forensics investigations.

Role of Mobile Phones and PDAs

Mobile phone technology has been (and is being) used, for example, by criminals—such as terrorists—to communicate undetected. Terrorists prevent their communications from being traced through simple practical means such as using prepaid (or stolen) mobile phones. In fact, the terrorists who bombed trains in Madrid in 2004 used this technique to organize their attacks. Members of organized crime groups have also been known to use numerous prepaid cell phones during their daily operations and discard them after use.[1] When mass retention of telecommunications (including fixed telephony and mobile phone) and electronic communications data (see Chapter 3) was debated in the European Parliament in 2005, members of this political organization vehemently argued that such a measure can be easily evaded by criminals. Indeed, one member of the European Parliament stated in respect of this measure that: "I think we are . . . chasing the wrong crooks, because if you are a crook who does not have the brains to use hotmail or prepaid mobile phone networks, then you are a stupid crook and we are really chasing the stupid crooks."[2]

Prepaid mobile phones do not require a contract for phone activation and the minutes for the phone can be purchased at most local convenience stores. Law enforcement agencies of countries in the European Union (EU) have noted the problems posed by the anonymity afforded by prepaid mobile phone cards. To remedy this situation, one EU member-state, France, no longer permits prepaid cards to be sold without proof of identification.[3] Other countries in the EU, such as Italy and Greece, have followed suit. Nevertheless, intelligence shows that terrorists have been able to conceal their true identities and thus avoid detection by law enforcement authorities by using passports and IDs—obtained by sympathizers or those that have been altered or counterfeited—to buy prepaid mobile phone cards. Moreover, terrorists may communicate undetected by acquiring phone cards from outside the EU and switching them regularly, using different mobile telephones from foreign suppliers, and using public telephone boxes.

Both mobile phones and PDAs have been used in crimes. Nowadays, the majority of PDAs function also as mobile phones. Consequently, their use in illicit activities has been increasing. Mobile phones and PDAs may be involved in an investigation in various ways.

They may be the target of an attack. Mobile phones and PDAs are increasingly becoming the targets of attacks by hackers and malicious software. For example, the Symbian operating system has been attacked by both worms and Trojan horses. One well-known worm is Cabir, which hit mobile phones in 2004. Cabir displayed a message on the mobile phones it infected, and required a user to accept it to use the phone. The Trojan horse known as Skulls also targeted the Symbian operating system; it prevented applications from running and replaced the icons of an application with a skull and crossbones.

The Windows CE operating systems was hit by a virus, Duts, that required a user to provide a positive response to a message displayed to use the system; the affirmative response then allowed Duts to spread to the files on mobile devices running Windows CE. Another virus that targeted these devices was the "Curse of Silence" virus, which sent a text message to unsuspecting users.[4] After users received this text message, they were no longer able to receive Short Message Service (SMS) or Multimedia Messaging Service (MMS).[5] Some models may be able to properly function after factory reset, but this depends on the operating system of the mobile phone.

Additionally, worms such as "ikee" and "Duh" have been used against mobile phones—specifically, against iPhones. They seek to build a botnet and steal users' data, such as banking information.[6]

Mobile phones and PDAs may be used as tools to commit a crime or incident. Criminals can clone mobile phones to networks and communicate at the expense of victims. Cloning occurs when the identity of a mobile phone is copied to the suspect's mobile phone. To clone an existing mobile phone, two pieces of data are stolen for use by the suspect: the mobile identification number (MIN) and the electronic serial number (ESN).[7] According to Denning and Baugh, in one case, the operatives of a Colombian drug cartel cloned the phone number of the U.S. Drug Enforcement Agency (DEA) and used it to make phones calls within Colombia.[8] Authorities made this discovery after reviewing their phone bills and finding an unusually large numbers of calls to Colombia.

Mobile phones have also been used as a tool for cyberbullying (see Chapter 6). In school settings, phone (or text) bullying may occur when one student who does not like the victim passes the victim's number to other bullies, who subsequently bombard the victim with humiliating text messages. In addition, mobile phones and PDAs (at least those that have messaging capabilities) have been used for sexting (also covered in Chapter 6).

In some instances, mobile phones have been used to detonate explosives, as in the following cases:

- *Bali bombings.* On October 12, 2002, a terrorist attack was executed in Bali by a militant Islamic group based in Southeast Asia, known as Jemaah Islamiyah.[9] Two nightclubs were targeted in the Bali area that were very popular with Australian tourists, Paddy's Pub and Sari Club. The two nightclubs were located opposite each other. The first bomb was detonated in Paddy's Pub by a suicide bomber wearing a vest filled with explosives. Nine people were killed by this bomb. Survivors of the bomb frantically exited Paddy's Pub, only to encounter a second bomb that had been placed in a vehicle parked in front of the Sari Club. This bomb was triggered by a mobile phone 15 seconds after the initial blast.[10]

- *Madrid bombings.* On March 11, 2004, bombs detonated on four commuter trains in Madrid, Spain, killing 191 people and injuring more than 1,800 individuals. The terrorists responsible for this attack detonated the bombs with mobile phones. This event represented the first coordinated terrorist attack conducted by Islamic extremists on European soil. These terrorists were inspired by al-Qaeda's cause, rather than explicitly directed by al-Qaeda personnel.[11]

- *London bombings.* On July 7, 2005, during morning rush hour in London, three suicide bombers (Shehzad Tanweer, Mohammad Sidique Khan, and Jermaine Lindsay) detonated bombs on their persons (less than one minute within one another) on three separate subway trains heading in different directions. Approximately one hour later, another suicide bomber (Hasib Hussain) detonated his explosive device on a bus. It was believed that mobile phones were used to set off these bombs. Like the Madrid bombers, these terrorists were inspired by al-Qaeda's cause, but not directed by this group. This event was the first coordinated suicide bombing by Islamic extremists on European soil.[12]

- *Bombing in a Saudi Arabian palace.* On August 28, 2009, an al-Qaeda suicide bomber, Abdullah Asieri, detonated explosives hidden in his rectum in an attempt to kill the head of Saudi Arabia's counterterrorism operations, Prince Mohammed Bin Nayef. Asieri had passed through two sets of airport security and had spent more than 30 hours with Saudi Arabian secret service agents before reaching the Prince. Once in the Prince's company, Aiseri informed him that more senior members of al-Qaeda wanted to surrender and convinced the Prince to speak with them on his mobile phone. The bomb is believed to have been triggered by a text message that was sent to Asieri's phone.[13]

Mobile phones and PDAs may provide evidence of a crime or incident. Today's PDAs and cell phones have vast storage capacities—and, therefore, may contain very valuable information to an investigation. The types of information normally retrieved from PDAs and mobile phones for use in an investigation include, but are not limited to:

- *PIM data.* **Personal information manager (PIM)** information includes:
  - A memo pad, personal notes, diary, and/or a journal
  - A calendar, date book, and/or events list, which contains appointments and reminders
  - A "to-do list," which records tasks that a user needs to accomplish
- *The numbers dialed, the numbers from which calls were received, missed calls, and the dates and times of these calls.* An investigator can also get this data from a service provider. This information has been quite useful in the investigation of terrorist attacks. For example, missed call data aided investigators in tracking down the perpetrators of the 2004 bombings in Madrid.
- *Contacts listed in the phone book.* These data may include names, e-mail addresses, home addresses, and phone numbers (work, home, and/or mobile).
- *Text messages.* Such messages may take multiple forms:
  - **Instant messages (IMs)**
  - SMS messages
  - MMS messages, which can include text and image, video, and/or sound
  - **Enhanced multimedia messages (EMS)**, which can be used, for example, to send ringtones
- *E-mail and Internet data.* E-mails sent and received can be stored in mobile phones and PDAs. Draft e-mails may also be found. Information on websites accessed can be obtained from these sources as well.
- *Image, sound, or audio files.* Photographs, audio recordings, and video clips can be stored on the memory card. **Memory cards** are storage media that allow users "to store additional files beyond the device's built-in capacity and provide another avenue for sharing data between compatible devices."[14]
- *Applications.* Examples include programs used to view and create documents, spreadsheets, and presentations.
- *Subscriber identifiers.* These identifiers are used to authenticate the user to the network and to verify the services tied to the account.
- *Other data.* The **personal identification number (PIN)** and financial information (e.g., credit and debit card numbers) of the mobile phone or PDA user can also be retrieved, including data from the user's voice mail account.

If a person (including an investigator) enters the wrong PIN three consecutive times, he or she will be locked out of the user's phone. To unlock access to the phone after incorrectly typing in the PIN in this manner, the investigator must type in the **personal unlock key (PUK)**. The PUK is unique to each **subscriber identity module (SIM)** card. The SIM card stores information identifying the subscriber to a particular network. An investigator must ensure that he or she does not enter the wrong PUK 10 times. If this occurs, the SIM card will be rejected. If the SIM card is rejected, the only way to use the phone is to request another SIM card. Certain phones may also ask for a second personal unlock key (**PUK2**) after a user has inserted the PUK as a means of enhancing security.

During the course of an investigation, service providers may be contacted for this information in an attempt to access a user's account. The **integrated circuit card identifier (ICCID)**, which is imprinted on the SIM card, can be used by the service provider operator to trace the SIM card back to the number that it was assigned to. The **mobile subscriber integrated services digital network number (MSISDN)** on the SIM card also provides the real number of a **Global System for Mobile Communications (GSM)** mobile phone.[15] The MSISDN consists of the country code, national destination code, and subscriber number, in that order.

SIM cards are used in GSM devices. Their equivalent in **Universal Mobile Telecommunications Systems (UMTS)** is the **Universal Integrated Circuit Card (UICC)**. UMTS runs the **Universal Subscriber Identity Module (USIM)** application, which is a component of the UMTS 3G network. The USIM stores data that identifies a mobile phone user as well as his or her subscriber data, contacts, preferences, and so on.

As mentioned earlier, **voice mail** access numbers and passwords can also be obtained from mobile phones and PDAs. Voice mail may contain information that is vital to an investigation. Messages in the voice mail system are held by the network, but the storage space for such messages is finite. Thus it is in the best interest of investigators to access the voice mail and review the messages as soon as they have the legal authority to do so.

Information retrieved from a mobile phone should be verified with the service provider, wherever possible. In particular, the SIM and the **international mobile equipment identifier (IMEI)**[16] are used to identify mobile phones and match them to subscribers. The SIM card can store phone numbers, call information, and text messages. If this information is deleted, in most cases it can be recovered from the SIM card using the appropriate forensics tools. An IMEI number may be requested when a service provider wants to determine whether a mobile phone has been stolen. However, IMEIs can easily be manipulated by users, and manufacturers might assign these numbers multiple times. Accordingly, the accurate identification of mobile terminals and subscribers based on these numbers might be difficult. Most service providers do not use IMEI numbers to identify mobile phone users, but instead use the **international mobile subscriber identity (IMSI)** number assigned by the provider and stored on the customer's chip (SIM) card.

Location data can also be retrieved from mobile devices. For instance, third-generation mobile phones can pinpoint a user's location to within a few meters, requiring only that the phone is switched on to identify this site. Additionally, the GPS functionality included in most mobile phones and PDAs enables the pinpointing of the location of the user to within a few feet. GPS navigation systems can record the user's home address, work address, and other areas to which the individual traveled. Moreover, Google offers mapping capabilities that allow the user of a mobile phone to pinpoint the locations of his or her friends.[17] Certain phones, such as the Motorola Droid, have a feature that enables other Droid users to find the exact locations of their Droid-user friends—as long as their phones are turned on. This capability could prove extremely useful in missing children cases, as long as the child's phone remains on long enough for law enforcement authorities to identify the location of the child. This feature can even be used when Motorola Droid cell phones have been stolen. On the downside, the Droid user tracking feature could also be used to stalk individuals.

Given that the majority of PDAs and mobile phones have digital picture and video capabilities, images or recordings of a crime, evidence, victims, or accomplices can be stored on it. One well-known example of this usage involved Robert P. Hanssen. Hanssen, an agent of the FBI who received a sentence of life in prison for selling secrets to Moscow, "used his Palm III [PDA] to keep track of his schedule to pass information to his Russian contacts."[8] Another example involved Julian Torres, who was part of an identity theft ring.[19] Evidence of his crimes was stored on his PDA, a Sony Clie Palm. Specifically, Torres had stored the names and personal information of his victims on his PDA. The Palm device also supports a list feature, which Torres had used to store his planned tasks, such as picking up supplies with which to make counterfeit checks. Furthermore, his Palm contained images of his accomplices.

Criminals may also rely on such devices to assist them in engaging in illicit activities. As Lee stated,

> Drug dealers use contact lists to track buyers and suppliers . . . , while drug makers, like those who run clandestine methamphetamine laboratories, use memos to keep recipes and ingredient lists. Pimps use the devices to keep track of clients, revenues and expenses. Smugglers and money launderers track their transactions on spreadsheets. Stalkers have been known to store their fantasies and victims' schedules on their [PDAs].[20]

Phone records may sometimes prove or disprove a suspect's testimony or alibi. Cell phones are constantly communicating with whichever signal tower is closest to them. Providers such as Sprint, Verizon, T-Mobile, and AT&T keep track of which phone numbers are communicating with every signal tower at any given time. This information can then be used to plot out the course and subsequent locations of a mobile device. Evidence of this type has been used in many criminal investigations. One notable case involved a professor from the University of Richmond who was murdered outside of his home. Piper Rountree, the ex-wife of Professor Frederic Jablin, was convicted of murdering him.[21] The most significant evidence against Rountree was her mobile phone records, which contradicted her alibi—namely, her contention that she was in Texas at the time of Jablin's murder in Virginia. Rountree claimed that she had spoken with her son shortly before the murder, using her mobile phone while in Texas. Her provider's records showed that the phone call was actually made from the same county in Virginia where Jablin resided.

Another example involved Darryl Littlejohn, a former bouncer at the Falls night club in New York City's Soho district.[22] Littlejohn was considered a suspect in Imette St. Guillen's rape and murder. He was last seen arguing with the young woman at 4 A.M. while leaving the nightclub. Littlejohn claimed that he was visiting his mother in a nursing home at the time of St. Guillen's murder. After obtaining his phone records from a communications service provider, an investigator found that several calls were placed near the site where St. Guillen's body was found on the day she was murdered.[23]

In another case, Mikhail Mallayev was hired by Dr. Mazoltuv Borukhova to kill her ex-husband, Daniel Malakov.[24] Mallayev had claimed that on the day of Malakov's murder, he was outside of New York.[25] However, he made a call from his mobile phone that day that disproved his alibi and placed him in New York on the day of the murder.

Yet another case concerned four teenagers who were involved in the murder of an immigrant, Abelino Mazaniego.[26] Mazaniego was beaten to death by three of the teenagers; the other teenager videotaped the beating with his mobile phone. This video was then circulated among teenagers within the town of Summit, New Jersey.[27]

Mobile phone data have proven critical in numerous other criminal prosecutions:

- *The murder of Veronica Guerin.* Veronica Guerin was a well-known Irish journalist who reported on gangs and drug dealers in Dublin. In 1996, she was murdered in her car by individuals who drove up alongside her vehicle in a motorcycle and shot her multiple times. Mobile phone call data proved critical in the criminal prosecutions of those responsible for Guerin's murder.[28]

- *Omagh bombing.* On August 15, 1998, the Real Irish Republican Army (Real IRA), a splinter terrorist group from the Provisional IRA, executed an attack in Omagh, Northern Ireland, that killed 29 individuals and injured 220.[29] The information retrieved from mobile phone data led to the criminal prosecutions of the perpetrators of the Omagh bombing.[30]

- *Soham murders.* In 2002, two 10-year-old girls, Holly Wells and Jessica Chapman, were murdered in Soham, United Kingdom.[31] Authorities were able to use mobile phone records to track the location of Jessica and Holly. The records indicated that right before Jessica's mobile phone was turned off on the night the pair was killed, she was close to the home of Ian Huntley, who was charged in the murders. Mobile phone records were also used to discredit the testimony of Huntley and his girlfriend, Maxine Carr. Huntley and Carr had stated that they were together, but mobile phone records revealed that Carr was actually miles away from Huntley on the night of the murders.

- *Murder of Avis Banks and her unborn child.* Keyton Pittman was engaged to Avis Banks, who was pregnant with his child. However, Pittman was also having an affair with Carla Hughes, a fellow teacher at his school. In November 2006, Hughes went to Banks' home, where she stabbed and shot Banks and her unborn child.[32] Her mobile phone records helped place Hughes at the crime scene during the murder. This, along with the other evidence, proved critical in her conviction for two counts of capital murder.[33]

Mobile Phones and PDAs Versus Other Electronic Devices

Many electronic devices, such as computers, laptops, mobile phones, and PDAs, are similar in that they read and write data using some kind of storage medium. Indeed, both mobile phones and PDAs contain memory similar to that of a hard drive of a computer, which provides for nonvolatile and volatile storage of data. Consider, for instance, the SIM and USIM smart cards contained in most mobile phones. These phones typically have a **microprocessor** and three types of memory: **random access memory (RAM)**, **read-only memory (ROM)**, and **electrically erasable programmable read-only memory (EEPROM)**.[34]

Memory cards could also be used to store the data from PDAs and mobile phones. For example, a micro secure digital (SD) memory card, which is a removable card, can be used to store data. This card can be inserted in a cell phone or PDA and can store numbers, files, and other data. Generally, these cards are used when an individual would like to store data whose size exceeds the device's built-in capacity for storage. In essence, **micro SD** cards are the equivalent of a flash drive that is used for a computer. However, micro SD cards do not have similar storage capabilities in respect to the amount of data that can be stored. These memory cards are diminutive in size—which means that an investigator might easily overlook them during his or her search.

Other types of memory cards that can be used in mobile phones and PDAs include the following devices:[35]

- **Multi-Media Card** (MMC): a form of nonvolatile computer storage
- **Memory sticks**: used to store data from portable electronic devices
- **TransFlash**: similar to a micro SD card (In fact, it was an earlier version of it.)
- **CompactFlash** (CF): provides mass storage for these devices

Fortunately, these memory cards are "normally formatted with a conventional filesystem (e.g., FAT) and can be treated similarly to a disk drive, imaged and analyzed using a conventional forensic tool with a compatible media adapter that supports an **integrated development environment (IDE) interface**."[36]

Forensically, these electronic devices are also similar because they rely on an operating system for their functioning. However, there are numerous makes and models of mobile phones and PDAs on the market, which differ significantly in respect to their specifications (e.g., processing capacity, storage size and other capabilities, such as phone, texting, and other features) and the types of operating systems they use (e.g. Macintosh, Linux, or a particular version of Windows). In addition, the operating systems of these devices "accommodate mobility aspects such as power management, specialized file systems, automatic file compression, and execute-in-place programs."[37]

Some of the most common operating systems of PDAs are identified here:

- *PalmOS.* Palms are PDAs that run on the Palm operating system. Sony and Palm are among the manufacturers that use this operating system.

- *Windows CE.* Pocket PCs are PDAs that use the Windows CE operating system. Usually, Compaq, Casio, and Hewlett-Packard PDAs run this operating system.

Other PDAs have proprietary operating systems. For example, BlackBerry PDAs run on the RIM BlackBerry operating system.

PDAs are primarily used as organizers and are intended to link up with computers. Given this interrelationship, the file structures of PDAs and computers are compatible. This similarity enables the investigator to easily access PDA files, because they are essentially computer files residing on a different device.

At the same time, PDAs differ from computers in some ways. Specifically, these devices often contain additional functions that allow them to operate as cell phones. A device that has both mobile phone and PDA capabilities is known as a **smartphone**. Depending on the make and model of the smartphone, it can either be a mobile phone with added PDA capabilities or a PDA with added mobile phone capabilities. Smartphones can, for example, sync with a user's personal and/or corporate e-mail provider.

The operating systems of smartphones vary. Like PDAs, smartphones can run on Windows CE and Palm OS. BlackBerry smartphones have their own proprietary operating system. Apple Computer's iPhone also has its own operating system. Moreover, the Motorola Droid runs on the Google Android 2.0 operating system, while the Symbian operating system is used for smartphones from Nokia, Samsung, and Panasonic (to name a few).

Some of the major differences between computer systems and mobile phones, PDAs, and smartphones are that the latter devices are "oriented toward mobility, they depend on battery power, emphasize wireless connectivity, and use specialized interfaces and media."[38] These devices also typically use volatile memory to store user data. As such, special care must be taken by the investigator to ensure that no loss of data occurs when examining their contents.

Similar approaches may be used when conducting investigations of these devices, yet all require different methods when it comes to the nitty-gritty of handling data within these devices. Specifically, different types of software and hardware are needed to effectively copy and analyze the data from each type of device. The method of data recovery from these devices depends on the operating system, and each operating system requires its own unique method of collecting and handling evidence. Therefore, each operating system requires specialized knowledge as to how the operating system and a particular device work, and how evidence can be extracted without altering its integrity.

Given that each manufacturer may use its own proprietary operating system, and that each device may operate in a completely different manner from other devices offered by the same manufacturer, it is nearly impossible to develop an all-inclusive forensics tool to deal with the many different makes and models of mobile phones, PDAs, and smartphones on the market from various manufacturers. "With the advent of new [devices] coming into the market at an exponential rate, as well as new companies coming into the market using a whole different blend of proprietary software, the problem has been even more compounded as time progresses."[39]

Existing forensics tools for mobile phones, PDAs, and smartphones are highly diverse. At one point, the National Institute of Standards Technology (NIST) undertook an extensive overview of forensics tools for both the mobile phones and PDAs. According to the authors of these analyses, "a considerable number of software tools and toolkits exist, but the range of devices over which they operate is typically narrowed to distinct platforms for a manufacturer's product line, a family of operating systems, or a type of hardware architecture."[40] Clearly, forensics tools need to adapt to the new mobile phones, PDAs, and smartphones that are constantly being developed. These tools must also be able to adapt to the upgrades of the operating systems of these devices. Mobile phones, PDAs, and smartphones may be offered without having the appropriate tools for retrieving data from them available on the market. To remedy this omission, it has been recommended that developers shorten "the latency period between the introduction of the [electronic device] and the time [when the] forensic software" for this device becomes available.[41]

Which Tools Can Be Used to Retrieve Evidence?

As stated earlier, the forensics tool that is used on mobile devices to obtain evidence depends on the make and model of the device and its operating system. These devices require different tools than those used on computer systems.[42] Some of the tools that have been successfully used in forensics investigations of PDAs, mobile phones, and smartphones are explored in further detail in this section.

PDA Forensics Tools

According to the NIST, PDA forensics tools are chosen for use in an investigation based on their ability to successfully meet the following demands:[43]

- Acquire the contents of the device
- Retrieve information associated with PIM applications (e.g., calendar, "to-do list")
- Locate graphic, text, video, and audio files
- Identify websites visited and obtain e-mails exchanged
- Find and display fields acquired from the device
- Locate data within compressed, archived, and formatted files

- Recover deleted, misnamed, and hidden files
- Retrieve files from a removable memory card
- Acquire data after a hard reset is performed
- Obtain the user's password to acquire the contents of the device in question

Ultimately, the data retrieved depend on the functions of the PDA. That is, if a PDA does not have Internet capabilities, some of the data listed above will obviously not be retrieved using the forensics tool.

The tools most commonly used to conduct an investigation involving PDAs are profiled here.

Encase

Encase is one of the most popular forensics tools, but is incompatible with PDAs running Pocket PC. Instead, it is used to acquire data from PDAs with Palm operating systems (OS). This tool also facilitates investigations of Linux-based PDAs. With Encase, "a complete physical bit-stream image of Palm OS devices is created and this bit-stream image is checked with the already obtained existing CRC (**Cyclical Redundancy Checksum**) values."[44] Encase images the device and provides a hash value for it. Accordingly, the evidence file that is created by this tool can be reviewed for potential evidence. A report of the analyses performed and the results of these analyses can also be provided using this tool. Furthermore, Encase has organizing and bookmarking capabilities; the latter may be used to highlight and store certain data for future reference.

Palm dd

The **Palm dd** (pdd) tool is used to acquire data from a Palm device with a Palm operating system that is running in console mode. This tool does not have report, bookmarking, and search capabilities. The files that are "created from pdd can be imported into a forensic tool, such as [Encase], to aid analysis; otherwise the default tool is a hex editor."[45] Although this tool can image the memory of the device, it does not provide hash values for the data that are acquired; as such, a separate procedure must be used to ensure the integrity of the data collection process.[46] As of January 2003, this forensics tool was no longer being updated, although Paraben has included a pdd engine in its PDA Seizure software.[47]

PDA Seizure

The **PDA Seizure** tool can be used to extract data from PDAs running the Windows CE, Pocket PC, and Palm operating systems. This tool can image RAM and ROM and works in a Windows environment. Unlike the pdd, PDA Seizure can provide hash values for acquired information. This toolkit is oriented toward PDAs without mobile phone functions, as it does not include features such as the ability to acquire SIM data. An investigator can use this tool to evaluate the contents of the PDA and provide a report on the results of his or her analyses. Similarly to Encase, PDA Seizure has organizing and book-marking capabilities.

POSE

If an investigator uses Palm Operating System Emulator (**POSE**), he or she can view the data in the same manner that the user of the PDA did. That is, the POSE interface allows an investigator to work on the device and access items (e.g., menus, calendars) on it as if he or she was working on the original device.

Pilot-link

Pilot-link is incompatible with PDAs running Pocket PC; rather, it is used to acquire data from PDAs with Palm OS. It was "developed for the Linux community to allow information to be transferred between Linux hosts and Palm OS devices. It runs on several other desktop operating systems besides Linux, including Windows and Mac OS."[48] Similarly to the PDA Seizure forensics tool, it physically acquires the contents of RAM and ROM.[49] The contents retrieved with this tool can be examined with Encase, POSE, or a hex editor.[50]

Duplicate Disk

The **Duplicate Disk** (dd) tool functions in a similar manner to pdd, in that it creates a bit-by-bit image of the contents of the PDA. It is compatible with a Linux-based PDA. If this forensics tool is used incorrectly, it could delete or overwrite parts of the content that are stored on the device.

Mobile Phone Forensics Tools

The information located on a mobile phone will depend on several factors:[51]

- Capabilities of the mobile phone
- Modifications made to the phone by a communications service provider
- Services that the user has subscribed to and utilizes
- Modifications made to the phone by the user

According to the NIST, mobile phone forensics tools are chosen for an investigation based on their ability to successfully meet the following demands:[52]

- Acquire the contents of the device
- Obtain subscriber data
- Recover location data (e.g., the place where the device was last used)
- Retrieve information associated with PIM applications (e.g., a phonebook)
- Find numbers dialed, numbers from which calls were received, missed calls, and deleted calls

- Locate SMS, MMS, and EMS messages sent, received, saved as drafts, and deleted
- Obtain information on Web applications (e.g., Internet sites browsed)
- Acquire sent, received, saved, and deleted e-mails and IMs
- Find graphic, text, video, and audio files
- Locate data within compressed, archived, and formatted files
- Recover deleted, misnamed, and hidden files
- Retrieve files from a removable memory card
- Acquire data after a hard reset is performed
- Recover data from a device after its power has been completely drained
- Locate the password for the device to access its contents

Some of the most widely used mobile phone forensics tools are described in this section.

Cell Seizure

Paraben's **Cell Seizure** product targets certain models of GSM, TDMA (time division multiple access), and CDMA (code-division multiple access) phones.[53] This forensics tool can acquire, search, bookmark, and examine data from the mobile phone device and report the results of the analyses performed.

.XRY

The **.XRY** forensics tool retrieves a wealth of stored data from certain models of GSM phones.[54] In addition, this tool creates an encrypted file of the data acquired from the mobile phone device and enables the investigator to inspect, search, and provide a report of the analyses conducted. An additional important feature of this tool is that it protects the data from accidental or intentional deletion and overwriting.

Oxygen Phone Manager

The **Oxygen Phone Manager** forensics tool is compatible with numerous mobile phones and smartphones (Nokia, Samsung, Siemens, Ericsson, and Panasonic models, to name a few) and is widely used by law enforcement and government agencies.[55] The Oxygen Forensic Suite 2010 can be used on more than 1550 mobile phones, smartphones, and PDAs with diverse operating systems.[56] An additional feature of this forensics tool is that it can extract geographic positioning data from the device.

MOBILedit!

The **MOBILedit!** forensics tool is also widely used by law enforcement, government agencies, and forensic investigators. It can acquire, search, and examine data from virtually all mobile phones on the market.[57] It also generates an extensive report of the results that can be electronically stored or printed.

SIM Cards

To extract SIM card information in a forensically sound manner, an investigator can use any of the following tools:[58]

- Mobile phone forensics tools
- SIM card readers
- Tools from the manufacturer

SIM Forensics Tools

Certain SIM tools have been designated for forensic purposes, such as Cell Seizure, GSM.XRY, and MOBILedit! Forensic. These tools can recover data from a variety of SIM cards, including those inserted in handsets. Other tools can analyze only external SIM cards. One of the latter tools is **SIMIS**, which can both securely acquire data from the SIM card and use cryptographic hashes to protect the integrity of the data.[59] Another tool, **ForensicSIM**, can clone the SIM card and examine it without altering the contents of the original device. All of these tools can acquire and report the results of analyses, and all can be used to examine the data on a SIM card.

One especially well-known SIM forensics tool is Paraben's **SIM Card Seizure**.[60] This tool can perform a complete acquisition and analysis of the data on a SIM card. It can also recover deleted data from the SIM card by searching unallocated space on this device. Finally, SIM Card Seizure can calculate hash values, perform search functions, and bookmark data.

SIM Card Readers

SIM card readers enable users to back up the data on their mobile phone devices. Numerous SIM cards readers are available on the market:[61]

- Dekart SIM reader
- SIMGuard
- SIMClone
- SIMScan
- SIMMaster
- SIMCopy
- SIM Tools

The integrity of the data that are extracted using these tools, however, cannot (and has not) been established.

Manufacturer-Provided Tools

A vast number of manufacturer tools are available that are "designed to back up, restore, synchronize, or transfer data to and from their phones and domestic computers."[62]

Investigators should keep in mind that such tools are not forensically sound and their use could alter the data in the device.

Popular Forensics Tools for Mobile Phones, PDAs, and Smartphones

The two most popular forensics tools geared toward mobile phones, PDAs, and smartphones are CellDEK and Device Seizure.

CellDEK

CellDEK is widely used in investigations, as it can extract a wealth of information from the majority of mobile phones, PDAs, and smartphones on the market, while preserving the integrity of the data. According the Logicube website, this tool is compatible with approximately 1700 mobile phones and PDAs (including BlackBerry PDAs).[63] It also supports iPhones, iPods, and handheld navigation devices (e.g., GPS devices from Garmin, TomTom, and iPAQ).

Device Seizure

Paraben's **Device Seizure** also acquires and examines data from PDAs, mobile phones, smartphones, and navigation devices. In particular, according to its website, it can be used on PDAs with various operating systems (e.g., BlackBerry, Windows CE, Symbian, and Pocket PC) and on more than 1000 models of mobile phones.[64] Similarly to CellDEK, this tool preserves the integrity of the evidence it extracts from these devices.

It is important to note that the tool an investigator chooses to use with mobile phones, PDAs, and similar devices must be specifically designated for forensic purposes. Those that have not been designated for forensic use should be avoided. Using these tools may render the evidence retrieved by them inadmissible in court.

Mobile Phone and PDA Investigations

In mobile phone and PDA investigations, the investigator must first locate and document all of the relevant evidence to the case. Specifically, the investigator must document the location, make, model, serial number, identifying marks (if any), and condition of the electronic device (on, off, standby) to be seized. In addition, if the device is on, the investigator must write down all of the information projected on the display of the electronic device and its current battery charge in his or her notebook. The screen of the mobile device (if on) and other related evidence should also be photographed.

Forensic protocol dictates that if mobile devices (mobile phones, smartphones, and PDAs) are discovered during an investigation, they must be left in the state that they were found. If the device is off, it should remain off. If the device is turned on, the data within it may be modified. If the device is found on, the device should remain on. These devices are powered by battery; they must be charged to ensure they remain on. As such, an investigator should ensure that the "on" state is maintained by powering the devices with, for example, a battery pack.

The investigator cannot know how long a device has been on and how long the battery will last without power. In fact, the battery life among mobile device varies according to their make and model. Consider two phones that Samsung has on the market: the Strive (SGH-a687) and the Propel Pro (SGH-i627). The former has less than five hours of talk time battery life, whereas the other one has approximately eight hours.[65] This duration, however, may be significantly less depending on the frequency with which the device is used. Additionally, the phone will consume more power if it is blocked from connection to the mobile network, because it will continually attempt to connect to it. Thus the battery can be depleted much faster if the device is isolated from the mobile phone network. Moreover, some mobile phones will reset or clear network data after a predetermined number of failed attempts to connect to the network—and these data may be critical to the investigation. Accordingly, investigators must familiarize themselves with mobile devices for the following purposes:

- Prevent the devices from losing power

- Help them better understand how to handle the devices

- Avert any data loss

The investigator must have external batteries with him or her that can be attached to the mobile device to prevent power loss. If the battery of the device is depleted or the investigator turns the device off, volatile data will be lost. Moreover, an investigator may be locked out of the phone and be required to enter a PIN (which he or she may not have or know) when the device is turned back on. Although other electronic devices such as computers must be powered down before transport, PDAs, mobile phones, and smartphones should remain on if they were found in this state.

If a mobile device such as a PDA is in a cradle (an instrument connected to the computer that allows data to be transferred from the PDA), a different procedure should be followed. In particular, the investigator must remove the connection of the device to the computer. By doing so, the investigator will prevent any further communication between the PDA and the computer, thereby blocking any data from being modified in the mobile device.

Investigators must remember that mobile devices that remain on are susceptible to damage. That is, data within these devices can be remotely modified and/or deleted if they are powered on. In fact, according to the National Institute of Justice,

> software is available for mobile and [smartphones] that can be activated remotely to render the device unusable and make the data it contains inaccessible if the phone is lost or stolen. This software can produce similar results if activated on a device seized by law enforcement. First responders should take precautions to prevent the loss of data on devices they seize as evidence.[66]

The same applies when an investigator is dealing with a PDA. Furthermore, if the device is turned on, newly transmitted messages may destroy existing messages that are stored on it. Forensic protocol dictates that Faraday bags be used to prevent any electronic signals from entering or exiting the mobile device.

When collecting mobile phone, PDA, or smartphone evidence, investigators should look for memory and SIM cards. The suspect may have hidden such items. Given their small size, these cards may be easily overlooked by investigators. In addition to those items, the cradle and all synchronization and power cables associated with the mobile device should be seized, labeled, packaged, and transported back to the forensic laboratory (if available). All peripheral devices, cloning equipment, and instruction and informational manuals related to the electronic devices (if present) should also be seized. All of the seized devices must be handled very carefully to avoid any destruction or tampering of the evidence.

Investigators should take special care when examining a mobile phone for evidence. Care must also be taken to ensure that a forensic tool does not alter the data in the device. Highly specialized tools are required to recover data from PDAs and their memory cards. In addition, specialized tools are required for recovering data from mobile phones, smartphones, and SIM cards. The tools used depend on the operating system of the mobile device. As noted in the previous section, the numerous makes and models of mobile devices have different operating systems. The investigator must ensure that the forensic tool that is used is documented in his or her notebook.

The evidence stored on the mobile device (PDAs, mobile phones, and smartphones) will depend on its make and model. All storage devices can have deleted files or fragments recovered using a forensic procedure. Recovery depends on the time a manufacturer allotted for the retention of this information on a specific device. For instance, iPhones tend to retain data longer than any device on the market. Other makes and models of mobile phones may store data for approximately 12 months before those data are over-written—though this storage time depends on the frequency with which the devices are used. Other forms of evidence may be accessible from the service provider. Again, the availability of these data depends on the retention practices of the specific service provider.

Special Issue: Deleted Text Messages

The ability to recover deleted text messages depends on the make, model, and service provider of a cell phone. The service provider stores text messages in different places. The identity of the service provider may provide an investigator with a general idea of where a message can be stored—for instance, in the handset or the SIM card (if one exists). From the make and model of the phone, an investigator can determine which type of information can be stored on the phone. If unsure of such information, the investigator can review informational booklets on the mobile phone, visit the phone company website to review the capabilities of the phone, and contact the manufacturer. If a particular mobile phone is supported by forensic software, an investigator can use the tool to acquire the data stored in it.

When conducting an investigation, the analyst must ensure that the mobile device is properly connected to a forensic workstation (i.e., computer) before acquiring data from it. For example, to acquire data from a PDA, the investigator must make sure that the device is powered on, in the appropriate cradle, and properly connected to the computer via a USB or serial port. The most important thing for an investigator to remember is to make sure that no matter which device is involved (e.g., PDA, mobile phone, smartphone), he or she must create a bit-by-bit copy of the contents of the device. The investigator must then verify that this bit-by-bit image is an exact copy of the contents of the device. As stated in Chapter 7, this is done by calculating the MD5 hash of both the original and the copy. If they match, then the image is an exact copy of the contents of the device.

The investigator must also make sure that no modifications are made to the original contents of the device; doing so may render the evidence inadmissible in court. For example, a device could be used to copy the SIM card while simultaneously preventing it from connecting to the mobile phone network. A search can then be performed securely (i.e., without threat of modification of the data) on the contents of the device and pertinent data to the investigation can be retrieved (if any). An investigator may also use software, such as ForensicSIM, to clone a mobile phone's or smartphone's SIM card. This cloned card can then be used to examine the contents of the device without worrying about the possibility of modifying any potential evidence.

Criminals have been known to destroy mobile devices to prevent investigators from retrieving data stored on them by breaking, burning, and submerging the device in water. However, even in these instances, an investigator can retrieve the data stored by the suspect's service provider.

Missing SIM Card

What should an investigator do if a SIM card is missing (and he or she cannot find it at the crime scene)? The specific steps to be taken will depend on the forensic tool used. Suppose an investigator is using the .XRY forensic tool. If a SIM card is missing from a phone, the following tasks should be performed:

1. The phone should be connected to the host workstation.

2. The last IMSI should be recovered.

3. The IMSI should be typed into the .XRY program and copied to a clone card.

4. After the insertion of the cloned card into the phone, data can be securely acquired from the device.

Depending on the device, data may also be archived online or on the suspect's desktop or workplace computer. Indeed, **Bluetooth** (i.e., short-range, wireless communication) technology affords users with the opportunity to synchronize and transfer the data on their mobile devices with their home and work computers and laptops. Bluetooth technology can also be used to synchronize and transfer the data on the user's mobile devices to other mobile devices. For example, a suspect might transfer incriminating evidence from one smartphone to another that he or she does not frequently use. These data will then be deleted from the smartphone that is used more often.

Furthermore, new makes and models of mobile devices are offering users the ability to store the contents of their devices on the Internet. For instance, Verizon's Kin One and Kin Two models enable users to sort, share, and store all of the contents of their phones online.[67] Palm Pre also requires users to create a Palm Profile (complete with username and password), which saves their data and applications online for backup purposes. One feature of the Palm Profile is that it allows a user to erase personal data remotely. With this feature, users may remotely access the data stored online to erase any incriminating evidence if they believe that they are under investigation. The availability of such areas of remote storage further complicates the work of a forensic investigator.

Investigators must stay current in the field. Failure to do so may cause them to overlook vital evidence. Accordingly, to do their job efficiently and effectively, investigators must keep abreast of changes in mobile device technology and its capabilities.

Kill Switches in Mobile Phones and Smartphones

Kill switches are no longer limited to automobiles, where they have long been used to prevent motor vehicle theft. Such techniques are now being used in communications technology, albeit for different reasons—for example, to remotely delete applications that violate the terms of service for users' phones. In addition, they may be used by individuals and service providers to wipe the contents of a mobile device when a phone has been lost or stolen. Unfortunately, kill switches have been used for nefarious purposes as well.

Microsoft has added to the remote access capabilities of its users' phones by introducing **Digital Manner Policies** (DMP)-enabled devices. DMP devices could set users' phones to vibrate when individuals enter certain buildings or board a flight. They could also prevent users from using certain services and applications on their phones, such as taking photographs, in specific facilities. With this technology, someone else will have control over what is done with a user's phone in specific designated areas.

1. What is your opinion of DMP?
2. Will these devices assist or hinder mobile device investigations? Why do you think this is the case?

Mobile Device Investigation

An individual is suspected of having engaged in drug trafficking. The phone found at the crime scene is an LG Vu Phone (**Figure 12–1**). The investigator is unfamiliar with the phone.

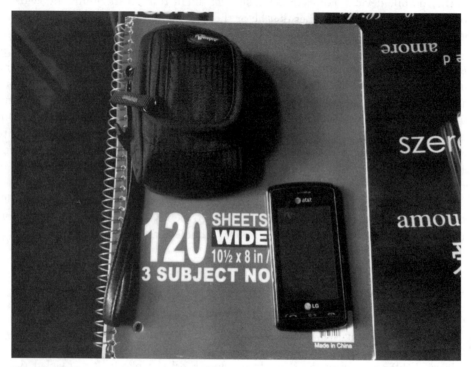

Figure 12–1

1. What is the first thing that the investigator should do?

2. What should he or she do next (i.e., how should the investigator conduct the investigation)?

3. Where would the investigator look for evidence?

4. How would the investigator retrieve this evidence?

5. Which obstacles might the investigator encounter?

Chapter Summary

Mobile devices, such as PDAs, mobile phones, and smartphones, have played a major role in investigations. These devices have been the targets of crimes, have been used to commit crimes, and have stored evidence of crimes. Mobile devices can be used to make phone calls and send and receive text messages. Many of these devices also have GPS navigation systems. In many ways, mobile devices behave like computers; for example, they can be used to store files, create and save images, browse the Internet, send e-mails, and record videos. Accordingly, mobile devices hold a wealth of data that may be used as evidence in an investigation.

Some differences between computers and mobile devices complicate investigations focusing on the latter devices. The procedures used to seize, transport, acquire, and examine the contents of mobile phones, PDAs, and smartphones will depend on the particular device. The tools used to acquire evidence depend on the operating systems of the mobile devices. The tool that is used to conduct the investigation on the contents of the mobile device must be forensically sound. Apart from the internal memory of the mobile device, data can also be stored on memory cards, SIM cards, websites, computers, and related devices. The exponentially growing number and variety of mobile devices on the market have made the job of an investigator increasingly difficult. To successfully conduct an investigation that involves mobile devices, an investigator must stay current in the field and keep up-to-date with changes in technology.

Key Terms

Bluetooth
Cell Seizure
CellDEK
CompactFlash
Cyclical Redundancy Checksum
Device Seizure
Digital Manner Policies
Duplicate Disk
Electrically erasable programmable read-only memory (EEPROM)
Enhanced multimedia messages (EMS)
ForensicSIM
Global System for Mobile Communications (GSM)
Instant messages (IMs)
Integrated circuit card identifier (ICCID)
Integrated development environment (IDE) interface
International mobile equipment identifier (IMEI)
International mobile subscriber identity (IMSI)
Memory card
Memory sticks
Micro SD
Microprocessor
Mobile subscriber integrated services digital network number (MSISDN)
MOBILedit!
Multi-Media Card
Oxygen Phone Manager
Palm dd
PDA Seizure
Personal identification number (PIN)
Personal information manager (PIM)
Personal unlock key (PUK)

Pilot-link

POSE

PUK2

Random access memory (RAM)

Read-only memory (ROM)

SIM card reader

SIM Card Seizure

SIMIS

Smartphone

Subscriber identity module (SIM)

TransFlash

Universal Integrated Circuit Card (UICC)

Universal Mobile Telecommunications
Systems (UMTS)

Universal Subscriber Identity Module
(USIM)

Voice mail

.XRY

Practical Exercise

Conduct an online search for a case (not described in this chapter) where a mobile device (PDA, mobile phone, or smartphone) was involved in a criminal or civil investigation.

1. What were the facts of the case?

2. What was the role of the device in the investigation?

Critical Thinking Questions

1. New mobile devices are constantly being introduced. How does this ever-changing market affect the work of a forensic investigator?

2. In this and previous chapters, different electronic devices were considered. What do these devices have in common, in a forensic sense? How are they different?

3. Is the forensic analysis of mobile devices easier or more complex than computer forensics analyses? Why?

Review Questions

1. What are the roles of mobile phones, smartphones, and PDAs in crime?

2. What problems do prepaid phones pose to investigators?

3. What is mobile phone cloning?

4. Which kinds of evidence can be retrieved from mobile phones and PDAs?

5. What are the similarities between computers and mobile devices?

6. What are the differences between computers and mobile devices?

7. Which types of evidence should PDA forensics tools extract from a device?

8. Which types of evidence should mobile phone forensics tools extract from a device?

9. How can SIM card information be extracted?

10. What should be done before a mobile device is transported?

11. What is Bluetooth technology? How does it affect investigations?

Footnotes

[1]Denning, D. E., & Baugh, W. E. (1999). Hiding crimes in cyberspace. *Information Communication and Society, 2*(3). Retrieved from http://all.net/books/iw/iwarstuff/www.infosoc.co.uk/00107/feature.htm

[2]European Parliament. (2005, December 13). *Data retention debate.*

[3]Gow, G. (2005, March 24). *Prepaid phone cards: The anonymity question.* Paper presented at Simon Fraser University. Retrieved from http://www.sfu.ca/cprost/prepaid/downloads.htm#Mar%2024%202005. Cited in Shields, P. (2006). Electronic networks, enhanced state surveillance, and the ironies of control. *Journal of Creative Communications, 1*(1), 21.

[4]Insight on mobile phone viruses and other malware. (2010, April 29). *Tech Guru.* Retrieved from http://www.techquark.com/2010/04/insight-on-mobile-phone-viruses-other.html

[5]Osada, J. (2009, January 14). Beware of the "curse of silence" mobile phone attack. Retrieved from http://www.asiaone.com/Digital/News/Story/A1Story20090114-114492.html

[6]Keizer, G. (2009, November 23). New iPhone worm steals online banking codes, builds botnet. *IT World* [online]. Retrieved from http://www.itworld.com/security/85761/new-iphone-worm-steals-online-banking-codes-builds-botnet?page=0%2C0

[7]Collins, M. (2000). Telecommunications crime: Part 3. *Computers and Security, 19,* 142.

[8]Denning, D. E., & Baugh, W. E. (1999). Hiding crimes in cyberspace. *Information Communication and Society, 2*(3). Retrieved from http://all.net/books/iw/iwarstuff/www.infosoc.co.uk/00107/feature.htm

[9]Council on Foreign Relations. (2009, June 19). Jemaah Islamiyah. Retrieved from http://www.cfr.org/publication/8948/jemaah_islamiyah_aka_jemaah_islamiah.html

[10]The Bali bombing plotters. (2010, March 10). *BBC News* [online]. Retrieved from http://news.bbc.co.uk/2/hi/asia-pacific/2385323.stm

[11]Rai, M. (2006). *7/7: The London bombings, Islam, and the Iraq War.* London: Pluto Press, p. 156.

[12]For further information on the London bombings, see: *Report of the official account of the bombings in London on 7th July 2005: Ordered by the House of Commons to be printed 11th May 2006* (HC 1087). London: The Stationery Office. Retrieved from http://www.official-documents.gov.uk/document/hc0506/hc10/1087/1087.pdf

[13]MacVicar, S. (2009, September 28). Al Qaeda bombers learn from drug smugglers. *CBS News* [online]. Retrieved from http://www.cbsnews.com/stories/2009/09/28/eveningnews/main5347847.shtml

[14]Jansen, W., & Ayers, R. (2005). An overview and analysis of PDA forensic tools. *Digital Investigation, 2,* 124.

[15]Phillips, E. (2002). Mobile phone: Friend or foe?. *Science and Justice, 42*(4), 226.

[16]A unique 15-digit serial number of the mobile device. This number is automatically transferred by the phone when the network asks for it. See EDRI. (n.d.). Technical questions on data retention. p. 1, Retrieved from http://www.edri.org/docs/Technical_Questions_on_Data_Retention_answers.pdf

[17]Barnard, A. (2009, July 6). Growing presence in the courtroom: Cellphone data as witness. *New York Times* [online]. Retrieved from http://www.nytimes.com/2009/07/06/nyregion/06cellphone.html?_r=1&adxnnl=1&adxnnlx=1280682046-V1yzsoQKG4t+Z3Z5Oenspg

[18]Lee, J. (2002, October 24). A palmtop for the prosecution. *New York Times* [online]. Retrieved from http://www.nytimes.com/2002/10/24/technology/circuits/24palm.html?pagewanted=all

[19]Clie used as evidence against criminals. (2002, October 28). Retrieved from http://www.pdablast.com/articles/2002/10/20021028-Clie-s-Used-As.html; Lee, J. (2002, October 24). A palmtop for the prosecution. *New York Times* [online]. Retrieved from http://www.nytimes.com/2002/10/24/technology/circuits/24palm.html?pagewanted=all

[20]Lee, J. (2002, October 24). A palmtop for the prosecution. *New York Times* [online]. Retrieved from http://www.nytimes.com/2002/10/24/technology/circuits/24palm.html?pagewanted=all

[21]Leung, R. (2005, May 20). Two wigs, a gun and a murder. *CBS News* [online]. Retrieved from http://www.cbsnews.com/stories/2005/05/19/48hours/main696659.shtml?tag=contentMain;contentBody

[22]Shifrel, S. (2009, June 3). Bouncer Darryl Littlejohn found guilty of murdering of Imette St. Guillen. *NY Daily News* [online]. Retrieved from http://www.nydailynews.com/news/ny_crime/2009/06/03_darryl_littlejohn.html

[23]Barnard, A. (2009, July 6). Growing presence in the courtroom: Cellphone data as witness. *New York Times* [online]. Retrieved from http://www.nytimes.com/2009/07/06/nyregion/06cellphone.html?_r=1&adxnnl=1&adxnnlx=1280682046-V1yzsoQKG4t+Z3Z5Oenspg

[24]Gendar, A., Bode, N., Belenkaya, V., & Standora L. (2008, February 8). Ex-wife of slain dentist Daniel Malakov arrested for murder plot connection. *New York Daily News* [online]. Retrieved from http://www.nydailynews.com/news/ny_crime/2008/02/08/2008-02-08_exwife_of_slain_dentist_daniel_malakov_a-1.html

[25]Barnard, A. (2009, July 6). Growing presence in the courtroom: Cellphone data as witness. *New York Times* [online]. Retrieved from http://www.nytimes.com/2009/07/06/nyregion/06cellphone.html?_r=1&adxnnl=1&adxnnlx=1280682046-V1yzsoQKG4t+Z3Z5Oenspg

[26]Henry, S. (2010, July 27). 3 N.J. Teens charged with immigrant death. *Time* [online]. Retrieved from http://www.time.com/time/nation/article/0,8599,2006922,00.html

[27]Black, C. (2010, July 28). Three N.J. teens fatally beat Salvadoran immigrant to death, show off video of the attack, say police. *CBS News* [online]. Retrieved from http://www.cbsnews.com/8301-504083_162-20011834-504083.html

[28]Lillington, K. (2007, March 2). Retention of mobile data a threat to everyone's privacy. *The Irish Times* [online]; Lillington, K. (2003, February 25). State secretly retaining phone data. *The Irish Times* [online]. Retrieved from http://radio-weblogs.com/0103966/stories/2003/02/25/stateSecretlyRetainingPhoneData.html

[29]Timeline: Omagh bombing. (2009, June 8). *The Guardian* [online]. Retrieved from http://www.guardian.co.uk/uk/2002/jul/26/northernireland. The Provisional IRA, which was founded in 1969, ended its armed struggle with the United Kingdom on April 10, 1998, with the Good Friday Agreement signed in Belfast, Northern Ireland. Those dissidents who believed that the struggle should not end formed the Real IRA. Other splinter groups have also been formed, such as the Continuity Irish Republican Army and the Irish National Liberation Army. For further information, see Gregory, K. (2010, March 26). Provisional Irish Republican Army. Council on Foreign Relations. Retrieved from http://www.cfr.org/publication/9240/provisional_irish_republican_army_ira_aka_pira_the_provos_glaigh_na_hireann_uk_separatists.html; Fletcher, H. (2008, May 21). IRA splinter groups. Council on Foreign Relations. Retrieved from http://www.cfr.org/publication/9239/ira_splinter_groups_uk_separatists.html

[30]Lillington, K. (2007, March 2). Retention of mobile data a threat to everyone's privacy. *The Irish Times* [online]; Lillington, K. (2003, February 25). State secretly retaining phone data. *The Irish*

Times [online]. Retrieved from http://radio-weblogs.com/0103966/stories/2003/02/25/stateSecretlyRetainingPhoneData.html

[31]Mathieson, S. A. (2004, January 29). Eyes on the child. *Guardian* [online]. Retrieved from http://www.guardian.co.uk/technology/2004/jan/29/childprotection.onlinesupplement; Summers, C. (2003, December 18). Mobile phones: The new fingerprints. *BBC News* [online]. Retrieved from http://news.bbc.co.uk/2/hi/uk_news/3303637.stm

[32]Katz, N. (2009, October 13). Carla Hughes verdict: Teacher found guilty of killing pregnant Avis Banks to steal her fiancé. *CBS News* [online]. Retrieved from http://www.cbsnews.com/8301-504083_162-5382677-504083.html

[33]Love triangle murder begins. (2009, October 6). *CNN Justice* [online]. Retrieved from http://insession.blogs.cnn.com/2009/10/06/love-triangle-murder-trial-begins/

[34]EEPROM provides nonvolatile storage of data. See Savoldi, A., & Gubian, P. (2007). SIM and USIM filesystem: A forensic perspective. *Proceedings of the 2007 ACM Symposium on Applied Computing.* Retrieved from http://pds3.egloos.com/pds/200705/25/00/sim_and_usim_filesystem_a_forensics_perspective.pdf

[35]Ayers, R., Jansen, W., Cilleros, N., & Daniellou, R. (2005, October). Cell phone forensic tools: An overview and analysis. National Institute of Standards and Technology. *NISTIR 7250*, pp. 6–7. Retrieved from http://csrc.nist.gov/publications/nistir/nistir-7250.pdf; Jansen, W., & Ayers, R. (2005). An overview and analysis of PDA forensic tools. National Institute of Standards and Technology, p. 7. Retrieved from http://csrc.nist.gov/groups/SNS/mobile_security/documents/mobile_forensics/ForensicArticle-DI-fin.pdf

[36]Ayers, R., Jansen, W., Cilleros, N., & Daniellou, R. (2005, October). Cell phone forensic tools: An overview and analysis. National Institute of Standards and Technology. *NISTIR 7250*, p. 6. Retrieved from http://csrc.nist.gov/publications/nistir/nistir-7250.pdf

[37]Jansen, W., & Ayers, R. (2005). An overview and analysis of PDA forensic tools. *Digital Investigation*, 2, 121.

[38]*Ibid.*

[39]Bhadsavle, N., & Wang J. A. (2008). Validating tools for cell phone forensics. *Technical Report* CISE-CSE-08-05, p. 2. Retrieved from http://it.spsu.edu/research/Tech_Report/Neil_Andy_CellPhoneForensics.pdf

[40]Jansen, W., & Ayers, R. (2005). An overview and analysis of PDA forensic tools. National Institute of Standards and Technology, p. 7. Retrieved from http://csrc.nist.gov/groups/SNS/mobile_security/documents/mobile_forensics/ForensicArticle-DI-fin.pdf

[41]Bhadsavle, N., & Wang J. A. (2008). Validating tools for cell phone forensics. *Technical Report* CISE-CSE-08-05, p. 2. Retrieved from http://it.spsu.edu/research/Tech_Report/Neil_Andy_CellPhoneForensics.pdf

[42]There are, however, some exceptions. For instance, different versions of Encase software can be used on mobile phones and PDAs with certain operating systems. For more information on the forensics tools for computer systems, such as FTK and Encase, see Chapter 7.

[43]Jansen, W., & Ayers, R. (2004, August). PDA forensic tools: An overview and analysis. National Institute of Standards and Technology. *NISTIR 7100*, pp. 31–34. Retrieved from http://www.csrc.nist.gov/publications/nistir/nistir-7100-PDAForensics.pdf

[44]Akkaladevi, S., Himabindu, K., & Luo X. (2008, March). Efficient forensic tools for handheld devices: A comprehensive perspective. *Proceedings of Southwest Decision Sciences Institute* (Houston, Texas), p. 352. Retrieved from http://www.swdsi.org/swdsi08/paper/SWDSI%20Proceedings%20Paper%20S406.pdf

[45]Jansen, W., & Ayers, R. (2004, August). PDA forensic tools: An overview and analysis. National Institute of Standards and Technology. *NISTIR 7100*, p. 8. Retrieved from http://www .csrc.nist.gov/publications/nistir/nistir-7100-PDAForensics.pdf

[46]Jansen, W., & Ayers, R. (2005). An overview and analysis of PDA forensic tools, p. 9. Retrieved from http://csrc.nist.gov/groups/SNS/mobile_security/documents/mobile_forensics/ForensicArticle -DI-fin.pdf

[47]Jansen, W., & Ayers, R. (2004, August). PDA forensic tools: An overview and analysis. National Institute of Standards and Technology. *NISTIR 7100*, p. 8. Retrieved from http://www.csrc .nist.gov/publications/nistir/nistir-7100-PDAForensics.pdf

[48]Ayers, R., Jansen, W., Cilleros, N., & Daniellou, R. (2005, October). Cell phone forensic tools: An overview and analysis. National Institute of Standards and Technology. *NISTIR 7250*, p. 11. Retrieved from http://csrc.nist.gov/publications/nistir/nistir-7250.pdf

[49]Jansen, W., & Ayers, R. (2005). An overview and analysis of PDA forensic tools. *Digital Investigation*, 2, 126.

[50]*Ibid.*

[51]Ayers, R., Jansen, W., Cilleros, N., & Daniellou, R. (2005, October). Cell phone forensic tools: An overview and analysis. National Institute of Standards and Technology. *NISTIR 7250*, p. 8. Retrieved from http://csrc.nist.gov/publications/nistir/nistir-7250.pdf

[52]*Ibid.*, pp. 22–25.

[53]Akkaladevi, S., Himabindu, K., & Luo X. (2008, March). Efficient forensic tools for handheld devices: A comprehensive perspective. *Proceedings of Southwest Decision Sciences Institute* (Houston, Texas), p. 354. Retrieved from http://www.swdsi.org/swdsi08/paper/SWDSI%20Proceedings%20 Paper%20S406.pdf

[54]Akkaladevi, S., Himabindu, K., & Luo X. (2008, March). Efficient forensic tools for handheld devices: A comprehensive perspective. *Proceedings of Southwest Decision Sciences Institute* (Houston, Texas), p. 354. Retrieved from http://www.swdsi.org/swdsi08/paper/SWDSI%20Proceedings%20 Paper%20S406.pdf

[55]Ayers, R., Jansen, W., Cilleros, N., & Daniellou, R. (2005, October). Cell phone forensic tools: An overview and analysis. National Institute of Standards and Technology. *NISTIR 7250*, p. 12. Retrieved from http://csrc.nist.gov/publications/nistir/nistir-7250.pdf

[56]For further information, see http://oxygen-forensic.com/en/features/#unique

[57]For more information on the capabilities of the MOBILedit! forensic tool, see http://www .mobiledit.com/products.asp

[58]Mellars, B. (2004). Forensic examination of mobile phones. *Digital Investigation*, 1, 269.

[59]Ayers, R., Jansen, W., Cilleros, N., & Daniellou, R. (2005, October). Cell phone forensic tools: An overview and analysis. National Institute of Standards and Technology. *NISTIR 7250*, p. 12. Retrieved from http://csrc.nist.gov/publications/nistir/nistir-7250.pdf

[60]For further information, see http://paraben-forensics.com/catalog/product_info.php?products_ id=289

[61]Mellars, B. (2004). Forensic examination of mobile phones. *Digital Investigation*, 1, 271.

[62]*Ibid.*, p. 269.

[63]For further information, see http://www.logicubeforensics.com/products/hd_duplication/ celldek.asp

[64]See the following website for further information: http://www.paraben-forensics.com/handheld_ forensics.html

[65]Review and compare information on Strive (SGH-a687) and Propel Pro (SGH-i627) from http://www.pcmag.com/article2/0,2817,2362704,00.asp and http://www.pcmag.com/article2/ 0,2817,2347209,00.asp, respectively.

[66]National Institute of Justice. (2008, April 14). Chapter 1. Electronic devices: Types, descriptions and potential evidence. In US Department of Justice, *Electronic crime scene investigation: A guide for first responders* (2nd ed.). Retrieved from http://www.ojp.usdoj.gov/nij/publications/ecrime-guide-219941/ch1-electronic-devices/handheld-devices.htm

[67]See the Verizon website: http://www.verizonwireless.com/b2c/index.html

Chapter 13

The Pretrial and Courtroom Experiences of a Computer Forensics Investigator

This chapter explores the challenges inherent in the computer forensics investigator's role of providing testimony in pretrial hearings and the courtroom. It also stresses the importance of the use of a defensible approach to evidence collection, transportation, extraction, and preservation by the investigator and his or her meticulous attention to chain of custody issues and the rules of evidence.

Pretrial Procedures

In pretrial procedures, a preliminary examination of the evidence may occur. However, the "court has an absolute right to refuse to decide the admissibility of evidence, allegedly violative of some ordinary rule of evidence, prior to trial."[1] Additionally, in these procedures, a preliminary examination of a potential **witness** to determine his or her suitability to serve as expert or technical witness can occur, as well as an examination of the defendant to determine his or her competence to stand trial. Furthermore, these procedures are essentially used to determine whether a case should be tried in criminal or civil court. Two key pretrial procedures in either criminal or civil cases are motions and pretrial hearings.

Motions

Motions are filed by either party in a criminal or civil trial. Many different types of motions could be filed in courts. Motions, which must be supported by fact, are usually filed by attorneys to raise an objection to the introduction of a piece of evidence. These requests ask that a judge rule on a specific matter such as a violation of constitutional rights (e.g., unreasonable search and seizure) or rules of evidence.

An example of a motion filed for a constitutional violation is a **motion to suppress evidence**. Motions to suppress evidence can, for instance, focus on evidence and testimony or confessions of a suspect that have been illegally obtained by law enforcement agencies. A confession is illegally obtained if the suspect was not informed of his or her rights via what is now known as the *Miranda* **warning** before being interviewed by the police. The *Miranda* warning, outlined in *Miranda v. Arizona*,[2] is as follows:

> You have the right to remain silent. Anything you say can and will be used against you in a court of law. You have the right to speak to an attorney, and to have an attorney present during any questioning. If you cannot afford a lawyer, one will be provided for you at government expense.

Other motions can focus on rules of evidence. Such motions may include motions in limine or motions for discovery. When filing a **motion in limine**, the attorney seeks a pretrial ruling on the admissibility of evidence. In *Harper v. Patterson*,[3] the court discussed two ways in which a motion in limine can be used:

> 1) The movant seeks, not a final ruling on the admissibility of evidence, but only to prevent the mention by anyone, during the trial, of a certain item of evidence or area of inquiry until its admissibility can be determined during the course of the trial outside the presence of the jury. 2) The movant seeks a ruling on the admissibility of evidence prior to the trial.

The **motion for discovery**, in contrast, seeks pretrial disclosure of evidence. In particular, each party requests evidence or information from the other party. According to Rule 26(b)(1) of the Federal Rules of Civil Procedure:

> Parties may obtain discovery regarding any nonprivileged matter that is relevant to any party's claim or defense—including the existence, description, nature, custody, condition, and location of any documents or other tangible things and the identity and location of persons who know of any discoverable matter. For good cause, the court may order discovery of any matter relevant to the subject matter involved in the action.

Moreover, if either party "discovers additional evidence or material before or during the trial," they "must promptly disclose its existence to the other party or the court if: (1) the evidence or material is subject to discovery or inspection under this rule; and (2) the other party previously requested, or the court ordered, its production."[4] **Discovery**, therefore, involves the exchanging of information concerning the evidence and witnesses that each party intends to present at the trial.

Electronic discovery may pose problems to either party because of the enormous volume of electronic data that can be stored in databases and backup tapes. The more data stored in the database, the more difficult (both in terms of time and effort) the search for the information requested. Additionally, data are stored in a raw format; it must be restored to a more-formatted version before it can become identifiable and thus be of any use to those requesting it. Therefore, the data in backup tapes may not be reasonably accessible and easily searchable. A party may not have to produce the data if the production of the desired electronic evidence within these backup tapes is an extremely burdensome task or quite costly,[5] although this party would have to prove that such was the case to avoid retrieving the information.

During the discovery process, if an individual fails to provide information to the opposing party, "a party may move for an order compelling disclosure or discovery. The motion must include a certification that the movant has in good faith conferred or attempted to confer with the person or party failing to make disclosure or discovery in an effort to obtain it without court action."[6] Sanctions are provided to those who fail to produce documents during discovery. This outcome occurred in *Anderson v. Crossroads Capital Partners, L.L.C.*,[7] where the plaintiff failed to produce the requested discovery. Upon reviewing the plaintiff's hard drive, a computer forensics expert discovered that the plaintiff had installed a data-wiping application after having agreed not to delete any existing documents within the computer.

Some data-wiping applications are touted online as being so powerful as to defeat court-validated computer forensics tools. One case in which a plaintiff used this type of data-wiping software (specifically, a product known as Evidence Eliminator) to delete and overwrite more than 15,000 files was *Kucala Enters., Ltd. v. Auto Wax Co.*[8] In this case, the court sanctioned the plaintiff for acting "willfully and with the purpose of destroying discovery by purchasing and then using Evidence Eliminator on his computer. Any reasonable person can deduce … that Evidence Eliminator is a product used to circumvent discovery."[9]

Two forms of discovery are interrogatories and depositions. **Interrogatories** involve written questions that are provided to a witness. A witness must, in turn, answer these questions in written form. A **deposition** is testimony that is taken outside of a court of law. As such, depositions do not involve a judge or jury. Instead, the attorneys of both parties are present during the testimony of the witness and the questioning of the potential witness under oath by either party. This process is recorded by transcript and/or video. A party may use all or part of the recorded deposition provided by a witness as evidence in a court of law.[10] Moreover, according to Rule 15(b)(1) of the Federal Rules of Criminal Procedure

> A party seeking to take a deposition must give every other party reasonable written notice of the deposition's date and location. The notice must state the name and address of each deponent. If requested by a party receiving the notice, the court may, for good cause, change the deposition's date or location.

Depositions are less common is criminal proceedings. They are most often used in civil proceedings.

Pretrial Hearings

Pretrial hearings both determine if a crime or civil wrong has been committed—depending on the court of law—and evaluate the evidence and testimony to be presented in court. In respect to the former task, in criminal court, evidence is presented to determine whether an actual crime has been committed; in civil court, evidence is presented to determine whether a civil wrong has been committed. If there is sufficient evidence of a crime or civil wrong, then the case can go to court (criminal and civil court, respectively).

In respect to the assessment of evidence and testimony, the information retrieved during discovery is evaluated to determine which evidence will be admitted in court. Evidence is examined by the opposing party's experts. The tools, techniques, and methodologies used to examine and extract electronic evidence are subject to evaluation in these proceedings. After the validity and reliability of the evidence has been established, it is introduced in court for consideration by the judge and/or jury. Witnesses are also presented and each party has the right to cross-examine the other's witnesses. Each party will seek to discredit the evidence and testimony provided by the opposing party.

Testimony and Rules of Evidence

Computer forensics investigators/specialists have come to play an extremely important role in terms of courtroom testimony owing to the many advances in computer technology and related devices. Computer forensics investigations have been conducted on computers, mobile devices, and other media, with the results of these investigations subsequently being presented as crucial evidence in the courtroom. A computer forensics specialist is responsible for the collection, documentation, preservation, interpretation, and analysis of the evidence contained within the suspect's computer(s) and related electronic devices. Accordingly, this individual is often called to testify about the evidence, including how it was discovered, created, and obtained. This individual also testifies as to any modifications that may have occurred (either intentionally or accidentally) to the evidence during the recovery process. Therefore, investigators are not only tasked with finding and recovering evidentiary data, but are also often required to present the information found during the investigation to a judge and/or jury.

Testimony is provided to either give an opinion so as to assist the court in understanding a piece of evidence or to establish a fact that is presented in court. In particular, a computer forensics specialist has one of two roles in pretrial and trial proceedings.

On the one hand, a computer forensics expert can take on the role of a technical or scientific witness. This type of witness is restricted to providing only the facts of the investigation that has been conducted. When giving this type of testimony, the witness presents the evidence and explains it. The witness also explains how the evidence was obtained. For example, an individual may testify as to how certain data were retrieved from a computer using a specific computer forensics tool. The individual is not authorized to provide any observations, opinions, or conclusions concerning the evidence.

On the other hand, a computer forensics investigator can take on the role of an **expert witness**. According to Rule 702 of the Federal Rules of Evidence,

> If scientific, technical, or other specialized knowledge will assist the trier of fact[11] to understand the evidence or to determine a fact in issue, a witness qualified as an expert by knowledge, skill, experience, training, or education, may testify thereto in the form of an opinion or otherwise, if (1) the testimony is based upon sufficient facts or data, (2) the testimony is the product of reliable principles and methods, and (3) the witness has applied the principles and methods reliably to the facts of the case.

Basically, in this role, the computer forensics specialist collects, analyzes, and evaluates evidence and forms an opinion on it. This opinion is then communicated to the court (judge and/or jury). Expert witnesses are the only witnesses in court who are permitted to give their opinion about the case without having been physically present during the crime or civil wrong or involved in it in any way after it occurred. This opinion, however, must be supported by facts. In particular, Rule 703 of the Federal Rules of Evidence states that

> the facts or data in the particular case upon which an expert bases an opinion or inference may be those perceived by or made known to the expert at or before the hearing. If of a type reasonably relied upon by experts in the particular field in forming opinions or inferences upon the subject, the facts or data need not be admissible in evidence in order for the opinion or inference to be admitted.

Moreover, Rule 705 of the Federal Rules of Evidence states that "the expert may testify in terms of opinion or inference and give reasons therefor[e] without first testifying to the underlying facts or data, unless the court requires otherwise. The expert may in any event be required to disclose the underlying facts or data on cross-examination." As such, unlike a technical or scientific witness, expert witnesses are allowed to provide testimony that includes their opinions of any observations made during the investigation.

A party is required to "disclose to the other parties the identity of any witness it may use at trial to present evidence under Federal Rule of Evidence 702, 703, or 705."[12] According to Rule 26(a)(2)(B) of the Federal Rules of Civil Procedure, the disclosure must be accompanied by a written report that is prepared and signed by the potential technical or expert witness, which contains the following information:

- All of the opinions to be expressed that are relevant to the case, the basis for these opinions, and any exhibits that may be used to summarize or support these opinions
- The individuals' qualifications, including any publications made that are relevant to the issue the individual is providing an opinion on
- The cases in which the individual served as an expert witness during pretrial and/or court proceedings

To establish the competency of a technical or expert witness to participate in a trial, a thorough examination of that person's background and credentials is required. This occurs through a process known as ***voir dire***.[13]

Voir Dire

Before a technical or expert witness can provide testimony, he or she must be qualified in the *voir dire* process. The statements made during the *voir dire* process are submitted under oath. To verify that a witness is qualified to provide testimony, the party that is seeking to hire the witness questions him or her. This process seeks to either validate or refute the background and credentials of the witness. The opposing party also has the opportunity to cross-examine the witness during this process to highlight any weaknesses in his or her education or experience.[14]

During this process, the opposing party's attorney may attack the credibility of the individual providing the testimony. Primarily, the attorney will attack the credentials, methods used, and any oral or written statements made by the witness. Often, the opposing party's attorney may attack the qualifications of the expert, as opposed to his or her methods or interpretation of the results. Certain guidelines are included in the Federal Rules of Evidence pertaining to how an individual's credibility may be attacked. In fact, Rule 607 of the Federal Rules of Evidence states that "the credibility of a witness may be attacked by any party, including the party calling the witness." For example, the credibility of the witness may be undermined by providing evidence in support of the untruthfulness of the character of the witness. However, evidence showing the truthful character of the witness may be admitted only after an attack of the nature previously described has occurred.[15]

During the *voir dire* process the following information about the witness must be verified:

- Name
- Title
- Employment history (positions held, length of positions, and duties in each position)
- Current occupation (or occupations) including the position (or positions) held, duties performed, and length of time in the current position (or positions)
- Any specializations in the current field of employment (e.g., in the field of security an individual may be specialized in protective services operations, whereas in the field of computer forensics an individual may be specialized in network forensics)
- Employment address
- Education, including any degrees held, in which subject(s), and from which college(s) and/or university (or universities)

- Licenses (from which states in the United States and in which fields)
- Specialized training
- Board certification as a specialist in the field
- Any teaching or lectures given, including the date and place of where teaching or lectures took place
- Publications in the individual's field of specialization
- Cases where the individual served as a technical or expert witness in pretrial or court proceedings
- Membership in professional societies, organizations, or associations, and which position an individual held (or holds) in them
- Honors, awards, or any other special achievements in the individual's field
- Consulting the individual may have provided to private and/or public agencies

During the *voir dire* process, it would be foolish to rely on the résumé or curriculum vitae (CV) provided by the potential expert witness for all of these data, because individuals frequently embellish the information provided in these documents. Instead, witnesses should be subjected to a thorough background investigation. One can begin this process by conducting research on the potential witness. Several resources are available to help attorneys find the right experts, evaluate their credibility, and assess the admissibility of their testimony. One such resource is the Technical Advisory Service for Attorneys (TASA). TASA serves legal professionals, insurance companies, businesses, and government agencies by providing them with experts to assist them in various areas of litigation, including case evaluation and testimony.

Every aspect of the background of the potential technical or expert witness should be reviewed—beginning with the individual's résumé or CV. All of the information included in the résumé or CV should be verified. An investigator should contact prior employers to determine, among other things:

- If the potential witness actually worked there and if so, the length of employment
- The position or duties performed by the individual during his or her employment

This investigation is important because some individuals may truthfully include previous workplaces in their résumé or CV but may exaggerate their responsibilities or positions in these workplaces. Both the educational institutions and the degrees earned should also be verified, by contacting the educational institution and enquiring as to the individual's educational background. Online searches can be conducted to verify an individual's education background as well. For instance, the National Student Clearinghouse website (http://studentclearinghouse.com/) can provide verification of degrees earned and enrollment in an educational institution.

Public records searches can also provide information about the individual's aliases, maiden name, relatives, criminal records, property, military service, address history, phone numbers, and e-mail address (or addresses). All licenses, professional certificates, and training listed for the witness should be checked as well, by contacting the institutions, agencies, or organizations that provided them. Professional license information may also be found on websites (free or with a fee) such as Black Book Online (http://blackbook online.info/).

Moreover, publications reported by the witness should be verified. If the individual has published an article, to search for this publication, the investigator can go directly to the publisher's website and browse for the article in question. If the publication is a book, an online search for it would suffice. Chances are such a search would produce numerous results. Alternatively, the publisher's website could be searched. Other essential sources with which to check published material include the U.S. Copyright Office, the Library of Congress catalog, LexisNexis, HeinOnline, and Westlaw. Additional sites where publications can be found include Amazon.com, Borders, Barnes and Nobles, and other online bookstores.

Insight into the personal and professional lives of witnesses is also sought. Some information on these individuals may be found on social and business networking sites (e.g., Twitter, Facebook, and MySpace). It is important to note that the use of an online search engine (e.g., Google or Bing) could produce information on many of the items mentioned previously (e.g., education, employment, publications). Moreover, interviews with potential witnesses should be conducted, both over the phone and in person, to gain insight into their background and character.

Prior cases in which expert witnesses participated should also be checked—for example, to reveal any existing biases or challenges to the testimony they had provided, their methods, or their qualifications as experts. The prior testimony (or testimonies) of a potential witness should also be reviewed to reveal that individual's strengths and weaknesses as a witness. Legal research websites and self-reports from technical and expert witnesses are inadequate tools for witness due diligence.[16] One very important, yet sometimes overlooked, informational resource that seeks to remedy this situation is the *Daubert Tracker*. According to its website, the primary purpose of this database is to provide information about challenged expert witnesses, their disciplines and areas of expertise, and the results of these challenges.[17] This database holds information on all reported and numerous unreported state and federal cases where the methodology or qualifications of a testifying scientific or expert witness have been challenged. One can easily search this website using the name of the technical or expert witness.

Attorneys should pay special attention to researching whether the potential witness has publications or has provided testimony that conflicts with the findings in the current case. The attorney should also research whether a prospective witness provides testimony for only one particular party (e.g., he or she always testifies for the defense in a criminal trial). This one-sidedness can illustrate bias. By comparison, impartial witnesses tend to testify on behalf of either party in criminal or civil courts.

The interrogation and background check to which the potential expert witness is subjected is "quite intense, because acceptance of an unqualified expert (or conversely, exclusion of a qualified expert) is considered grounds for overturning the verdict in a higher court."[18] A prime example of what happens when the qualifications of an expert are not checked can be found in the case of Fred Zain.[19] Zain was employed as a forensic expert and testified in hundreds of criminal cases. Zain, however, was never qualified to perform forensic work. His educational transcripts revealed that he was a mediocre student and had failed chemistry. Despite his poor academic background, he was employed as a chemist at the Department of Natural Resources after completing his education. His employment background as a chemist was one of the main reasons why his lab chief at the time, Ray Barber, claimed that he felt that he did not need to check Zain's background before hiring him. After reviewing 189 cases in which Zain provided a report and/or testimony, a court warned that "any testimony or documentary evidence offered by Zain at any time should be deemed invalid, unreliable and inadmissible."[20]

There have been many instances where witnesses have lied about their education and qualifications. A case in point is James Earl Edmiston. Edmiston included false information about his educational history in his résumé and communicated this false information to individuals who questioned him when he was being qualified as an expert witness.[21] He claimed in his résumé that he had attended certain educational institutions and received degrees from them that were, in reality, not even offered by those institutions.

Another expert witness who lied about her credentials was Carolyn Ridling.[22] In 2008, Ridling lied under oath by claiming that she was certified as a Sexual Assault Nurse Examiner by the Texas Attorney General's office. She had been certified as a Sexual Assault Nurse Examiner at one point, but her certification had expired on April 18, 2004. As a result of Ridling's fabrication, the cases in which she gave testimony after the date that her certificate expired were reevaluated.

Yet another case involved Saami Shaibani, who significantly embellished his education and work experience on his résumé.[23] The numerous murder cases he testified in as an expert witness, which led to the convictions of the defendants on trial, are now in jeopardy of being overturned. In fact, the outcome in one of the cases in which Shaibani testified has already been overturned because he had lied about his qualifications as an expert: The murder conviction of Douglas Plude, who was serving a life sentence for poisoning his wife and drowning her in a toilet, was overturned by a Wisconsin Supreme Court.[24]

Authentication of Evidence

Evidence must be authenticated before it is submitted in court. According to Rule 901(a) of the Federal Rules of Evidence, "the requirement of authentication or identification as a condition precedent to admissibility is satisfied by evidence sufficient to support a finding that the matter in question is what its proponent claims."[25] One way in which evidence can be authenticated is through testimony by a technical or expert witness "that a matter is what it is claimed to be."[26] This testimony does not need to be provided by an expert on the subject matter. For instance, in *United States v. Whitaker*,[27] the court held that the authentication requirement may be met "through the testimony of a witness who was present and observed the procedure by which the documents were obtained from [the] Defendant's computers."

Nevertheless, to testify about the authenticity of evidence, a technical or expert witness must be knowledgeable of the issue at hand. According to Rule 602 of the Federal Rules of Evidence, "a witness may not testify to a matter unless evidence is introduced sufficient to support a finding that the witness has personal knowledge of the matter." Specifically, to testify as experts, witnesses must possess specialized knowledge and experience with which to explain evidence and certain events in relation to the crime. However, there is no rule as to the level of knowledge required to qualify a witness as a technical or expert witness in the field. For example, in *United States v. Scott-Emuakpor*,[28] the court held that to be considered an expert witness in computer forensics, knowledge of how to develop a sophisticated software program is not required. Instead, the court stated that the expert should possess the skills needed to find evidence on a hard or Zip drive. Therefore, to provide testimony as a computer forensics witness, knowledge of electronic evidence recovery is required, but an investigator does not need to be trained as a computer forensics investigator. Thus an individual who is skilled in computer forensics but has not had formal training can still qualify as an expert. The witness' knowledge of the matter at hand, however, will play a role in determining the weight given to and credibility of his or her testimony in a court of law. The task of finding the right witness is challenging because the field of computer forensics is constantly evolving. It is paramount that computer forensics specialists stay current in the field.

Based on the authenticity of the information provided by the potential technical or expert witness and a thorough review of his or her background and qualifications, a decision is made as to whether an individual is competent to serve as a technical or expert witness. This decision is then communicated to the court by the hiring attorney.

The Role of the Computer Forensics Investigator in Pretrial Proceedings and Court

A computer forensics investigator should always anticipate that his or her case may go to court. Accordingly, every step of the investigation must be documented. A computer forensics investigator can be questioned on all phases of his or her investigation and each step of the examination of the electronic evidence. This is why the previous chapters in this book have focused on (and stressed) the use of lawful procedures for conducting investigations and the selection of forensically sound tools for extracting evidence. An investigator's examination of seized evidence must be verifiable and repeatable.

A computer forensics investigator may be called as a technical or expert witness even before the trial—for example, in a pretrial hearing or to give a deposition. As a technical or expert witness, one must be prepared to provide definitions of terms and to explain the following points.[29]

The investigator must provide his or her qualifications. Computer forensics witnesses must be prepared to answer questions about their work experience (current position and employment history), educational background, training, licenses, certificates, membership in professional organizations, awards, publications, and testimony provided in other cases. The individual's qualifications as a technical or expert witness are extremely important to the case. The computer forensics investigator needs to be viewed as a credible witness to ensure that the validity and reliability of the electronic evidence and its handling are upheld in court.

The investigator must explain how things work. The investigator must be able to explain his or her actions in layperson terms. More specifically, when the investigator is called as a witness, he or she must be able to explain complex terms and processes in a simple and easy-to-understand manner. Given the complexity of the recovery of electronic evidence from computers and related devices, expert and technical witnesses are expected to testify about the data and the manner in which those data were recovered. Because the judge or the members of a jury may not be well versed in technology, computers, electronic devices, and computer forensics, an expert must be able to interpret the evidence for the court. Accordingly, in pretrial and courtroom proceedings, a computer forensics witness may be asked to show how the evidence was extracted from a computer or other electronic device. Indeed, many individuals—even computer users—do not really understand computer forensics processes such as data analysis and the acquisition of data using forensics tools. For example, some individuals may not know that when a user deletes data it can be recovered. To clarify these points, a computer forensics witness would be called upon to thoroughly explain how he or she recovered the deleted information. In doing so, it is important for the testimony of the computer forensics witness to turn this complex information into easy-to-understand, yet precise language that judges and jurors can comprehend.

Essentially, the testimony of the computer forensics expert is expected to shed light on electronic evidence presented in court. The judge or jurors may not be able to fully appreciate the significance of certain evidence if they are not familiar with computers and related technology. Testimony is used to provide the supplemental information necessary to help the judge and jury comprehend the importance of certain evidence. Being able to explain how a drive was imaged and the pertinent evidence recovered in a simple and concise way could be the deciding factor as to how a judge or jury decides the case. Without such testimony, a judge or jury might misinterpret or misunderstand the evidence, ultimately coming to an erroneous conclusion about its significance and relevance to the case.

The investigator must explain why a specific forensics tool was used to extract evidence and/or why another tool was not used instead. The technical or expert witness must be able to justify his or her use of a particular computer forensics tool. As a general rule, computer forensics specialists should use a computer forensics tool whose use has previously been upheld in court. New computer forensics tools that have recently become available on the market should be avoided, because their validity and reliability as appropriate computer forensics tools could be called into question in pretrial or courtroom proceedings. If the validity and reliability of the tool cannot be established, then the evidence retrieved using this tool would be inadmissible. To avoid rendering evidence inadmissible in court, the computer forensics tool that is chosen must have been successfully admitted and upheld in previous court proceedings.

The investigator must explain how an examination of evidence is completed with a particular forensics tool (i.e., the one used in the investigation). A computer forensics technical or expert witness must have knowledge of the forensic protocols used to locate evidence and protect the data on the computer (or computers) from which the evidence is extracted. The correct procedures for extracting the electronic evidence must also be followed to ensure the admissibility of the evidence in court. The investigator must be able to explain to the judge or jury the method and tools used to extract the evidence. This testimony is provided so that the judge and jury will not question the validity and reliability of the process through which the evidence was retrieved.

The witness must also be able to demonstrate that the evidence was recovered in a manner that did not jeopardize its integrity. Indeed, computer forensics witnesses must show that the evidence presented is authentic and has not been altered in any way. For instance, in child pornography cases, a digital image expert is often called to testify about the authenticity of an image, including whether it has been manipulated. In particular, the court has stated that the evidence presented must be "in substantially the same condition as when the crime was committed."[30] Accordingly, witnesses may testify that the evidence presented is in the same condition as it was when it was found by law enforcement agents.[31]

The investigator must account for the accuracy and interpretation of results. The accuracy and interpretation of the results of computer forensics analyses are often challenged in pretrial and courtroom proceedings. Computer forensics witnesses are called to testify in these proceedings to prove the accuracy of the results and to support their interpretation of the results of their analysis of the electronic evidence. For the most part, after his or her analysis of the evidence, the computer forensics specialist "forms an opinion based on what he or she considers as the most probable explanation of the results. During his or her testimony, this individual must be careful to express an expert opinion that has a solid foundation based on test results."[32]

Additionally, the computer forensics witness must be able to demonstrate to the judge or jury that the data presented in court are what they are purported to be and that the tools used to retrieve the data functioned as designed, thereby producing an accurate result. The witness interprets the data to provide an accurate picture of the evidence and its relation to the crime. The witness must also demonstrate why his or her results are important to the case. In fact, the results of the investigator's analysis of the evidence are often used to link a particular suspect to the crime or exclude that person as a possible suspect.

The investigator must explain any gaps that may exist in the chain of custody. The manner in which the computer or related electronic devices and the electronic data within these devices were collected will be scrutinized by the opposing party to ensure that proper procedures for handling the evidence were followed and that the chain of custody was maintained. If procedures were not followed (e.g., data were not properly extracted) or the chain of custody was not maintained, the evidence (more often than not) will be deemed inadmissible in court (see Chapter 8). The attorney for the opposing party in pretrial proceedings and/or in the courtroom will seek to discredit the computer forensics expert or find problems with his or her chain of custody.

As part of his or her testimony, an investigator may be required to provide detailed explanations of the entire investigative process—that is, the identification, collection, documentation, tagging, bagging, transport, and preservation of evidence. Preparation is key to providing technical or expert testimony that will stand up in court. The investigator should keep a journal of the exact steps taken to analyze the evidence and maintain its integrity.

The attorney will prepare the witness for trial by, among other things:[33]

- Discussing the witness' testimony with him or her
- Reviewing the witness' notes and reports and highlighting any inconsistencies between those documents and the witness' oral testimony of the events, evidence, procedures used, observations, or opinions
- Discussing the possible line of questioning that the witness should be prepared for from both parties
- Rehearsing witness testimony through, for example, role playing

If the technical or expert witness has never testified in court before this case, then the attorney must explain the pretrial and courtroom proceedings to this individual. When testifying, a computer forensics witness must follow appropriate courtroom etiquette and maintain a high degree of professionalism. Casual dress in such proceedings should be avoided. The appropriate attire for pretrial and courtroom proceedings should be communicated to the witness by the hiring attorney. The witness must be confident when expressing his or her opinion or testifying on the facts of the case. Witnesses should request that questions be clarified if they do not understand what an attorney is asking. The answers to the questions posed by attorneys in either pretrial or courtroom proceedings should be truthful and carefully formulated. False testimony will result in a witness being disqualified, rendering his or her testimony inadmissible.

Are Computer Forensics Analysts Required by Law to Testify?

Many analysts provide reports of their analyses for use in pretrial and courtroom proceedings. In the past, few were called to testify about their findings. If they were called, it was at the discretion of the attorney of either party. This testimony by the analysts, however, was not required by law.

This situation changed in 2009. Specifically, prior to the court ruling in *Melendez-Diaz v. Massachusetts*,[34] forensic analysts' reports could be admitted as evidence without the expert being required to testify. In *Melendez-Diaz*, the drug analysis certificates of the bags of cocaine prepared by a lab technician were presented as evidence against the defendant in court. Luis Melendez-Diaz appealed this decision, arguing that the drug analysis certificates presented as evidence in court violated his constitutional rights. In particular, Melendez-Diaz claimed that the admittance of this evidence violated his Sixth Amendment rights. The **Sixth Amendment** to the U.S. Constitution makes the following statement:

> In all criminal prosecutions, the accused shall enjoy the right to a speedy and public trial, by an impartial jury of the State and district wherein the crime shall have been committed, which district shall have been previously ascertained by law, and to be informed of the nature and cause of the accusation; to be confronted with the witnesses against him; to have compulsory process for obtaining witnesses in his favor, and to have the Assistance of Counsel for his defense.

The Sixth Amendment contains several criminal trial rights. The one applicable to the *Melendez-Diaz* case is the right of a defendant to confront the witnesses against him or her (known as the **confrontation clause**). Previous court cases have established that "the need for confrontation is particularly important where the evidence is testimonial, because of the opportunity for observation of the demeanor" of the witness.[35] Indeed, rulings in cases such as *Crawford v. Washington*[36] have held that testimonial evidence cannot be introduced in court unless the defense has had an opportunity to cross-examine the witness providing the evidence in question. In *Crawford*, the court explicitly stated that the Framers of the U.S. Constitution "would not have allowed admission of testimonial statements of a witness who did not appear at trial unless he was unavailable to testify, and the defendant had had a prior opportunity for cross-examination."[37] This court further noted that the ultimate goal of the confrontation clause of the Sixth Amendment

> is to ensure reliability of evidence, but it is a procedural rather than a substantive guarantee. It commands, not that evidence be reliable, but that reliability be assessed in a particular manner: by testing in the crucible of cross-examination . . . Dispensing with confrontation because testimony is obviously reliable is akin to dispensing with jury trial because a defendant is obviously guilty. This is not what the Sixth Amendment prescribes.[38]

Accordingly, if the evidence introduced in proceedings is considered testimonial, the declarant must be called to testify. Failure to do so violates the defendant's rights. In *Melendez-Diaz*, the U.S. Supreme Court held that the lab report of the forensic analyst was testimonial evidence. As such, the Court ruled that the introduction of these documents without affording the defense the opportunity to cross-examine the forensic analyst in court violated the defendant's Sixth Amendment right to confront the witnesses against him. As a result of the ruling in *Melendez-Diaz*, a number of cases involving similar confrontation clause challenges to forensic evidence were vacated and remanded for reconsideration.[39]

The dissenting justices in the *Melendez-Diaz* case argued that the Sixth Amendment does not require analysts to provide testimony in court, citing numerous cases to support this conclusion.[40] Despite rulings in prior cases that established that analysts do not need to provide testimony to support their conclusions, the majority of the Court in *Melendez-Diaz* believed otherwise. Consequently, analysts are now required to testify in court or at least provide the defense with an opportunity to cross-examine them in proceedings.

Several different types of individuals may be involved in the computer forensics process:

- The investigator in the field who identifies, collects, labels, packages, and transports the evidence

- The analysts responsible for reviewing the evidence, interpreting it, and drawing conclusions about its relevance and significance to the case

- The computer technicians, network administrators, and other practitioners who may handle the evidence

Each individual must document and report his or her role in the investigatory process. As a result of the *Melendez-Diaz* case, these individuals can be called to testify if their reports are admitted as evidence.

Chapter Summary

Pretrial proceedings are held to determine if there is sufficient evidence with which to go to court. Several events can take place before a trial, including discovery, hearing of motions, and depositions. The computer forensics investigator or specialist who identifies, analyzes, and preserves the electronic evidence may be called to pretrial hearings or subsequent court proceedings to testify about his or her actions and findings. The accuracy of the results and the examiner's interpretation of the results may be called into question during these proceedings.

The computer forensics investigator has one of two roles in a pretrial or courtroom proceeding—as a technical witness or as an expert witness. As a technical witness, an individual can testify only as to the facts of the case, evidence, and procedures used. In contrast, as an expert witness, the individual can provide an opinion based on the investigation conducted and the observations he or she made. Computer forensics witnesses are used to help the court understand computer forensics, the process used to obtain electronic evidence, and the significance of this evidence to the case.

When evidence is collected and presented in pretrial and courtroom proceedings, a computer forensics technical or expert witness may be required to authenticate that evidence and explain its relevance to the judge and jury. The witness must be able to explain how the evidence was identified, extracted, and preserved. This step is intended to ensure that the judge and jury will not question the integrity of the process. To be effective, a witness must be able to clearly articulate his or her findings to the court.

Forensic analysts' reports are considered testimonial evidence under a recent ruling of the U.S. Supreme Court. Computer forensics reports cannot be admitted as evidence in pretrial or court proceedings without providing the defendant with the opportunity to cross-examine the analyst or investigator who authored the report. This testimony was mandated after the decision in *Melendez-Diaz*. If such information is admitted without allowing a defendant to cross-examine the analyst or investigator, the defendant's Sixth Amendment rights are violated. In particular, the right of a defendant to confront the witnesses against him or her is violated.

Key Terms

Confrontation clause	Motion for discovery
Deposition	Motion in limine
Discovery	Motion to suppress evidence
Electronic discovery	Pretrial hearing
Expert witness	Sixth Amendment
Interrogatories	Trier of fact
Miranda warning	*Voir dire*
Motion	Witness

Critical Thinking Question

In your opinion, how important is the testimony of a computer forensics witness to a case?

Review Questions

1. What are two types of pretrial proceedings?

2. What is a motion?

3. What happens during discovery?

4. What are some problems posed by electronic discovery?

5. What are two forms of discovery?

6. What are the two roles of a computer forensics specialist in pretrial or courtroom proceedings?

7. Why should the qualifications of potential witnesses be evaluated? What happens when attorneys fail to do so?

8. Which information about potential witnesses should be verified before making a decision as to their competency to participate in a trial? How can this be accomplished?

9. Why do computer forensics investigators testify?

10. Which types of questions should a computer forensics witness be prepared to answer in pretrial or courtroom proceedings?

11. Is a computer forensics analyst required to testify in court?

Footnotes

[1]*Harper v. Patterson*, 270 Ga. App. 437, 441, 606 S. E. 2d 887, 892 (2004). See also *Telcom Cost Consulting, Inc. v. Warren*, 621 S. E. 2d at 866.

[2]384 U.S. 436 (1966).

[3]270 Ga. App. 437, 441, 606 S. E. 2d 887, 892 (2004). See also *Telcom Cost Consulting, Inc. v. Warren*, 621 S. E. 2d at 866.

[4]Rule 16(c) of the Federal Rules of Criminal Procedure.

[5]See Federal Rules of Civil Procedure, Rule 26(b)(2)(B).

[6]Rule 37(a)(1) of the Federal Rules of Civil Procedure.

[7]2004 WL 256512 (D. Minn. Feb. 10, 2004).

[8]56 Fed. R. Serv. 3d 487 (N.D. Ill. 2003).

[9]*Kucala Enters., Ltd. v. Auto Wax Co.*, 56 Fed. R. Serv. 3d 487, 494 (N.D. Ill. 2003).

[10]See Rule 15(f) of the Federal Rules of Criminal Procedure.

[11]The trier of fact (i.e., the finder of fact) is the person or persons (i.e., judge and/or jury) who determine the facts of a case.

[12]Rule 26(a)(2)(A) of the Federal Rules of Civil Procedure.

[13]*Voir dire* is French. It means "to speak the truth."

[14]Girard, J. E. (2008). *Criminalistics: Forensic science and crime.* Sudbury, MA: Jones and Bartlett, p. 52.

[15]Rule 608(a), Federal Rules of Evidence.

[16]Due diligence refers to systematic research conducted to acquire valid and reliable information with which to make an informed decision on the issue at hand.

[17]See the *Daubert Tracker* website: http://www.dauberttracker.com/index.cfm

[18]Girard, J. E. (2008). *Criminalistics: Forensic science and crime.* Sudbury, MA: Jones and Bartlett, p. 52.

[19]Berry, S. M. (n.d.). When experts lie. *Truth in Justice.* Retrieved from http://www.truthinjustice.org/expertslie.htm

[20]Clines, F. X. (2001, September 5). Work by expert witness is now on trial. *New York Times* [online]. Retrieved from http://www.nytimes.com/2001/09/05/us/work-by-expert-witness-is-now-on-trial.html

[21]Ellis, J. (2007, June 30). Ex-witness gets 21 months in prison: Trial expert lied on resume and to investigators. *The Fresno Bee* [online]. Retrieved from http://www.fresnobee.com/2007/07/28/97923/ex-witness-gets-21-months-in-prison.html

[22]Garsee, R. (2008, February 27). Sex crimes expert lied on the stand. *KTEN News* [online]. Retrieved from http://www.kten.com/global/story.asp?s=7937161&ClientType=Printable

[23]Nationwide murder convictions in doubt due to "expert" lies. (2008, June 26). *Fox News* [online]. Retrieved from http://www.foxnews.com/story/0,2933,372466,00.html

[24]Foley, R. J. (2008, June 26). Expert's lies jeopardize murder convictions. *USA Today* [online]. Retrieved from http://www.usatoday.com/news/nation/2008-06-26-2301015538_x.htm

[25]Rule 901(a) of the Federal Rules of Evidence.

[26]Rule 901(b) of the Federal Rules of Evidence.

[27]127 F. 3d 595, 601 (7th Cir. 1997).

[28]2000 U.S. Dist. LEXIS 3118, 3133 (W.D. Mich. Jan. 25, 2000).

[29]This list is by no means exhaustive.

[30]*Gallego v. United States*, 276 F. 2d 914, 917 (9th Cir. 1960).

[31]*Wilson v. State*, 983 P. 2d 448, 462 (Okla. Crim. App. 1998).

[32]Girard, J. E. (2008). *Criminalistics: Forensic science and crime*. Sudbury, MA: Jones and Bartlett, p. 53.

[33]Wydick, R. C. (1995, September). The ethics of witness coaching. *Cardozo Law Review, 17*, 4–5.

[34]129 S. CT. 2527 (2009).

[35]*People v. Arreola*, 875 P.2d 736, 744 (Cal. 1994); *People v. Johnson* 18 Cal. Rptr. 3d 230, 233 (Ct. App. 2004).

[36]541 U.S. 36 (2004).

[37]*Crawford v. Washington*, 541 U.S. 36, 53–54 (2004).

[38]*Crawford v. Washington*, 541 U.S. 36, 61–62 (2004).

[39]*People v. Barba*, 2007 WL 4125230 (Cal. App. 2d Dist.), vacated, 129 S. Ct. 2857 (2009); *Ohio v. Crager*, 116 Ohio St. 3d 369 (Ohio 2007), vacated, 129 S. Ct. 2856 (2009); *Commonwealth v. Rivera*, 70 Mass. App. Ct. 1116 (Mass. App. Ct. 2007), vacated, 129 S. Ct. 2857 (2009); *Commonwealth v. Morales*, 71 Mass. App. Ct. 587 (Mass. App. Ct. 2008), vacated, 129 S. Ct. 2858 (2009); *Commonwealth v. Pimentel*, 2008 WL 108762 (Mass. App. Ct. 2008), vacated, 129 S. Ct. 2857 (2009). See also Froehlich, D. M. (2010, February). The impact of *Melendez-Diaz v. Massachusetts* on admissibility of forensic test results at courts-martial. *Army Lawyer*, p. 36.

[40]Such as *Sherman v. Scott*, 62 F. 3d 136, 139-142 (CA5 1995); *Minner v. Kerby*, 30 F. 3d 1311, 1313-1315 (CA10 1994); *United States v. Baker*, 855 F. 2d 1353, 1359-1360 (CA8 1988); *Reardon v. Manson*, 806 F. 2d 39 (CA2 1986); *Kay v. United States*, 255 F. 2d 476, 480-481 (CA4 1958). See also *Manocchio v. Moran*, 919 F. 2d 770, 777-782 (CA1 1990).

Glossary

Access control list: Rules that define network security policies and govern the rights and privileges of users of a specific system.

Access control systems: Systems used to restrict access to protected network resources.

Acquisition: The process of evidence retrieval in computer forensics investigations—from the search for the evidence to its collection and documentation.

Adam Walsh Protection and Safety Act of 2006: Law designed to protect child against sexual predators, passed in response to the abduction and murder of Adam Walsh.

Address Resolution Protocol (ARP): It is a protocol that is used to map an IP address to the unique physical address of a computer—the MAC address.

Administrative law: Law that focuses on the exercise of government authority by the executive branch and its agencies.

Adware: A computer program used to track users' online activities and deliver targeted pop-up ads to users.

African Network Information Center (AfriNIC): The registry of Internet number resources for Africa.

Age Discrimination in Employment Act of 1967: Law that prohibits an employer from discriminating against an individual 40 years or older on the basis of his or her age.

AMBER alert: A warning system developed to give notice to the public of recent child abductions.

American Registry for Internet Numbers (ARIN): The registry responsible for assigning and registering IP addresses in the North American region.

American Society for Industrial Security: An organization that is dedicated to increasing the efficiency of security professionals. It was founded in 1955.

American Standard Code for Information Interchange (ASCII): A standardized character set used in computers.

Americans with Disabilities Act of 1990: Law that prohibits employers from discriminating against an individual based on his or her disability.

Anonymity: The ability of individuals to conduct their lives without making their activities known to others.

Answering machine: A telecommunications device that may include forensic evidence in the form of recorded voice messages (current or deleted), missed calls, caller identification information, and the last number called or dialed on the device.

Application log: A type of event log that contains the events that are logged by programs and applications.

Applications and services log: An event log in Windows 7 that stores events from a single application or component.

ARP poisoning: A type of attack in which the offender causes the network devices to update their ARP tables with the false information, so that the offender could redirect traffic and launch attacks.

Asia Pacific Network Information Center (APNIC): The registry responsible for IP address allocation and registration in the Asia-Pacific region.

Attorney–client privilege: Protection of the communications between a client and an attorney concerning a legal matter.

Auction fraud: A type of fraud in which the seller of an item at an auction engages in some sort of deception (e.g., not delivering the item after receiving payment, rigging the bidding) to defraud the would-be buyer.

Authentication: A process whereby the original data are compared to a copy of the data to determine if the two are exactly the same.

Authentication-Results **field:** A header field that decides and/or makes a recommendation to the user as to the validity of the origin of the message and the integrity of its content.

Avoidance: A strategy to subvert surveillance in which individuals' actions are displaced to "times, places and means in which the identified surveillance is presumed to be absent."

Backdoor: A means by which a developer or user reenter a computer system at a later date and time, without going through the normal security processes.

Bad clusters: Areas of a computer's operating system where data are stored, but that are not accessed and thus are overlooked by the operating system.

Bagging: Packaging of evidence prior to its transport to a forensic laboratory.

Bank Secrecy Act of 1970: Law that requires financial institutions to report instances of money laundering and to take an active role in detecting and preventing money laundering.

Batch mode: Data processing that occurs all at once.

Berne Convention: An international agreement governing copyright protection of literary and artistic works.

Bid rigging: A type of corporate crime in which purchasers collude to manipulate the price of an item, unbeknownst to the seller.

Bitstream copy: A bit-by-bit copy of a hard drive that does not alter the original hard drive in any way.

Blocking moves: A strategy to subvert surveillance in which an individual physically blocks access to the communication or renders it unusable.

Bluetooth: A short-range, wireless communication technology.

Bootleg: Reproduction of someone else's recording, marketed under a different label.

Border search doctrine: The right of a sovereign nation to regulate the entry and exit of individuals at its borders and the conditions under which such entry and exit may occur.

Botherder: A person who controls a botnet.

Botnet: A network of zombie (bot-infected) computers; a network of computers that are controlled remotely without the knowledge of the computer user.

Bribery: "The offering or receiving of money, goods, services, information, or anything else of value for the purpose of influencing public officials to act in a particular way."

Bro Intrusion Detection System: Software that passively monitors network traffic for suspicious activity.

Cable Communications Privacy Act of 1984: Law that restricts cable operators from using their systems to collect personally identifiable information and from disclosing such data inappropriately.

Campus area network: A computer network that connects computer systems in a particular area.

Campus Sex Crimes Prevention Act of 2000: Law that requires sex offenders who enroll or work at a community college, college, university, or trade school to notify local law enforcement agencies of their presence.

CAN-SPAM Act of 2003: The Controlling the Assault of Non-Solicited Pornography and Marketing Act of 2003 sought to regulate unsolicited commercial e-mails by requiring such messages to include opt-out instructions for recipients and prohibiting the use of deceptive subject lines and false headers in these messages. This Act additionally provides severe penalties for violations of its provisions.

Carey-Winick approach: A search protocol in which investigators limit searches by searching the computer using filenames, directories, and a sector-by-sector search of the hard drive.

Carnivore: A U.S. government-controlled software program that, when installed at facilities of Internet service providers, can monitor all the traffic moving through them.

"Catch it as you can" tools: Network forensics analysis tools that capture and store packets that traverse certain points, and then analyze the recorded traffic in batch mode.

Cell Seizure: A forensics tool used with certain models of GSM, TDMA, and CDMA phones.

CellDEK: A forensics tool that can extract information from the majority of mobile phones, PDAs, and smartphones on the market, while preserving the integrity of the data.

Chain of custody: The process by which investigators preserve the crime (or incident) scene and evidence throughout the life cycle of a case.

Child exploitation: Child exploitation includes the manufacture, possession, and distribution of child pornography, child sexual molestation, child sexual abuse, enticement of children for sexual acts, child prostitution, and child sex tourism.

Child Online Protection Act of 1998: Law (declared unconstitutional in 2007) that tried to control the dissemination of information to children by prohibiting websites from knowingly making available harmful material to minors.

Child pornography: Child pornography involves the visual depiction of any kind of a minor engaging in sexual activity.

Child Pornography Protection Act of 1996: Law that explicitly criminalizes the exploitation of children online.

Child Protection and Obscenity Enforcement Act of 1988: Law that made it illegal for an individual to use a computer to depict or advertise child pornography.

Child Protection and Sexual Predator Punishment Act of 1998: Law that requires communications service providers that become aware of child pornography to report it.

Child Victim Identification Program: A U.S. organization that houses a large database of child pornography images for the purpose of identifying victims of child exploitation and abuse.

Children's Online Privacy Protection Act of 1998: Law that applies to children younger than 13 years of age and protects them from the collection and misuse of their personal information by commercial websites.

Chilling effect: "When individuals otherwise interested in engaging in a lawful activity are deterred from doing so in light of perceived or actual government regulation of that activity."

Ciphertext: An encrypted message.

Circumstantial evidence: Evidence that allows someone to infer the truth of a given fact.

Civil law: Law that governs the relationships between private parties, including both individuals and organizations.

Client/server e-mail: An e-mail system in which the client is the computer that sends or receives messages, and the server stores any messages received until they are retrieved by the user.

Cluster: A fixed-length block of data on a computer that is used to store files.

Code of Federal Regulations: The codification of rules and regulations under the Securities Exchange Act.

Communications Act of 1934: Law that regulates interstate and foreign communication by wire or radio.

Communications Assistance for Law Enforcment Act of 1994: Law that requires telecommunications and electronic communications service providers to ensure that government entities, pursuant to a lawful authorization (e.g., a search warrant), have access to all wire and electronic communications and call-identifying information.

Communications Decency Act of 1996: Law (eventually struck down by the U.S. Supreme Court) that was designed to protect children from exposure to indecent material.

CompactFlash: A memory card that provides mass storage for mobile phones and PDAs.

Computer forensics: A branch of forensic science that focuses on criminal procedure law and evidence as applied to computers and related devices.

Computer Matching and Privacy Protection Act of 1988: Law outlining a computer-supported process in which personal data from records about individuals held by government agencies are compared for the purpose of determining eligibility for federal benefits programs.

Computer Security Act of 1987: Law that sought to improve privacy protections and establish minimum acceptable security practices for federal computer systems that contain sensitive information.

Computer virus: A software program that is designed to spread itself to other computers and to damage or disrupt a computer (e.g., interrupting communications by overwhelming a computer's resources).

Confrontation clause: The right of a defendant to confront the witnesses against him or her, as specified in the Sixth Amendment.

Consent search: A type of search that can occur without a warrant and without probable cause if an individual who has authority over the place or items to be searched has consented to the search.

Content data: The spoken words in a conversation or the words written in a message.

***Content-type* field:** A header field that indicates the type of data included in an e-mail message, such as text, image, audio, video and multipart.

Controlled substance: A drug whose manufacture, distribution, possession, and use are tightly regulated by the government.

Convention on Cybercrime: A Council of Europe document that contains provisions for the collection of communications data, including procedures for the preservation of telecommunications and electronic communications traffic data.

Cookie: A file created by a website that is stored on a user's computer hard drive when he or she visits that particular website.

Coppolino standard: Guideline stating that a novel test may be admitted as evidence in court if its validity can be proven, even if the general scientific community is unfamiliar with it.

Copyright: A legal document that gives to the copyright holder the right to copy, reproduce, and sell works or authorship such as sound recordings, musical, literary, dramatic, architectural, pantomimes, choreographic, pictorial, graphic, sculptural works, and motion pictures and other audiovisual works.

Copyright Act of 1976: Law that protected copyright holders; it was amended by the Digital Millennium Copyright Act of 1998.

Copyright infringement: The unlawful copying of movies, TV shows, music, software, literature, and videogames; also known as piracy.

Copyright Infringement Liability Limitation Act of 1998: Part of the Digital Millennium Copyright Act of 1998, which provides immunity to Internet service providers for software piracy violations under certain exceptions.

Corporate and Criminal Fraud Accountability of 2002: Title VIII of the Sarbanes-Oxley Act, which identifies criminal penalties for altering documents.

Corporate Fraud Accountability of 2002: Title IX of the Sarbanes-Oxley Act, which concerns the actions of "tampering with a record or otherwise impeding an official proceeding."

Counterfeit: An item represented to be the genuine articles of a particular manufacturer when, in truth, it is not.

Cracking: An attempt to gain unauthorized access to a computer system so as to commit another crime, such as destroying information contained in that system.

Credit card fraud: A crime in which an individual obtains, uses, sells, or buys another individual's credit card information.

Criminal law: Law dealing with public offenses—that is, actions that are harmful to society as a whole.

Cyberbullying: The use of telecommunications and electronic communications technologies to harass, insult, or humiliate a child, pre-teen, or teen.

Cybercrime: The use of the Internet, computers, and related technologies in the commission of a crime.

Cyberextortion: A crime in which someone, via e-mail, threatens to attack an individual, business, or organization if money is not provided to prevent the attack.

Cyberharassment: Use of the Internet, e-mail, or other forms of communication to intentionally annoy, attack, alarm or otherwise bother another person.

Cyberprostitution: Nonmarital sexual services offered through the Internet.

Cyberslacking: Misuse of computer privileges by employees that results in financial loss, reduction in productivity, wasted resources, and unnecessary business interruptions for their employer.

Cybersmear investment scheme: A type of financial fraud in which the offender tries to lower the price of a stock by circulating false rumors about it.

Cyberstalking: A type of cybercrime in which an individual repeatedly harasses or threatens another person through the use of the Internet, e-mail, or other electronic communications devices.

Cyberterrorism: The politically motivated use of computers as weapons or as targets, by subnational groups or clandestine agents intent on violence, to influence an audience or cause a government to change its policies.

Cyclical Redundancy Checksum: A hash function that is used to detect errors in transmission or duplication of data.

Data manipulation: The misuse of a computer system by altering data within it without proper authorization. Such information is altered for the offender's material gain.

Data mining: The process of sorting and sifting through vast quantities of data to find valuable information.

Dating scam: A type of fraud in which an offender poses as an attractive person on an online dating site, develops an online relationship with the victim, and then seeks money from the person because of a supposed "emergency."

Daubert standard: Guideline stating that a novel test may be admitted in court if it satisfies criteria related to testing, peer review, the potential rate of error, and acceptability.

Decryption: The transformation of ciphertext (encrypted text) into plaintext (unencrypted text).

Defamation: An offense in which a person damages an individual's reputation "by making public statements that are both false and malicious."

Deleted files: Deleted files are those files that have been removed or erased from a computer system and/or related device. Depending on the manner in which a user removed or erased the file, some data may still reside on the hard drive of a computer.

Demonstrative evidence: Evidence that cannot independently prove a fact, but rather is used to explain other evidence; illustrate, demonstrate, or recreate an event; or show a situation similar to something being presented in the case.

Denial of service: An attempt to prevent users of a particular service from effectively using that service. Typically, a network server is bombarded with authentication requests; the attack overwhelms the resources of the target computers, causing them to deny server access to other computers making legitimate requests.

Deposition: Testimony that is taken outside of a court of law.

Device Seizure: A forensics tool that acquires and examines data from PDAs, mobile phones, smartphones, and navigation devices.

Digital camera: A non-film-based camera that may contain evidence in the form of images, sounds, and date and timestamps.

Digital Manner Policies: A Microsoft system that allows someone else to control what is done with a user's mobile phone in specific designated areas.

Digital Millennium Copyright Act of 1998: Law that amended the U.S. Copyright Act of 1976 to address digital information. It was passed to implement two 1996 World Intellectual Property Organization (WIPO) treaties: the Copyright Treaty and the Performances and Phonograms Treaty.

Direct evidence: Evidence that establishes a fact; it includes eyewitness testimony and confessions.

Directive 95/46/EC: The European Union's Data Protection Directive, which was created to protect individuals with regard to the processing of their personal data and the free movement of such data, in accordance with European Community law, and required communications service providers to erase data or make the data anonymous after they were no longer needed for business purposes.

Directive 2002/58/EC: The European Union's Electronic Communications Directive, which translated the principles set out in Directive 95/46/EC into specific rules for the protection of privacy and personal data in all forms of communications, including electronic communications.

Directive 2006/24/EC: The European Union's Data Retention Directive, which makes compulsory the a priori retention of traffic, location, and other related data of individuals to make sure that such information is available to authorities.

Discovery: An exchange of information concerning the evidence and witnesses that each party intends to present at a trial.

Discrimination: Illegal (negative) favoritism based on age, sex, race, national origin, religion, disability, and pregnancy.

Distributed denial of service: An attempt to gain control over multiple computers and then use them to launch an attack against a specific target or targets.

Documentary evidence: Any kind of writing, video, or sound recording material whose authenticity needs to be established if it is introduced as evidence in a court of law.

Documentation: An accurate, thorough description of a crime scene and all evidence found there.

Domain name: An identifying label by which computers are familiarly known on the Internet.

Domain Name System (DNS): A hierarchical, distributed database of unique IP addresses that correspond with domain names.

DRAM: Dynamic random access memory.

Dru's law: Law that called for establishment of a national sex offender registry.

Duplicate Disk: A forensics tool that creates a bit-by-bit image of the contents of a Linux-based PDA.

Dynamic Host Configuration Protocol (DHCP): It is a protocol that allows a server to dynamically assign IP addresses to networked computers.

Dynamic IP address: A temporary IP address that is assigned from a pool of available IP addresses when the user logs on and is released when the user logs off.

Economic espionage: The theft of trade secret for the benefit of a foreign government, foreign instrumentality, or foreign agent.

E-fense Helix3Pro: A computer forensics tool that is designed to ensure that data are not altered during the imaging process.

E-fense Live Response: A computer forensics tool that is designed to collect nonvolatile and volatile data from live running systems.

e-Government Act of 2002: Law that requires government agencies to perform privacy impact assessments on computer and information technology systems that are designed to collect, maintain, and disseminate information about individuals in identifiable form.

EHLO: Extended Hello; a means by which a sending server identifies itself when it initiates the transfer of e-mail to a receiving server.

Electrically erasable programmable read-only memory (EEPROM): Memory that provides nonvolatile storage of data.

Electronic Communications Privacy Act of 1986: Law that governs the privacy and collection, access, disclosure, and interception of content and traffic data related to electronic communications.

Electronic communications services: Under 18 U.S.C. § 2510, "any service which provides to users thereof the ability to send or receive wire or electronic communications."

Electronic discovery: An exchange of information concerning the electronic evidence and experts that each party intends to present at a trial.

Electronic evidence: Any type of information that can be extracted from computer systems or other digital devices and that can be used to prove or disprove an offense or policy violation.

Electronic Freedom of Information Act Amendments of 1996: Law that allows requesters to obtain electronic government records on specific matters (e.g., a document created by a federal agency).

E-mail header: The part of an e-mail that provides identifying information for both sender and receiver of the message.

Embezzlement: A type of theft in which an employee of a company transfers the company's financial assets to his or her own account, often via computer.

Encase: A computer forensics tool that is widely used by law enforcement agencies.

Encryption: A means of physically blocking third-party access to a file, either by using a password or by rendering the file or aspects of the file unusable.

Enemy combatant: An enemy combatant is an individual that can be detained for the duration of an armed conflict pursuant to laws and customs of war. This term has been used to describe individuals that are unlawful combatants (e.g., terrorists). These individuals do not receive the entitlement of a prisoner of war status and do not fall under the protection of the Geneva Convention, which is reserved for lawful combatants.

Enhanced multimedia messages (EMS): A form of text message communication on PDAs and mobile phones that may include text and image, video, and/or sound.

Enron scandal: A 2001 scandal in which the energy trading firm's personnel showed shareholders that Enron was making a profit when the company was actually billions of dollars in debt.

Equal Employment Opportunity Commission: U.S. government agency that deals with charges of discrimination.

Equal Pay Act of 1963: Law that mandated equal pay for equal work, regardless of gender.

ESMTPA: Extended SMTP with SMTP Authentication; a protocol used to transmit e-mail.

Evaluation: The stage of a computer forensics investigation in which the data retrieved during the investigation are analyzed to establish their significance and relevance to the case at hand.

Event log: A software capability that automatically records events that occur within a computer to provide an audit trail that can be used to monitor, understand, and diagnose activities and problems within the system.

Evidence log: A record of all pertinent facts related to evidence found at a crime scene.

Exclusionary rule: A rule that makes evidence that was obtained in violation of the Fourth Amendment generally inadmissible in court.

Executive Order 13133: An executive order that established the Working Group on Unlawful Conduct on the Internet to evaluate the extent of cybercrime and the tools and capabilities of law enforcement to deal with this threat.

Executive Order 13181: An executive order that protects consumers' health information by preventing the disclosure to and use of this information by government agencies.

Exigent circumstances: A type of warrantless search in which evidence is seized because its destruction is imminent and there is probable cause to believe that the item seized constitutes evidence of criminal activity.

Expert witness: A specialist who collects, analyzes, and evaluates evidence; forms an opinion on it; and then communicates this opinion during a court proceeding.

Extortion: Obtaining something of value from another person without his/her consent by threatening to use force, jeopardizing their livelihood, or tarnishing their reputation.

Fail-open mode: A mode in which switches begin to function as hubs and send the incoming traffic through all of their ports.

Fair Labor Standards Act of 1938: Law that established minimum wages and the 40-hour work week.

Fair use doctrine: An exception to the copyright infringement rules that allows limited use or reproduction of copyrighted material.

False Claims Act of 1863: Law that allows a whistleblower to file a civil lawsuit against a federal contractor on behalf of the government.

False negative: An outcome in which an intrusion detection system misses an actual intrusion or attack.

False positive: An outcome in which an intrusion detection system reads an activity as an attack or intrusion when it is not; a false alarm.

Family and Medical Leave Act of 1993: Law that provides employees with unpaid leave (maximum of 12 work weeks) to tend to family or medical emergencies.

Family Education Rights and Privacy Act of 1974: Law that protects the confidentiality of student records by requiring consent by the student or parent before disclosing personally identifiable information about a student to a third party.

Family Entertainment and Copyright Act of 2005: Law that makes it a criminal offense to use a camcorder or other similar recording device to record a movie playing in a movie theater.

Faraday bag: A container that prevents messages from being sent or received by electronic devices.

FAT: A type of file allocation table that is used in the Windows 3.1, 95, 98, and ME operating systems.

FAT32: A type of file allocation table that is used in the Windows 3.1, 95, 98, and ME operating systems.

Fax machine: A type of peripheral device that may hold computer forensics evidence.

Federal Information Security Management Act of 2002: The primary legislation that now governs the information security of U.S. government systems.

Federal Property and Administrative Services Act of 1949: Law that established the General Services Administration (GSA), which is responsible for the development of policy for and management of government property and records.

Fifth Amendment: The provision of the U.S. Constitution dealing with the right to avoid self-incrimination.

File carving: A means of searching for specific files in a hard drive based on the header, footer, and other identifiers within the file.

File server: A computer that handles requests from other computers on the network for data that are stored on one or more of the server's hard drives.

File slack: On a computer, unused space in a cluster owing to a file's data not filling out the entire cluster; also called slack space.

Firewall: A computer security program that blocks incoming network traffic based on certain predetermined criteria.

First Amendment: The provision of the U.S. Constitution dealing with freedom of religion, freedom of the press, the right to assemble, and the right to petition the government for a redress of grievances.

Fixed telephony: Fixed telephony consists of a telecommunications device that transmits speech by electronic signals and is connected to a fixed landline.

Foreign Corrupt Practices Act of 1977: Law that prohibits individuals from paying foreign government officials in an attempt to "obtain or retain business" within these countries.

Forensic laboratory: A laboratory dedicated to the analysis of evidence from crimes.

Forensics toolkit: A kit carried by an investigator to collect evidence, which includes documentation tools, disassembly and removal tools, package and transport supplies, and other essential items.

ForensicSIM: A forensics tool that can clone a SIM card and examine it without altering the contents of the original device.

Fourth Amendment: The provision of the U.S. Constitution dealing with illegal search and seizure—that is, with the right to privacy.

Fraud: Falsely representing a fact, either by conduct or by words or writing, to induce a person to rely on the misrepresentation and surrender something of value.

Freedom of Information Act of 1966: Law that allows requesters to obtain government records on specific matters (e.g., a document created by a federal agency).

Frye standard: Guideline stating that forensics tools, techniques, procedures, and evidence are admissible in court if they have a "general acceptance" within the scientific community, as proven by scientific papers and books that have been written about the testing procedures.

FTK: Forensic Toolkit; a computer forensics tool that can create images of hard drives, analyze the registry, scan slack space for file fragments, inspect e-mails, identify steganography, crack passwords, and files.

Global positioning system (GPS): A system used to pinpoint a geographic location.

Global System for Mobile Communications (GSM): A globally accepted standard for digital mobile telephony systems.

Good faith exception: An exception to the exclusionary rule that states illegally seized evidence is admissible in court if a law enforcement agent acted in good faith, believing that he or she was acting according to a valid search warrant that is later found defective.

Government e-mail scam: A type of fraud involving the distribution of malicious software through what appear to be legitimate government e-mails.

Gramm-Leach-Bliley Act: The Financial Modernization Act of 1999, whose primary purpose is to protect the financial information of individuals that is held by financial institutions.

Hacking: Unauthorized intrusion into computers.

Health Insurance Portability and Accountability Act of 1996: Law that governs the privacy of health information in electronic form.

Hearsay evidence: "An oral or written assertion or . . . nonverbal conduct of a person, if it is intended by the person as an assertion," other than the individual who makes the statement, "while testifying at the trial or hearing, offered in evidence to prove the truth of the matter asserted" (according to the Federal Rules of Evidence).

Hidden partition: A means of hiding drives or files on a computer.

Honeynet: A collection of honeypots that is used to mimic a more complex environment.

Honeypot: An intrusion detection/prevention system that lures offenders away from valuable network resources and can help capture new and unknown attacks.

HTTP: Hypertext Transfer Protocol; a protocol used to transmit files, including Web pages, across the Internet or a network.

http sessions: Computer interactions that are documented through data about the requests and transfers of files across the network.

Hub: A device that connects computers and sends data (packets) between the networked computers through all of its ports.

Identification: The stage of a computer forensics investigation in which an investigator explains and documents the origin of the evidence and its significance.

Identification fraud: The manufacture, transfer, possession, and/or use either of the identification documents of another person (or persons) or false identification documents in order to induce an individual to rely on this misrepresentation and surrender something of value.

Identity theft: The theft of an individual's identity by obtaining his or her personal information or by accessing an individual's account with the intention of using it to commit illegal activities.

Identity Theft and Assumption Deterrence Act of 1998: Law that made identity theft an offense in the United States.

Illegal gambling: Transmission of bets or wagers via the Internet, which is prohibited in the United States.

Illegal prescription drug sales: Online sales of prescription-only drugs either without a prescription or with an illegitimate prescription.

ILook: A government-use-only computer forensics tool that has imaging, advanced e-mail analysis, and data salvaging capabilities.

Imaging: The process by which a duplicate copy of the entire hard drive of a computer is created.

IMAP: Internet Message Access Protocol; a protocol used to retrieve and read e-mails.

Inevitable discovery exception: An exception to the exclusionary rule that allows evidence that has been illegally obtained to be introduced in court if it would have inevitably been discovered through lawful means.

Input scam: A type of corporate crime in which data within a computer database are altered, deleted, or fabricated.

Insider trading: A type of corporate crime in which director, officer, or shareholder of a corporation uses information that has not been available to the general public to trade shares of the company.

Instant messages (IMs): A form of text message communication on PDAs and mobile phones.

Integrated circuit card identifier (ICCID): A number imprinted on a SIM card, that can be used by the service provider operator to trace the SIM card.

Integrated development environment (IDE) interface: A set of programs that are run from a single user interface.

Intellectual property: Intangible property that the law grants ownership rights to.

Intellectual property theft: The unlawful possession, use, distribution, and sale of the intangible property (e.g., copyrights, trademarks, patents, and trade secrets) of another individual or organization.

International mobile equipment identifier (IMEI): An identifier used to identify a mobile phone and match it to a subscriber.

International mobile subscriber identity (IMSI): A number assigned to a mobile phone by the provider that is stored on the customer's chip (SIM) card.

Internet Assigned Numbers Authority (IANA): The organization responsible for coordinating the general pool of IP addresses and providing them to Regional Internet Registries.

Internet Protocol (IP) address: A unique identifier assigned to a computer by the Internet service provider when it connects to the Internet.

Interrogatories: Written questions that are provided to a witness in a trial; the witness must answer these questions in written form.

Interstate Stalking Act of 1996: Law that forbids the use of interstate or foreign commerce—including use of the Internet—to harass or threaten another person, that person's spouse or partner, or that person's immediate family.

Intrusion detection system: A computer security system that is intended to detect attacks on systems and the unauthorized use of such systems, networks, and related resources.

Intrusion protection system: A computer security system that analyzes network traffic in real time to avert a potential attack or intrusion; it drops a packet if it observes suspicious activity.

Invasion of privacy: A wrongful intrusion into the private life of a person by the government or other individuals.

Investment Advisers Act of 1940: Law that makes it illegal for investment advisers to defraud their investors.

Investment fraud: A type of fraudulent behavior involving market manipulation.

Jacob Wetterling Crimes Against Children and Sexually Violent Offender Act of 1994: Law that was passed as part of the Federal Violent Crime Control and Law Enforcement Act of 1994 and requires states to implement a sex offender and crimes against children registry.

Keeping the Internet Devoid of Sexual Predators Act of 2008: Part of the Violent Crime Control and Law Enforcement Act of 1994, which required states to implement a sex offender registry.

Keylogger: A type of spyware that records every keystroke of the user and reports this information back to its source.

Keyword searching: Keyword searching is used by a computer user when he or she seeks to locate a specific file (or files) on the computer that contains one or more of the words specified by the user.

Labeling: The process of affixing documentation to a piece of evidence prior to its transport to a forensic laboratory.

Latin American and Caribbean Internet Addresses Registry (LACNIC): The

regional Internet registry for part of the Southern Hemisphere.

Live analysis: A type of computer forensics analysis that retrieves data from a running system.

Local area network: The simplest type of computer network, which connects computers within a small area and provides these systems with the ability to share resources.

Location data: Data about the locations from which a particular communication was made.

Logic bomb: A type of malware that is triggered by an event such as a predetermined date and time or running of a particular program a specific number of times, at which point it executes a previously dormant program.

Lottery scam: A type of fraud in which the offender solicits money from a victim by claiming that the victim won a prize, typically a foreign lottery.

MAC address: Media access control address; a unique identifier that is used to connect a computer to a network.

MAC address spoofing: A type of computer attack in which the offender poses as an approved network device.

MAC flooding: A type of computer attack in which numerous ARP requests are sent requesting multiple MAC addresses, which eventually overwhelms the resources of network switches.

Malware: Malicious software that causes damage to a computer or invades a computer to steal information from it; includes computer viruses, worms, Trojan horses, spyware, and botnets.

Man-in-the-middle attack: A type of computer attack in which the offender hijacks a TCP connection between a client and a server, and eavesdrops on their communication.

MD5 hash algorithm: A method of determining whether a copy is an exact copy of the original computer drive.

Memory card: A storage medium that allows users "to store additional files beyond the device's built-in capacity and provide another avenue for sharing data between compatible devices."

Memory sticks: A type of memory used to store data from portable electronic devices.

Message ID **field:** A header field consisting of the name of the server and a unique string that the sending e-mail server assigned to the message.

Metadata: Data about data.

Metropolitan area network: A wide area network that is restricted to a particular city or town.

Micro SD: A micro secure digital memory card; a removable card that can be used to store data on a PDA or mobile phone.

Microprocessor: An integrated circuit that functions as a central processing unit (CPU) by performing the basic functions and operations of a computer.

MIME-Version **field:** A header field indicating that an e-mail was composed in compliance with RFC 1341, MIME.

Miranda **warning:** Notification of a suspect of his or her rights before the individual is interviewed by the police.

Mobile phone: An electronic device that connects to a wireless communications network to enable telecommunications within a defined area. Also known as *cell phone* or *cellular phone*.

Mobile subscriber integrated services digital network number (MSISDN): Data on a SIM card that provides the real number of a GSM-enabled mobile phone.

MOBILedit!: A forensics tool that can acquire, search, examine, and report on data from virtually all mobile phones on the market.

Money laundering: "The process by which criminals hide, disguise, and legitimize their ill-gotten gains."

Motion: In a court proceeding, a request that a judge rule on a specific, narrowly defined matter.

Motion for discovery: An attorney's request for pretrial disclosure of evidence.

Motion in limine: An attorney's request for a pretrial ruling on the admissibility of evidence.

Motion to suppress evidence: An attorney's request that a judge rule certain evidence inadmissible because it was illegally obtained by law enforcement agencies.

Multi-Media Card: A form of nonvolatile computer storage for mobile phones and PDAs.

Multimedia Messaging Service (MMS): Messages containing a combination of text, images, videos, and sound.

Multipurpose Internet Mail Extensions (MIME): A protocol that standardizes the header and body of e-mails.

National Bureau of Standards Act of 1901: Law that established the National Bureau of Standards, which was responsible for maintaining and developing the national standards for measurement and providing the private industry and government with the methods with which to make these measurements consistent with the existing standards of accuracy and uniformity. Today, this agency is known as the National Institute of Standards and Technology (NIST).

National Center for Missing and Exploited Children: The National Center from Missing and Exploited Children provides assistance in recovering missing children and informs the public about the ways that children can protect themselves from exploitation, molestation, and abduction.

National Institute of Standards and Technology: The U.S. agency that is responsible for developing standards and guidelines for government computer systems.

National Organ Transplant Act of 1984: Law that prohibits the selling of human organs or body parts.

National Sex Offender Registry: A website that provides investigators with information on sexual predators.

Negligent hiring: Failure to conduct a diligent inquiry into the background of a prospective employee and his or her suitability to the position in question before deciding to hire that person.

Negligent retention: Failure to take action against an employee after being made aware that the employee in question poses a risk to the company and the employees within the company.

NetDetector: A network forensics analysis tool that engages in continuous, real-time surveillance of a network.

NetIntercept: A network forensics analysis tool that captures and archives network data for analysis at a later date and time, and is used to detect spoofing.

NetWitness: A network forensics analysis tool that proactively detects threats and alerts authorities as to their presence; it also captures network traffic data and provides real-time analysis of it.

Network administrator: The individual responsible for managing a server-based network.

Network forensics: The use of scientifically proven techniques to investigate computer networks.

Network forensics analysis tool: A tool that is used to capture, record, and analyze computer network traffic, or is intended for use in a networked computer environment.

Network interface card: A network interface card is a device that is installed in or attached to a computer system, which enables the computer to transmit and receive data on a network.

Network traffic: Data in a network.

Networked computer: A computer that is connected to one or more computers in a manner that allows them to share data, software, and hardware.

NetworkMiner: A network forensics analysis tool (sniffer) that is used to capture network traffic.

NFTS: New Technology File System; the file allocation table used in the Windows NT, 2000, XP, 2003, 2008, Vista, and 7 operating systems.

Nigerian scam: Also known as a 419 scam; a type of fraud in which the offender claims that the victim has received an inheritance from a long-lost relative from Nigeria or claims to be a Nigerian government, business, or banking official.

No Electronic Theft Act of 1997: Law that specifies an individual does not have to distribute or reproduce copies of copyrighted works for profit to be considered to have violated the laws against copyright infringement.

No-knock warrant: A warrant that exempts law enforcement personnel from having to knock and announce their presence before executing a search.

Noncompete agreement: An agreement that employees will not work with a company's competitors within a particular geographical region for a reasonable period of time (no greater than two years) after leaving the company.

Non-content data: Communication data such as the telephone numbers dialed, customer information (name and address), and e-mail addresses of the message sender and recipient.

Nonsolicitation agreement: An agreement that prohibits an individual from trying to "steal" the employees or customers of his or her former employer.

Nonvolatile data: Data that are stored and preserved in the hard drive when the computer is powered off.

Notes: A document created by an investigator that includes anything relevant to the investigation at hand.

NVRAM: Nonvolatile random access memory.

Occupational Safety and Health Act of 1970: Law that mandates that companies ensure safe working conditions for their employees.

Octet: Four groups of 8 bits, for a total of 32 bits.

Omnibus Crime Control and Safe Streets Act of 1968: The Wiretap Statute; law that protected the privacy of only the content of telecommunications.

OmniPeek: A network forensics analysis tool that allows an investigator or network administrator to see every part of the network in real time, and capture and store network traffic.

Online sales fraud: Online sales of counterfeit consumer goods or nonexistent services.

Organized crime: Significant criminal activities perpetrated by three or more persons in a group that has some type of formalized structure and whose primary objective is monetary gain.

Output scam: A type of corporate crime in which an individual seeks to use company-owned data for personal gain.

Oxygen Phone Manager: A forensics tool that is compatible with numerous mobile phones and smartphones.

Packet: A unit of data transmitted over a network.

Pager: A type of handheld computing device that alerts the user of an incoming message.

Palm dd: A forensics tool that is used to acquire data from a Palm device with a Palm operating system that is running in console mode.

Particularity: In regard to a search warrant, the idea that the search warrant needs to specify the place that will be searched and the items that will be seized.

Password cracking: A type of software used to figure out someone's password and gain access to his or her computer system.

Patent: A document that provides an individual with the exclusive right to make, use, offer to sell, or sell his or her invention for a specific period of time.

PDA Seizure: A forensics tool that can be used to extract data from PDAs running the Windows CE, Pocket PC, and Palm operating systems.

Peer-to-peer networking configuration: A network in which each peer (or computer) manages authentication and access to its own resources.

Peripheral devices: Devices that are not essential parts of a computer system, such as scanners, copiers, printers, and fax machines.

Personal digital assistant (PDA): A type of handheld computing device once limited to use as a personal information organizer, but now often capable of many functions.

Personal identification number (PIN): A number that the user must input to access a mobile phone, PDA, or other account.

Personal information manager (PIM): A device that offers personal-management functions such as a memo pad, calendar or date book, and to-do list.

Personal unlock key (PUK): A number unique to a SIM card that may be used to access a mobile phone or PDA after the user is locked out of the device due to incorrect PIN entry attempts.

Pharming: A type of scam in which the offender creates a website that looks identical to an authentic website; the mirror website carries a malicious payload.

Phishing: A type of cybercrime in which individuals deceive others by posing as legitimately established enterprises so as to steal communication users' valuable personal information, such as account data and credit card information.

Phone phreaking: The illegal practice of tricking the telephone system to allow someone to make free phone calls.

Photograph: A type of documentation that provides a visual depiction of the crime scene, taken to preserve the crime scene and evidence.

Photograph log: Documentation of photographs taken at a crime, including information about the equipment, personnel, and circumstances of the photography.

Physical evidence: Tangible (real) evidence that may be used to corroborate suspect, witness, and victim statements; to link a suspect to a crime scene or crime; to link a suspect to a victim; or to rule out a person as a possible suspect.

Piggybacking: A strategy to subvert surveillance in which information that needs to be passed on to a third party undetected is attached to a legitimate object and transmitted over the Internet.

Pilot-link: A forensics tool that is used to acquire data from PDAs with Palm OS.

PING command: A strategy to validate an IP address found in an e-mail header that checks the accessibility of the IP address.

Pizzini: Small slips of paper—either tiny handwritten or typewritten notes—used for high-level communications to evade fax or phone taps.

Plain view doctrine: A type of warrantless search in which law enforcement agents are permitted to seize evidence (not specified in a search warrant) that is in plain view, whose incriminating nature is immediately apparent to the officers.

Plaintext: An unencrypted message.

Ponzi scheme: A type of fraud in which investors are promised high returns for little to no risk. The money provided by new investors is used to pay existing investors, but none of the money is actually invested anywhere.

POP3: Post Office Protocol 3; a protocol used to read e-mail.

Port scanner: A probe that makes requests to a computer system's ports to see which one responds or is open to the request.

POSE: Palm Operating System Emulator; a tool that allows an investigator to view, work on, and access items (e.g., menus, calendars) on a device as if he or she was working on the original PDA.

Pregnancy Discrimination Act of 1978: Law that prohibits employers from discriminating against pregnant women.

Presentation: The stage of a computer forensics investigation in which data pertinent to the case that were found during the investigation are reported.

Pretexting: Seeking to collect information about an individual using false pretenses.

Pretrial hearing: A court proceeding that both determines if a crime or civil wrong has been committed and evaluates the evidence and testimony to be presented in court.

Pretty Good Privacy: Software for encrypting files and messages.

Price fixing: The illegal collaboration of two or more parties to fix prices of particular goods or services.

Printer: A type of peripheral device that may hold computer forensics evidence.

Privacy: A multivalent social and legal concept, defined in ways ranging from the right to be let alone, to the capacity to keep certain things secret, to the right to control other individuals' access to oneself and information about oneself.

Privacy Act of 1974: Law that protects individuals' privacy by regulating the collection, maintenance, use, and dissemination of information by government agencies. It also seeks to limit the collection of personal data by government agencies.

Privacy Protection Act of 1980: Law that protects journalists, publishers, and other such individuals by not requiring them to turn over work that falls under the First Amendment protections to law enforcement agencies.

Privileged data: Personal, sensitive, confidential, and proprietary information that is protected from search and seizure, such as medical data, attorney–client communications, trade secrets, and journalism-related information from confidential sources.

Probable cause: A situation in which "the known facts and circumstances are sufficient to warrant a man of reasonable prudence in the belief that contraband or evidence of a crime will be found"; used as the basis for issuing a search warrant.

ProDiscover: A network forensics tool that can remotely acquire drive images in a live environment and securely investigate and extract network evidence from running systems.

Prosecutorial Remedies and Other Tools to End the Exploitation of Children Today Act of 2003: Law that requires Internet service providers to report child pornography violations to law enforcement officials, and prohibits the use of misleading domain names with the intent to deceive a minor into viewing online material that is harmful to minors.

Proxy server: A server that acts as an intermediary for client requests for resources from other servers.

PUK2: A second personal unlock key that a mobile phone requires to access the device after a user has inserted the first personal unlock key; used as a means of enhancing security.

Pump and dump investment scheme: A type of financial fraud in which the offender tries to increase the price of a stock by circulating false information about it.

PyFlag: An advanced network forensics analysis tool that can be used to analyze disk images and large volumes of log files, and can reconstruct Web pages viewed by a user.

Pyramid scheme: A type of financial fraud that requires new investors to become involved in marketing a product and recruiting other investors.

Random access memory (RAM): Computer memory where the operating system, application programs and other data that is being accessed or used is kept while the device is on. The contents of RAM are lost when the device is powered down.

Read-only memory (ROM): Computer memory that permanently stores instructions that the computer uses during the boot process. The contents of ROM cannot be altered.

***Received* field:** A header field that provides information on the recipient of an e-mail, and sometimes on the sender of the e-mail.

***Received-SPF* field:** A header field used for spam filtering purposes.

Regional Internet Registries (RIRs): Organizations that administer and register IP addresses within a defined region.

Regional Internet Registry for Europe (RIPE): The registry that assigns and registers IP addresses from Europe, the Middle East, and parts of Central Asia.

Regulation of Investigatory Powers Act of 2000: U.K. law that requires all Internet service providers in the United Kingdom to "install black boxes on their network to monitor all data as it passes and subsequently feed it to a central processing location controlled by the United Kingdom's security service MI-5."

Reliability: A characteristic of a technology or method in which the same results are obtained when using "the same method on identical test items in the same laboratory by the same operator using the same equipment within short intervals of time" (National Institute of Standards and Technology).

Remote computing services: Under 18 U.S.C. § 2711(2), "the provision to the public of computer storage or processing services by means of an electronic communications system."

Rental and real estate scam: A type of fraud involving rental properties and real estate.

Report: A narrative of what happened at the crime scene and how the investigation of the scene was conducted.

Reproducibility: A characteristic of a technology or method in which the same results are obtained when using "the same method on identical test items in different laboratories with different operators using different equipment (National Institute of Standards and Technology).

***Return-Path* field:** A header field that indicates where the e-mail should be sent back to if it fails to reach the recipient.

Right to Financial Privacy Act of 1978: Law that requires the notification of a subject if his or her financial records are sought and provides individuals with the ability to challenge the disclosure in court.

Rootkit: A computer program that is installed after an offender has gained full access to a system, and that helps hide the attacker's intrusion and access or modification of data.

Router: A device that connects two networks and routes data between them.

Sarbanes-Oxley Act of 2002: Law that sought to improve the accuracy and reliability of corporate disclosures so as to protect investors from fraudulent business practices.

SCADA systems: Supervisory Control and Data Acquisition systems; the computer systems of choice for most of the critical infrastructure industries in the United States.

Scanner: A type of peripheral device that may hold computer forensics evidence.

Scareware: Pop-up security warnings falsely informing a user that his or her computer has been infected with numerous computer viruses, with the goal of persuading the user to buy fake software and/or to download malicious software on the user's computer.

Search and seizure: A search and seizure is a legal procedure whereby law enforcement agents conduct an examination of a premises and forcibly take property of potential evidentiary value from a person who is suspected of violating or has violated the law.

Search incident to arrest: A type of warrantless search that occurs upon the arrest of an individual and encompasses a search of the arrestee and the areas under his or her immediate control.

Search protocol: A document that describes what is being searched for in a computer and which methodology will be used.

Search warrant: A court order allowing government personnel to enter a premises, search for the objects named in the warrant, and seize them.

Securing Adolescents from Exploitation Online Act of 2007: Law that expanded the definition of an Internet service provider to include wireless hot spots such as libraries, hotels, and municipalities, and required such entities to notify law enforcement officials about child pornography violations.

Securities Act of 1933: Law that covers securities fraud.

Securities and Exchange Commission: U.S. government agency that polices the financial industry.

Securities Exchange Act of 1934: Law that covers securities fraud.

Security log: An event log that records all login attempts (both valid and invalid) and the creation, opening, or deletion of files, programs, or other objects by a computer user.

Server-based network configuration: A network in which different servers manage authentication and access to resources from a database (directory).

Setup log: An event log that provides data on applications that are installed on a computer.

Sex Offender Registration and Notification Act of 2006: Law that created a national sex offender registry.

Sexting: The act of sending, receiving, or forwarding sexually explicit messages, photos, or images via cell phone, computer, or other digital device.

Sexual Exploitation of Children Act of 1978: Law that made it illegal for someone to manufacture and commercially distribute obscene materials that involve minors younger than 16 years old.

Sexual harassment: Unwelcome sexual advances, requests for sexual favors, and other verbal or physical conduct of a sexual nature that affect an individual's employment, unreasonably interfere with an individual's work performance, or create an intimidating, hostile, or offensive work environment.

Sexual predator: A term used to describe a person that commits a sex crime (or crimes), has been convicted of or plead guilty to committing a sex crime (or crimes), and who is likely to commit additional sex crimes in the future.

SHA hash algorithm: A method of determining whether a copy is an exact copy of the original computer drive.

Sherman Antitrust Act of 1890: Law that prohibits price fixing.

Short Message Service (SMS): Text messages.

SilentRunner: A network forensics analysis tool that can capture, analyze, and visualize computer network activity by uncovering misuse, intrusion attempts, and abnormal activity.

SIM card reader: A device that enable users to back up the data on their mobile phone devices.

SIM Card Seizure: A forensics tool that can perform a complete acquisition and analysis of the data on a SIM card.

SIMIS: A forensics tool that can both securely acquire data from an external SIM card and use cryptographic hashes to protect the integrity of the data.

Sixth Amendment: The provision of the U.S. Constitution dealing with criminal trial rights.

Sketch: An illustration made by an investigator to provide accurate dimensions of the crime scene and the location of the evidence in relation to that scene.

Slack space: Another term for *file slack*.

Smartphone: A device that has both mobile phone and PDA capabilities.

SMTP: Simple Mail Transfer Protocol; the protocol used to send e-mail across the Internet or across a network.

Sniffer: A type of software that is used to monitor and analyze networks, but can also be used to collect individuals' usernames, passwords, and other personal information.

Snort: A widely used intrusion detection system.

Spam: Junk e-mail that fills up individuals' e-mail accounts and may be used to commit online fraud.

Spyware: A type of malware that enables the remote monitoring of a computer user's activities or information on an individual's computer where this software has been installed. It may also secretly gather information on users without their knowledge and relay it to interested third parties.

SRAM: Static random access memory.

Stand-alone computer: A computer that is not connected to another computer or network.

Static IP address: The permanent address of a specific computer on the Internet.

Steganography: An information hiding technique that makes data and messages invisible by hiding them in various files.

"Stop, look and listen" tools: Network forensics analysis tools that analyze packets in a rudimentary manner and store only certain types of data for further analysis.

Stored Wire and Electronic Communications and Transactional Records Act of 1996: Law that requires the preservation and disclosure of customer communications records to government entities.

Subpoena: An order that compels someone to appear for a hearing and testify on a suspected violation or that requests an organization provide a government agency with records, documents, books, or papers to assist its personnel in their investigation and prosecution of a case.

Subpoena *duces tecum*: A subpoena used to command the production of evidence.

Subscriber identity module (SIM): A card that stores information identifying the subscriber to a particular network (e.g., a mobile phone or PDA user).

Switch: A more efficient and intelligent version of a hub that sends each packet only to the port that it is supposed to travel through.

SYN flood attack: A type of denial of service attack in which the attacker floods a network server with phony authentication requests so that individuals with legitimate requests are prevented from accessing it; ultimately the server shuts down.

Systems log: An event log that provides information on Windows system components.

Tagging: Labeling of evidence prior to its transport to a forensic laboratory.

Tax evasion: An illegal attempt to avoid paying taxes.

Taxpayer Browsing Protection Act of 1997: Law that criminalized federal and state employees' unauthorized browsing of tax-payer information.

TCP handshake: The process used to set up a TCP connection between two computers.

Telecommunications Act of 1996: Law that requires telecommunications companies to obtain the consent of the customer before using customer proprietary network information.

Telecommunications devices: Fixed telephony, mobile phones, and answering machines.

Telecommunications fraud: The theft of telecommunications services or the use of telecommunications services to commit other forms of fraud.

Telephone Records and Privacy Protection Act of 2006: Law that prohibits the use of pretexting to buy, sell, or obtain individuals' telephone records.

Terrorism: The unlawful use of force or violence against persons and/or property in the pursuit of political, religious, ideological or individual goals in order to intimidate or demoralize a government and its population.

Testimony: A type of evidence that consists of witnesses speaking under oath, including eyewitness and expert testimony.

Throughput scam: A type of corporate crime in which individuals commit crimes against business computers.

Title VII of the Civil Rights Act of 1964: Law that prohibits employers from discriminating against a potential employee on the basis of race, gender, religion, or national origin.

Tor: An anonymous Internet communication system that provides individuals (and organizations) with the ability to share information and communicate over public networks without compromising their privacy.

Tracert: Trace route; a tool that can determine the route a packet traveled across a computer network.

Trade secret: A type of intellectual property consisting of specialized financial, business, scientific, technical, economic, or engineering information.

Trade secret theft: A type of corporate crime in which the offender steals intellectual property consisting of specialized financial, business, scientific, technical, economic, or engineering information.

Trademark: A document that protects a commercial identity or brand.

Trademark Counterfeiting Act of 1984: Law that makes it a criminal offense for an individual to traffic in counterfeit goods or services.

Traffic data: Data about a communication, including the type, source, destination, and time/length of communication.

Trafficking of counterfeit goods: The sale of counterfeit goods in a manner intended to cause confusion, to cause mistake, or to deceive the buyer.

TransFlash: A type of memory card that is similar to a micro SD card; it is used in mobile phones and PDAs.

Transmission Control Protocol (TCP): The language that computers use on the Internet to communicate with one another.

Trier of fact: The finder of fact; the person or persons (i.e., judge and/or jury) who determine the facts of a case in a court proceeding.

Trojan horse: A type of malware that tricks the computer user into thinking that it is legitimate software, but actually contains hidden functions. When the computer user downloads and installs the program, these hidden functions are executed along with the software.

True negative: An outcome in which an intrusion detection system determines that network activity is normal and does not raise an alarm.

True positive: An outcome in which an intrusion detection system detects that an actual attack or intrusion has occurred, and authorities are alerted as to its presence.

Tyco International scandal: A 2002 scandal in which Tyco executives siphoned millions of dollars from the company through fraudulent stock sales and unauthorized business loans.

Unallocated space: Space available on a computer because it was never used or because the information in it was deleted.

Uniformed Services Employment and Reemployment Rights Act of 1994: Law that protects individuals from workplace discrimination because of their military service.

U.K. Regulation of Investigatory Powers Act of 2000: U.K. law that grants law enforcement agencies the power to compel suspects to provide their decryption keys.

Universal Integrated Circuit Card (UICC): The equivalent of a SIM card, but geared toward UMTS devices.

Universal Mobile Telecommunications Systems (UMTS): A standard for mobile phones, PDAs, and similar devices.

Universal Serial Bus (USB): An interface that is commonly used to connect peripheral devices to computers.

Universal Subscriber Identity Module (USIM): A component of the UMTS 3G network that stores data that identifies a mobile phone user as well as his or her subscriber data, contacts, and preferences.

USA Patriot Act of 2001: Law that sought to deter and punish terrorist acts in the United States and worldwide and to enhance law enforcement surveillance powers by easing the restrictions on government and law enforcement access to communications data.

U.S. Constitution: The U.S. Constitution establishes the basic principles of the laws of the United States of America that determine the powers and duties of the government and guarantees certain rights for the American people.

Validity: The extent to which measures indicate what they are intended to measure.

Vandalism: Defacement of someone else's property, either physical (e.g., graffiti on walls) or virtual (e.g., unauthorized access to a website and the alteration or replacement of its content).

Victims of Child Abuse Act of 1990: Law that sought to improve the investigation and prosecution of child abuse cases.

Video: A type of documentation that provides a visual depiction of the crime to those involved in the case and may reveal vital clues that investigators initially missed while at the crime scene.

Violent Crime Control and Law Enforcement Act of 1994: The most comprehensive law covering crime and law enforcement in U.S. history. It covered varied subjects and a broad spectrum of crimes (several of them new). Some of the provisions in this Act sought to increase police presence, improve cooperative efforts between law enforcement agencies and communities, and expand the powers of law enforcement agencies to deal with criminal aliens.

Virus: See *computer virus.*

Virus protection scam: A type of fraud in which the offender tries to get a computer user to buy fake software and/or to download malicious software on the user's computer.

Vishing: Voice phishing; a type of fraud that occurs via voice or text messages sent to cell phones.

Voice mail: Audio messages stored on fixed and mobile phones.

***Voir dire*:** French for "to speak the truth"; a legal proceeding that establishes the competency of a technical or expert witness to participate in a trial.

Volatile data: Data that disappear when the computer is powered off.

Warrantless search: A search conducted without a warrant that is considered valid because it falls under one of the exceptions to the warrant requirement (e.g., stop-and-frisk procedures, open fields, automobile exceptions, search incident to arrest, exigent circumstances, plain view, consent searches, and border searches).

Web-based e-mail: An e-mail system in which e-mail accounts are accessed through a Web browser and e-mails are stored in the e-mail service provider's server.

Web defacement: A type of attack in which one or more individuals alter or replace the content of a website without authorization.

Whistleblower: An individual who exposes a person, group, or business for engaging in illegal activities.

WHOIS: A query tool that allows a user to find out the contact and location information of the owner of an IP address.

Wide area network: A network that connects computers in different locations.

WIPO Copyright Treaty of 1996: The World Intellectual Property Organization's international treaty designed to control access to copyrighted works.

WIPO Performances and Phonograms Treaty of 1996: The World Intellectual Property Organization's international treaty designed to protect recordings of artists' performances.

Wireless Communications and Public Safety Act of 1999: Law that required mobile phones to have GPS tracking capabilities.

Wireshark: A sniffer program that captures network traffic in real time and records it.

Witness: A person who gives testimony in a court proceeding.

Work-at-home scam: A type of fraud in which the offender advertises opportunities for individuals to create their own home business or advertises positions in which employers are seeking individuals for at-home work.

Workplace injury: An injury that arises out of and in the course of employment.

Workplace violence: A type of violence that occurs between employees on the job or when outsiders seek to harm an employee or employees within the company.

World Intellectual Property Organization: International organization dedicated to protecting the rights of intellectual property owners.

Worldcom scandal: A 2002 scandal in which, by falsifying business records, Worldcom employees were able to show billions in profits, even though the company was actually in debt.

Worm: A type of malware that does not need to piggyback on a file or program to replicate itself or require user activity to make copies of itself and spread.

Write blocker: A device that blocks the wires that would otherwise communicate data to be written to a hard drive or other data source.

Wrongful termination: The termination of employment in violation of the law or the terms of the employee's contract.

***X-Mailer* field:** A header field that specifies the e-mail system used to send a message.

***X-Originating-IP* field:** A header field that reveals the IP address of the computer from which an e-mail was originally sent.

.XRY: A forensics tool that retrieves stored data from certain models of GSM phones.

INDEX